EVOLUTION AND TRENDS OF SUSTAINABLE APPROACHES

EVOLUTION AND TRENDS OF SUSTAINABLE APPROACHES

LATEST DEVELOPMENT AND INNOVATIONS IN SCIENCE AND TECHNOLOGY APPLICATIONS

Edited by

DANIEL ALEJANDRO ROSSIT

Engineering Department, Universidad Nacional del Sur, Bahía Blanca, Argentina

CHAUDHERY MUSTANSAR HUSSAIN

Department of Chemistry and Environmental Sciences, New Jersey Institute of Technology, Newark, NJ, United States

ELSEVIER

Elsevier
Radarweg 29, PO Box 211, 1000 AE Amsterdam, Netherlands
125 London Wall, London EC2Y 5AS, United Kingdom
50 Hampshire Street, 5th Floor, Cambridge, MA 02139, United States

ISBN: 978-0-443-21651-0

For information on all Elsevier publications visit our website at https://www.elsevier.com/books-and-journals

Publisher: Candice Janco
Acquisitions Editor: Jessica Mack
Editorial Project Manager: Akanksha Marwa
Production Project Manager: Bharatwaj Varatharajan
Cover Designer: Christian J. Bilbow

Typeset by TNQ Technologies

Contents

Part 2

Sustainable operations management

6. Maintenance in operational condition and obsolescence management in a sustainable operations management context

Pascal Vrignat, Mariem Besbes, Marc Zolghadri, and Frédéric Kratz

7. Circular economy strategies for improving furniture management in educational institutions

Pablo Maya-Duque, Brigitte Roldán-Ramírez, Betzy Cárdenas-Ibáñez, and Luisa Jiménez-Ramírez

8. A multiobjective maximal covering/p-dispersion model for reverse vending machine location

Sebastián Bedoya, Jaime Luis Mejía, Pablo Maya-Duque, and Juan G. Villegas

9. Closed-loop supply chain and extended producer responsibility (EPR): A literature review

Milagros Marizcurrena, Victoria Morás, Guillermo Ulery, Daniel Rossit, and Pedro Piñeyro

10. Solving a routing problem with time windows for an Argentinian case study in solid waste management

Diego Gabriel Rossit, Adrián Andrés Toncovich, and Franco Bautista Heine

11. A knowledge roadmap for digitally enabled sustainable construction and building supply chain management

Vidyasagar Gembali, Aalok Kumar, and P.R.S. Sarma

Part 3

Sustainability and technology

Contributors

Sebastián Bedoya Departamento de Ingeniería Industrial, Facultad de Ingeniería, Grupo Analytics and Research for Decision Making, Universidad de Antioquia, Medellín, Colombia

Mariem Besbes Quartz Laboratory, ISAE-Supméca, Saint-Ouen, France

David Castelo Branco Energy Planning Program, Federal University of Rio de Janeiro, Rio de Janeiro, Brazil

Betzy Cárdenas-Ibáñez Departamento de Ingeniería Industrial, Facultad de Ingeniería, Grupo ALIADO, Universidad deAntioquia, Medellín, Colombia

Javier Carmona-Murillo Departamento de Ingeniería de Sistemas Informáticos y Telemáticos, Centro Universitario de Mérida, Universidad de Extremadura, Mérida, Spain

Harpreet Kaur Channi Department of Electrical Engineering, Chandigarh University, Mohali, Punjab, India

Andrea Teresa Espinoza Pérez Industrial Engineering Department, Faculty of Engineering, University of Santiago of Chile (USACH), Santiago, Chile; Program for the Development of Sustainable Production Systems (PDSPS), Industrial Engineering Department, Faculty of Engineering, University of Santiago of Chile (USACH), Santiago, Chile

Katianny Gomes Santana Estival Department of Administrative and Accounting Sciences, Universidade Estadual de Santa Cruz, Ilhéus, Brazil

Jesús Galeano-Brajones Departamento de Ingeniería de Sistemas Informáticos y Telemáticos, Centro Universitario de Mérida, Universidad de Extremadura, Mérida, Spain

René André Garrido Lazo Program for the Development of Sustainable Production Systems (PDSPS), Industrial Engineering Department, Faculty of Engineering, University of Santiago of Chile (USACH), Santiago, Chile; Facultad de Ingeniería, Departamento de Geografía, Universidad de Santiago de Chile, Santiago, Chile

Vidyasagar Gembali Indian Institute of Management, Visakhapatnam, Andhra Pradesh, India

Virginia I. González Universidad del Norte, Barranquilla, Colombia

Franco Bautista Heine Department of Engineering, Universidad Nacional del Sur, Bahía Blanca, Argentina

César Augusto Henao Universidad del Norte, Barranquilla, Colombia

Chaudhery Mustansar Hussain Department of Chemistry and Environmental Science, New Jersey Institute of Technology, Newark, NJ, United States

Luisa Jiménez-Ramírez Departamento de Ingeniería Industrial, Facultad de Ingeniería, Grupo ALIADO, Universidad deAntioquia, Medellín, Colombia

Frédéric Kratz PRISME Laboratory, INSA Centre-Val de Loire, Bourges, France

Aalok Kumar Indian Institute of Management, Visakhapatnam, Andhra Pradesh, India

Raman Kumar Department of Mechanical and Production Engineering, Guru Nanak Dev Engineering College, Ludhiana, Punjab, India

João Gabriel Lassio Energy Planning Program, Federal University of Rio de Janeiro, Rio de

Janeiro, Brazil; Cepel - Electrical Energy Research Center, Rio de Janeiro, Brazil

Francisco Luna-Valero Departamento Lenguajes y Ciencias de la Computación, Universidad de Málaga, Málaga, Spain; ITIS Software, Universidad de Málaga, Málaga, Spain

Alessandra Magrini Energy Planning Program, Federal University of Rio de Janeiro, Rio de Janeiro, Brazil

Aline Patrícia Mano Department of Engineering and Computing, Universidade Estadual de Santa Cruz, Ilhéus, Brazil

Milagros Marizcurrena Universidad de la República, Montevideo, Uruguay

Yunesky Masip Macía Escuela de Ingeniería Mecánica, Pontificia Universidad Católica de Valparaíso, Quilpué, Chile

Denise Matos Energy Planning Program, Federal University of Rio de Janeiro, Rio de Janeiro, Brazil; Cepel - Electrical Energy Research Center, Rio de Janeiro, Brazil

Pablo Maya-Duque Departamento de Ingeniería Industrial, Facultad de Ingeniería, Grupo ALIADO, Universidad deAntioquia, Medellín, Colombia; Departamento de Ingeniería Industrial, Facultad de Ingeniería, Grupo Analytics and Research for Decision Making, Universidad de Antioquia, Medellín, Colombia

Jaime Luis Mejía Departamento de Ingeniería Industrial, Facultad de Ingeniería, Grupo Analytics and Research for Decision Making, Universidad de Antioquia, Medellín, Colombia

Marcelo Mena Carrasco Escuela de Ingeniería Bioquímica, Centro de Acción Climática, Quilpué, Chile

Victoria Morás Universidad de la República, Montevideo, Uruguay

Amaro Olímpio Pereira, Jr. Energy Planning Program, Federal University of Rio de Janeiro, Rio de Janeiro, Brazil

Pedro Piñeyro Universidad de la República, Montevideo, Uruguay

Andrés Felipe Porto Department of Industrial Engineering, Institución Universitaria Americana, Barranquilla, Colombia; Universitat Politècnica de Catalunya, Barcelona, Spain

Umabharati Rawat Indian Institute of Technology Roorkee, Roorkee, Uttarakhand, India

Pablo Rodríguez Machuca Facultad de Ingeniería, Departamento de Geografía, Universidad de Santiago de Chile, Santiago, Chile

Ángel Alexander Rodríguez Soto Escuela de Ingeniería Mecánica, Pontificia Universidad Católica de Valparaíso, Quilpué, Chile

Brigitte Roldán-Ramírez Departamento de Ingeniería Industrial, Facultad de Ingeniería, Grupo ALIADO, Universidad deAntioquia, Medellín, Colombia

Diego Gabriel Rossit Department of Engineering, INMABB, Universidad Nacional del Sur (UNS)-CONICET, Bahía Blanca, Argentina

Daniel Rossit Department of Engineering, INMABB, Universidad Nacional del Sur (UNS)-CONICET, Bahía Blanca, Argentina

Pavlo Santander-Tapia Industrial Engineering Department, Faculty of Engineering, University of Santiago of Chile (USACH), Santiago, Chile; Program for the Development of Sustainable Production Systems (PDSPS), Industrial Engineering Department, Faculty of Engineering, University of Santiago of Chile (USACH), Santiago, Chile

Rafael Bertoldo dos Santos Broto Incubadora Universitária, Universidade Estadual de Santa Cruz, Ilhéus, Brazil

P.R.S. Sarma Indian Institute of Management, Visakhapatnam, Andhra Pradesh, India

Daniel Serafini Departamento de Física, Universidad de Santiago de Chile, Santiago, Chile

Adrián Andrés Toncovich Department of Engineering, Universidad Nacional del Sur, Bahía Blanca, Argentina

Cibelle Trama Energy Planning Program, Federal University of Rio de Janeiro, Rio de Janeiro, Brazil; Cepel - Electrical Energy Research Center, Rio de Janeiro, Brazil

Guillermo Ulery Universidad de la República, Montevideo, Uruguay

Natalia Yulissa Velastín Osorio Industrial Engineering Department, Faculty of Engineering, University of Santiago of Chile (USACH), Santiago, Chile

Juan G. Villegas Departamento de Ingeniería Industrial, Facultad de Ingeniería, Grupo Analytics and Research for Decision Making, Universidad de Antioquia, Medellín, Colombia

Pascal Vrignat PRISME Laboratory, Orléans University, Orléans, France

Marc Zolghadri Quartz Laboratory, ISAE-Supméca, Saint-Ouen, France

CHAPTER

1

Introduction to—evolution and trends of sustainability

Daniel Rossit[1] *and Chaudhery Mustansar Hussain*[2]

[1]Department of Engineering, INMABB, Universidad Nacional del Sur (UNS)-CONICET, Bahía Blanca, Argentina; [2]Department of Chemistry and Environmental Science, New Jersey Institute of Technology, Newark, NJ, United States

This book aims to provide different trends and approaches within the sustainability framework to evaluate their impact and offer possible solutions to the problems facing the global sustainability paradigm. Sustainability assessment approaches support different levels of both decision-making and policy processes, thereby improving the management of natural and human systems. Additionally, there are many different approaches to quantifying and estimating sustainability. Among the most notable are sustainability indicators (SI), as they are widely used to measure and communicate progress toward sustainable development. Additionally, another option for assessing sustainability based on the industry's life cycle is the life cycle sustainability assessment (LCSA), which balances the three dimensions of sustainability (environmental, social, and economic). This book brings together different facets and approaches of sustainable production, as well as the impact of new technologies and different climate phenomena on the processes. In the different parts of this book, there is a comprehensive and orderly approach to current problems.

Starting with Part 1, a series of problems associated with sustainable development are presented, analyzing different issues, within which the social aspect stands out. A very valuable factor incorporated into the UN Sustainable Development Goals agenda is the consideration of the human factor as an intrinsic element of development. That is, sustainable development implies the development of a more just and inclusive society, with pillars of this development being the consideration of the need for work and the development of economic activities that do not undermine the development possibilities of future generations. Chapters 2 and 3 address the life cycle analysis of energy systems and how this analysis can provide useful information for decision makers. In this sense, energy studies and associated topics are central, as they are a transversal input to all productive sectors. Chapter 4 studies how new technologies and, particularly, the post-COVID-19 pandemic era affect the labor market. This chapter

Evolution and Trends of Sustainable Approaches
https://doi.org/10.1016/B978-0-443-21651-0.00014-0

studies the new characteristics of the labor market, where labor flexibility has permeated countless work activities that were unthinkable in previous stages. Likewise, these new flexibilities, in terms of industrial activities, represent new challenges in terms of human capital management. Another key aspect of achieving sustainable human development is the emergence of new economic systems that allow the development of regional economies, which must be integrated into a globalized and competitive market. For this purpose, Chapter 5 analyzes how small businesses with an environmentally friendly business model can emerge and develop as sources of work and economic drivers of regional economies. Within the chapter, different case studies of ventures are presented, and they study what were the main strengths of these ventures that allowed them to consolidate, as well as the main weaknesses in their growth.

In the second part of the book, operations management is analyzed in depth within the framework of sustainable development. This part presents a compendium of central themes to understand operations management as a tool to promote sustainable industrial development. These topics include maintenance management, product and obsolescence management, extended producer responsibility, and logistics operations. Regarding maintenance management, Chapter 6 presents a very interesting study on the approach to this management, where a paradigm is proposed that allows including the obsolescence of assets and spare parts, to ensure levels of operability. In Chapter 7, the obsolescence criterion is extended to other consumer products and not production assets; in this case, strategies are considered to implement a circular economy paradigm in the use of consumer products, considering a strategy of product as a service. To do this, the furniture within educational institutions is taken as a case study and a plan is developed to manage this furniture through a circular economy paradigm. Later, in Chapter 8, the concept of extended producer responsibility is introduced, where the producer must be responsible for the final impact that their products will have on the environment once their useful life has ended. In this particular case, the authors study strategies and models to manage the recovery of finished products, so that the correct disposal of these products is facilitated for the consumer. Chapter 9 analyzes the literature related to reverse logistics and remanufacturing processes associated with extended producer responsibility. This literature review presents a compendium and in-depth analysis of the literature, showing different models (mainly mathematical programming) that addressed different case studies and the particularities of each case. In Chapter 10, optimization tools for efficient energy management in very high-density energy systems are proposed. Finally, in Chapter 11, the last of Part 3 of the book, an innovative approach is addressed in the management of waste that has already taken the path of final disposal. These wastes, impossible to avoid in any human process, are important to address appropriately and efficiently, so that the management of operations, mainly logistics, has as little impact as possible.

In the final part of the book, Part 3, innovative and cutting-edge technologies that contribute to sustainable development are presented. Within these technologies, given the book's focus on operations management, digital technologies that allow information and operational orders to be managed in a more agile and direct way are of special interest. However, this part also incorporates studies in innovative energy management both at the basic technology and management levels. Chapter 12 analyzes how the digitalization of supply chains impacts in terms of sustainability, that is, how digital technologies contribute to improving the sustainable development of activities. As a particular case, they consider the construction supply chain. In Chapter 13, a more general study than the previous one is presented, and it analyzes how industrial logistics processes are enhanced by the incorporation of cyber-physical systems. These technologies, typical of Industry 4.0, allow physical systems to be integrated with digital systems directly, and for both systems to behave as a single system. Specifically, the chapter addresses a case study of the implementation of these technologies, and how they contribute to improving awareness about sustainable development. Chapter 14 presents a holistic approach to how digital technologies based on Industry 4.0 impact sustainable development, and how, through the adoption of these technologies, production processes can contribute to improving the sustainable development of society as a whole. Finally, the last Chapter 15 presents an interesting and innovative study in the energy sector considering hydrogen as a source and means of energy storage. This technology has acquired notable weight in recent years due to its potential zero impact on the environment, as it does not generate greenhouse gases. This chapter analyzes what the value chain of said technology is like and the management tools that are used, to indicate which are still the gaps that traditional value chain management tools fail to meet in desired levels of performance.

PART 1

Approaches to sustainable solutions

CHAPTER

2

Integrating life cycle sustainability assessment and multicriteria decision-making methods

João Gabriel Lassio[1,2], *David Castelo Branco*[1], *Alessandra Magrini*[1] *and Denise Matos*[1,2]

[1]Energy Planning Program, Federal University of Rio de Janeiro, Rio de Janeiro, Brazil; [2]Cepel - Electrical Energy Research Center, Rio de Janeiro, Brazil

1. Introduction

The sustainability paradigm has assumed a prominent role on the international agenda, primarily driven by the global climate crisis, increasing public awareness, and the most recent international agreements and commitments, such as the 2030 Agenda for Sustainable Development. This inaugural international consensus on addressing global challenges corresponds to an ambitious framework focused on 17 Sustainable Development Goals (SDGs) prioritizing people and the planet (Sanyé-Mengual and Sala, 2022; UN, 2015).

In this context, life cycle assessment (LCA) emerges as an interesting tool to actively support the environmental ambitions of the SDGs (Sanyé-Mengual and Sala, 2022). This technique falls under the umbrella of environmental assessment tools that aim to guide the latest policies and corporate decision making toward environmentally friendly products and practices. Only by a life cycle thinking lens is it possible to comprehensively observe the environmental implications associated with all stages of the value chain of produced goods and services. Hence, LCA facilitates recognizing and avoiding environmental tradeoffs among various geographical areas, life cycle stages, and impact categories (Moltesen and Bjørn, 2018; Owsianiak et al., 2018).

However, the single focus of LCA on the environmental dimension of sustainability neglects the economic and social aspects linked with the SDG agenda (UN, 2015). Consequently, it is imperative for life cycle thinking to acknowledge that sustainability encompasses more than just environmental considerations. It should also incorporate social issues, such as

Evolution and Trends of Sustainable Approaches
https://doi.org/10.1016/B978-0-443-21651-0.00003-6

human health, labor conditions, and community impacts, as well as economic concerns, including cost-effectiveness and resource efficiency (Zanni et al., 2020).

This implicit call for aligning LCA's scope with the multidimensional sustainability goals more effectively requires a broader outlook that also encompasses social and economic dimensions, such as the perspective offered by the life cycle sustainability assessment (LCSA). This comprehensive approach is often associated with the triple bottom line (TBL) concept (Elkington, 1998), providing a more holistic understanding of a product or service's overall sustainability performance. Accordingly, LCSA enhances decision making by offering a complete picture of the sustainability impacts and enabling stakeholders to identify burden-shifting and synergies between different sustainability dimensions (Guinée, 2016; Kloepffer, 2008; UNEP/SETAC Life Cycle Initiative, 2011).

Nevertheless, unlike LCA, LCSA still needs an internationally standardized or widely accepted methodological structure (Guinée, 2016; Kalbar and Das, 2020; Kloepffer, 2008). As a result, conducting an LCSA poses several methodological and practical challenges. Some of the most common concerns include the need for more input data, methodological proposals, and effective strategies for reporting results; the integration of environmental, social, and economic indicators to translate the product or service sustainability degree; and the subjectivity inherent to the sustainability concept and its qualitative metrics, especially in the social dimension (Guinée et al., 2011; Kalbar and Das, 2020; Lassio et al., 2021). In response to this situation, considerable efforts have been undertaken to enhance the LCSA framework, and some authors have advocated using multicriteria decision-making (MCDM) methods (Guinée et al., 2011; Kalbar and Das, 2020).

In light of this preamble, this chapter focuses on integrating MCDM methods and LCSA. Firstly, it discusses the challenges in applying LCSA and provides an overview of MCDM methods based on their theoretical references and practical studies in the literature. Then, it delves into using the analytic hierarchy process (AHP) and fuzzy logic in combination with LCSA, evaluating their effectiveness in addressing current LCSA obstacles and proposing a framework for their integration with LCSA.

2. Materials and methods

2.1 Life cycle sustainability assessment

Following the global debate about sustainable development, the LCA scope has expanded to encompass the social and economic domains, leading to the development of LCSA. While this comprehensive approach is suitable for integrating environmental, social, and economic aspects into sustainability-oriented decision-making processes, the need for a standardized and widely accepted methodology for its application remains (UNEP/SETAC Life Cycle Initiative, 2011).

One prominent resource currently available to guide the application of LCSA is the research project "Coordination Action for Innovation in Life Cycle Analysis for Sustainability" (CALCAS). CALCAS offers an integrated framework and roadmap for LCSA, guiding the incorporation of sustainability's environmental, social, and economic dimensions

(CALCAS, 2009; Guinée, 2016). Furthermore, a noteworthy collaboration between the United Nations Environment Program (UNEP) and the Society of Environmental Toxicology and Chemistry (SETAC) has resulted in a report titled "Toward a Life Cycle Sustainability Assessment." This publication illustrates the utilization and integration of LCA, life cycle costing (LCC), and social life cycle assessment (SLCA) (Eq. 2.1) as starting points for the LCSA application (UNEP/SETAC Life Cycle Initiative, 2011).

$$\mathrm{LCSA} = \mathrm{LCA} + \mathrm{SLCA} + \mathrm{LCC} \tag{2.1}$$

LCA is used to evaluate the environmental impacts associated with the entire life cycle of a product or service, starting from the extraction and processing of raw materials to waste recycling and final waste disposal. This environmental assessment tool adheres to standardized guidelines outlined in ISO 14040 (ISO, 2006) and ISO 14044 (ISO, 2006), which establish a four-step methodology for its application: (i) goal and scope definition, (ii) life cycle inventory (LCI), (iii) life cycle impact assessment (LCIA), and (iv) interpretation. Presently, various LCI databases and LCIA methods are available to analyze elemental flows and their corresponding environmental impacts across different categories. Ecoinvent (Wernet et al., 2016) stands out as the most comprehensive and widely employed LCI database (Bjørn et al., 2018). Additionally, notable LCIA methods include Eco-Indicator 99 (Goedkoop and Spriensma, 2000), ReCiPe 2016 (Huijbregts et al., 2017), and USEtox (Rosenbaum et al., 2008).

SLCA is a relatively new methodology compared to LCA; as such, its methodological development, practical implementation, and consensus are still in the early stages. This technique aims to evaluate the impacts of a product or service's life cycle on the well-being of stakeholders (UNEP, 2020). To guide practitioners in SLCA studies, UNEP/SETAC provides a framework that outlines key elements and offers specific guidance (UNEP/SETAC Life Cycle Initiative, 2009; UNEP, 2020). This international collaborative effort also defines both generic and specific social indicators and proposes potential data sources for their collection (UNEP, 2021). In the realm of SLCA studies, two central databases stand out: the Social Hotspot Database (SHDB) (Benoit-Norris et al., 2012) and the Product Social Impact Life Cycle Assessment (PSILCA) database (Ciroth and Eisfeldt, 2016). These databases compile pertinent social information and exclusively address social performance at the country or sector level (Kühnen and Hahn, 2017; Toniolo et al., 2020).

Finally, LCC is related to the first mention of the life cycle thinking at the end of the 1950s when Novick (1959) analyzed investments in military weaponry through a lifecycle-based method. Like LCA and SLCA, LCC enables the comprehensive evaluation of a product's costs throughout its life cycle, considering its interactions within the economic domain (Rödger et al., 2018). This allows for more informed decision making and supports identifying cost-saving opportunities (Hunkeler et al., 2008; Toniolo et al., 2020). However, even though the industry already uses it, it still lacks an agreed methodology consistent with the LCSA. What comes closest to this is the methodology provided by Hunkeler et al. (2008) that aligns LCC with the LCA framework, ensuring a cohesive analysis of economic and environmental aspects in lifecycle-based analysis.

To establish an overarching LCSA, it is imperative that the perspectives and aims of these three key techniques, namely LCA, SLCA, and LCC, are aligned (UNEP/SETAC Life Cycle Initiative, 2011). While the ISO 14040 (ISO, 2006) series predominantly focuses on environmental aspects and does not encompass social and economic concerns, it can serve as a valuable reference framework for implementing LCSA. Nevertheless, a growing body of literature highlights the challenges encountered in this process. In this context, two prominent issues arise: the integration of indicators and the subjectivity inherent to sustainability, as well as its qualitative metrics, especially social ones (Guinée, 2016). In light of this, the integration of LCSA with MCDM methods has emerged as an approach to overcome these challenges (Kalbar and Das, 2020).

2.2 Multicriteria decision making

2.2.1 Brief overview of MCDM methods in the sustainable context

In the decision-making process, it is common to consider various criteria that may potentially conflict with one another (Kalbar and Das, 2020). Within this context, MCDM methods are effective tools for dealing with this problem as they encompass many data, relationships, and objectives typically inherent in a particular real-world problem (Munda, 2005). These methods can be categorized into two main groups: (i) *multiattribute decision making* (MADM) and (ii) *multiobjective decision making* (MODM). MADM methods are employed when confronting decision-making problems that involve a finite number of alternatives (discrete variables). In contrast, MODM methods are specifically suited for problems encompassing an infinite range of alternatives (continuous variables) (Hwang and Yoon, 1981). It is worth noting that there are other classifications, including those based on different schools of thought, such as the French and American approaches.

To support sustainability-oriented decision making, MADM methods have gained prominence (Kalbar and Das, 2020; Kazimieras Zavadskas et al., 2019; Wang et al., 2009). Within this group, there are various MCDM methods. Selecting the most suitable method requires a thorough analysis of the convenience of adopting a compensatory approach (Gomes and Gomes, 2019; Munda, 2005). The reasoning behind this is that using a compensatory approach allows for formulating a measure of global merit for the analyzed alternatives. Conversely, a noncompensatory approach ranks the alternatives in relative terms without indicating their global merits. Compensatory approaches have gained preference due to the benefits of a single synthesis measure, such as simplifying systems and communicating complex information.

In this context, the AHP method is currently the most employed MADM method. On the other hand, fuzzy logic has been increasingly applied and shown a potential to support sustainability-oriented decision making (Martín-Gamboa et al., 2017; Wang et al., 2009; Wulf et al., 2019; Zanghelini et al., 2018). In both cases, decision makers must be aware of the burden-shifting (or trade-off) between different indicators due to the aggregation process. Note that it allows interchange between natural and anthropogenic capital, implying a weak sustainability approach. It is crucial to acknowledge that compensatory MADM methods may lead to imbalanced actions. While these actions may exhibit exceptional performance in specific areas, they may also encounter shortcomings in other aspects (Gomes and Gomes, 2019; Kalbar and Das, 2020).

2.2.2 *Analytic hierarchy process*

In 1990, Saaty (1990) devised the AHP to enable the consideration of both quantitative and qualitative data when evaluating criteria on an absolute scale. This methodology entails decomposing a complex decision-making scenario into smaller subproblems structured hierarchically (Fig. 2.1).

The AHP method application involves four main steps: (i) hierarchical structuring, (ii) pairwise comparison, (iii) priority weights derivation, and (iv) alternatives evaluation. Note that the first step is the most important in decision making. In this step, the decision maker defines the overall goal and organizes the problem concerning selecting the alternative that best meets the set of criteria in several hierarchical levels (Saaty, 1990).

As illustrated in Fig. 2.1, the simplest way to structure a decision problem consists of a three-level hierarchy. The first level is assigned to an element corresponding to the overall goal; the second level comprises the evaluation criteria; and the third includes the alternatives to be evaluated. Then, by performing pairwise comparisons at each level of this structure using a 1–9 scale, it becomes possible to obtain the ranking of alternatives (Saaty, 1990).

2.2.3 *Fuzzy logic*

To enhance the flexibility of associating elements with sets, Zadeh (1965) introduced the concept of the grade of membership (μ_A) for each element (x) with respect to a specific set (A). To this end, a membership function ($\mu_A(x)$) associates a real number from the interval [0,1] to an element (x) of the universe of discourse (U) (Gomes and Gomes, 2019).

In short, U corresponds to the space where set A's elements are defined, while the membership function ($\mu_A(x)$) determines the extent to which one of these elements belongs to the set. Thus, the function $\mu_A(x) : U \rightarrow [0, 1]$ associates each element of U with a membership degree (Gomes and Gomes, 2019; Ortega, 2001). A fuzzy set A can then be defined as an ordered set of pairs consisting of a generic element x_i ($i = 1, 2, 3...n$) and its corresponding grade of membership $\mu_A(x)$, according to Eq. (2.2) (Munda, 1995):

$$A = \{x_i, \mu_A(x_i) | x_i \in U\} \tag{2.2}$$

Every membership function must encompass a width referred to as the support set (S_A) (Eq. 2.3). This set includes the elements from the universe of discourse (U) that exhibit a nonzero degree of membership to the fuzzy set (Fig. 2.2) (Shepard, 2005).

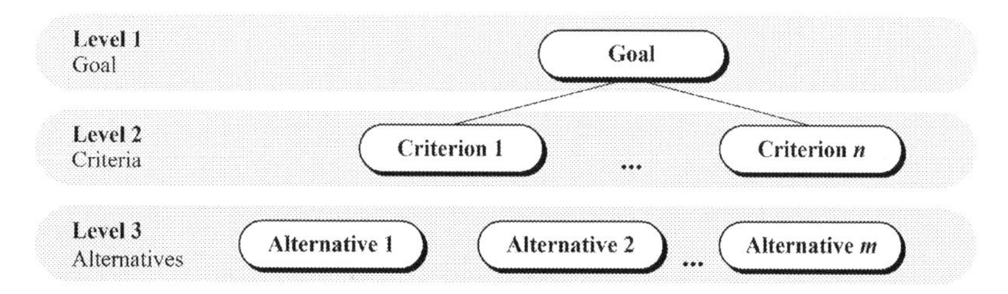

FIGURE 2.1 A generic hierarchy model of AHP. *Credit: Elaborated by the authors.*

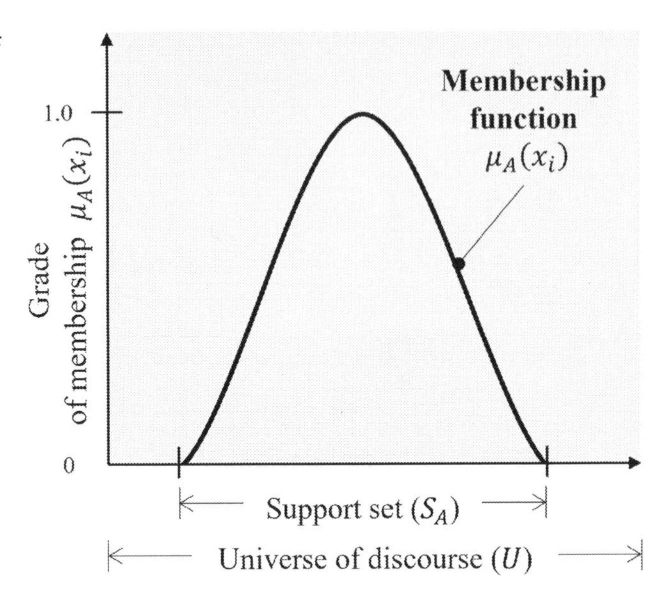

FIGURE 2.2 A generic fuzzy set. *Credit: Elaborated by the authors.*

$$S_A = \{x \in U | \mu_A(x) > 0\} \tag{2.3}$$

When using fuzzy logic, there are three primary steps to follow: (i) fuzzification, (ii) fuzzy inference, and (iii) defuzzification. In summary, the first step involves linking the input data (x) for a particular criterion to linguistic variables (or term sets) through membership functions ($\mu_A(x)$). In the subsequent step, these term sets are associated with a value range (S_A) of a fuzzy set (A), by establishing rules that rely on conditional IF-THEN propositions. Finally, in the third step, the resulting linguistic variables are converted into numerical values to obtain the output data.

3. Results and discussion

3.1 Literature review

3.1.1 Main LCSA challenges

Implementing life cycle-based analyses that go beyond the environmental dimension and include the social and economic dimensions has been challenging for researchers in this field. Numerous studies have reported the emergence of various obstacles when conducting LCSA. By giving greater recognition to LCSA complexity and delving into the difficulties of its application, Guinée (2016) draws attention to the lack of practical examples; the need to avoid double counting of issues addressed by the LCA, SLCA, and LCC; and the difficulty both in defining sustainability indicators and in communicating the results clearly and transparently. These challenges generally revolve around two central issues: the subjectivity inherent in sustainability and the integration of the LCA, SLCA, and LCC tools.

Concerning the subjectivity inherent in sustainability, it is important to emphasize that decision making guided by sustainability involves the consideration of multiple objectives, which are often conflicting and intangible. In addition, not all sustainability-related issues, especially social ones, can be measured quantitatively, implying the need to use qualitative metrics (Bitter et al., 2017). Such metrics generally rely on statements expressed in natural language that contain ambiguous and imprecise linguistic terms. For example, social acceptance of a given energy project is *low* since its interference in the landscape is *high*. Although the concepts of *low* and *high* are a consensus, the magnitude each one attributes to these terms is different. As a result, decision making guided by sustainability is subject to being modeled through vague and confusing information (Shepard, 2005).

One consequence of this situation is the challenge of selecting suitable and quantifiable sustainability indicators that align with a tool integrating diverse disciplines, such as LCSA. This challenge is particularly exacerbated when considering the social dimension, which encompasses subjective issues and indicators requiring access to confidential and restricted organizational information. In this context, the absence of a standardized and widely accepted methodological framework for conducting LCSA accentuates the risk of conducting unbalanced analyses. This risk primarily manifests through an underrepresentation of the social dimension and in the occurrence of eventually redundant indicators, which can lead to inaccurate and misleading conclusions.

In parallel, integrating and harmonizing the LCA, SLCA, and LCC results pose significant obstacles in applying the LCSA (Guinée, 2016; Kalbar and Das, 2020; Zanni et al., 2020). As these tools can consist of several and often conflicting criteria, it is difficult for decision makers to determine an alternative that meets or maximizes sustainability (Demirtas, 2013). Thus, rationally selecting alternatives will typically require the transformation of the various environmental, social, and economic indicators into a global sustainability index and other possible intermediate measures (Wang et al., 2009).

3.1.2 MCDM supporting LCSA application

As seen in the previous sections, performing an LCSA requires the systematic application of the trio LCA, SLCA, and LCC. However, integrating these tools and combining their results still need standardized approaches and consensus within the scientific literature. This poses challenges for researchers who have utilized MCDM methods to overcome them (Fig. 2.3). This section explores how the AHP method and Fuzzy Logic have provided ways to conduct previous LCSA studies, seeking to offer valuable insights into their application.

3.1.2.1 AHP

The AHP method has been used in LCSA studies to integrate the results of the LCIA step into more straightforward measures and when there is a need to consider the (often conflicting) preferences of various decision makers. Next, several studies employing the AHP are reviewed, emphasizing its application in addressing the complexity of decisions involving multiple criteria. The idea is to highlight its usefulness in supporting sustainability analyses based on the life cycle approach.

Sonetti and Lombardi (2020) integrate the LCA methodology with the AHP method to assess the sustainability performance of three types of building roofs. While the LCA

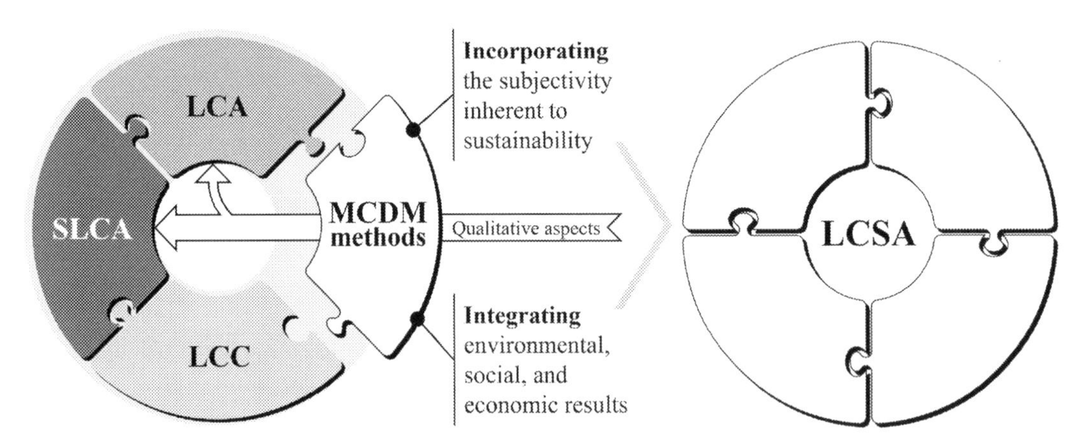

FIGURE 2.3 MCDM methods supporting LCSA application. *Credit: Elaborated by the authors.*

provides thermal performance and potential environmental impacts, the AHP method enables the assessment of roof types' social and aesthetic performance, as well as the attribution of weights to the criteria through interviews with designers.

Ghazvinei and colleagues (2017) propose combining the LCA methodology with the AHP method to select the most appropriate solid waste management system for a university campus. For this purpose, LCA is used to assess the potential environmental impacts of different solid waste management options (recycling, incineration, and landfill), taking into account environmental and economic criteria. At the same time, the AHP method is used to complement this assessment according to environmental and economic data based on expert opinion.

According to the literature, the most frequently studied subject is the evaluation of energy system aspects (Lindfors, 2021). Within this field, Karger and Hennings (2009) investigate the advantages and disadvantages of decentralized electricity generation in Germany. Their study encompasses four potential future electricity supply scenarios, evaluated based on a hierarchically structured set of sustainability criteria. Amer and Daim (2011) undertake a pioneering investigation into Pakistan's energy sector by employing the AHP method to assess various renewable energy options for electricity generation. Specifically, they explore the sustainability aspects of wind, solar photovoltaic, solar thermal, and biomass energy sources. In another study, Demirtas (2013) evaluates five electricity generation technologies in terms of environmental, social, economic, and technical dimensions. The author determines the selection criteria weights by pairwise comparison matrices of the AHP method. Ahmad and Tahar (2014) propose a methodology based on the AHP method to evaluate and rank different renewable sources for electricity generation. Finally, Corona (2016) suggests using the AHP method to weigh the multiple criteria for selecting the most sustainable concentrating solar power (CSP) technology.

3.1.2.2 Fuzzy logic

Concerning the field of sustainability, it is remarkable the space that Fuzzy Logic has been gaining within the framework of tools aimed at sustainability evaluation

(Andriantiatsaholiniaina et al., 2004; Phillis and Andriantiatsaholiniaina, 2001; Wang et al., 2009). In the lifecycle-based literature, the use of fuzzy logic has been focused on treating the outputs of the LCIA step. Afrinaldi and Zhang (2014) propose a methodology for carrying out the optional elements of normalization and aggregation within the LCIA step. In this approach, the performances according to the impact categories are normalized based on emission reduction targets. Then, they are grouped to obtain a global index. From the same perspective, Güereca and colleagues (2007) developed a method that links the results of the LCIA step with an inventory of emissions and sustainability goals, allowing them to rank different alternatives in order of preference.

Faced with the need to have not only in-depth environmental knowledge but also extremely precise data, González and colleagues (2002) provide a methodology that integrates fuzzy logic with LCA. Their main argument is to make the application of LCA more accessible in small and medium-sized companies. From the ecodesign point of view, Herva and colleagues (2012) propose an integration method based on fuzzy logic that incorporates three environmental assessment methodologies: LCA, ecological footprint (EF), and environmental risk assessment (ERA). The study's objective is to obtain a final index to classify different product options from an environmental perspective named fuzzy ecodesign (FEcoDI). The method is applied in a case study comparing two types of plastic water bottles produced from polyethylene terephthalate (PET) and polyvinyl chloride (PVC).

In energy sector applications, Kouloumpis and Azapagic (2018) propose a model that integrates the LCA, SLCA, and LCC with fuzzy logic to handle inaccurate information. In another study, Hemdi and colleagues (2013) provide a method to support sustainability-oriented decision making, enabling the modeling of complex systems with dynamic behavior and inputs that are not well defined. Concerning renewable energy sources, Bitter and colleagues (2017) assess a wind farm located in Germany using a set of 24 indicators distributed across seven dimensions of sustainability. For this purpose, the authors use fuzzy logic to integrate these indicators, resulting in dimension-specific indices and an overall sustainability index.

3.2 Proposal for LCSA integration

3.2.1 AHP

In LCSA studies, the AHP method can help prioritize and evaluate alternatives based on multiple criteria and their relative importance. In this process, defining the specific goal of the LCSA that will guide the AHP application is the first step. The LCSA goal generally is selecting the most sustainable alternative for a product, service, technology, etc. Then, the key criteria relevant to the LCSA goal must be identified. Such criteria should capture the various desirable dimensions of sustainability. LCSA studies typically follow the TBL concept (Elkington, 1998) and consider environmental, social, and economic dimensions.

With this information, LCSA practitioners can build the AHP hierarchical structure, organizing the goal, criteria, and subcriteria. Its first level should include the overall LCSA goal. For example, "selecting the most sustainable product or service." The second level may encompass the three sustainability dimensions: environmental, social, and economic. Then, each criterion is subdivided into subcriteria in the third level of this hierarchical structure.

It means that each sustainability dimension is divided into its respective issues. For example, the environmental dimension may contain the following issues: global warming, land use, water consumption, and so on. Finally, the fourth level relates to the alternatives analyzed. Fig. 2.4 illustrates an AHP hierarchy structure that can be used in LCSA studies. Note that it visually represents the relationships between different elements.

Following the guidelines presented earlier, the AHP application in LCSA studies proceeds with pairwise comparisons, priority weights derivation, and alternatives evaluation. Since this method requires dimensionless input parameters, normalizing procedures are mandatory for its application. In this context, internal normalization is the most suitable approach to transform LCA, SLCA, and LCC results (expressed in different units) into a dimensionless scale (Pizzol et al., 2017).

Once the dimensionless input parameters are obtained, the alternatives can be evaluated against each criterion based on their performance. In each sustainability subcriterion (level 3), the normalized alternatives' performances are multiplied by the corresponding subcriterion weight to obtain weighted scores. Then, these weighted scores of subcriteria belonging to the same sustainability dimension are aggregated to determine the alternatives' performances on the environmental, social, and economic dimensions (level 2). Finally, these performances are also multiplied by the corresponding dimensions weights and then aggregated to obtain each alternative's overall sustainability performance or preference (level 1) (see Fig. 2.4). Both sets of weights are obtained through pairwise comparisons based on expert judgments.

3.2.2 *Fuzzy logic*

Similar to the AHP method, defining the sustainability criteria of LCSA is one of the first steps in combining it with fuzzy logic. Since each sustainability criterion (m) is particular and has a unique scale, numeric range, and subcategories that hold value, it is crucial to define its measurement unit and universe of discourse (U_M). The latter is central to the fuzzy inference engine since it corresponds to the valid range of input values (Shepard, 2005).

To effectively convey sustainability criteria, establishing linguistic variables (or term sets) is imperative. As previously stated, *low*, *medium*, and *high* can be employed to generally describe the level of environmental, social, and economic impacts. For each one of these

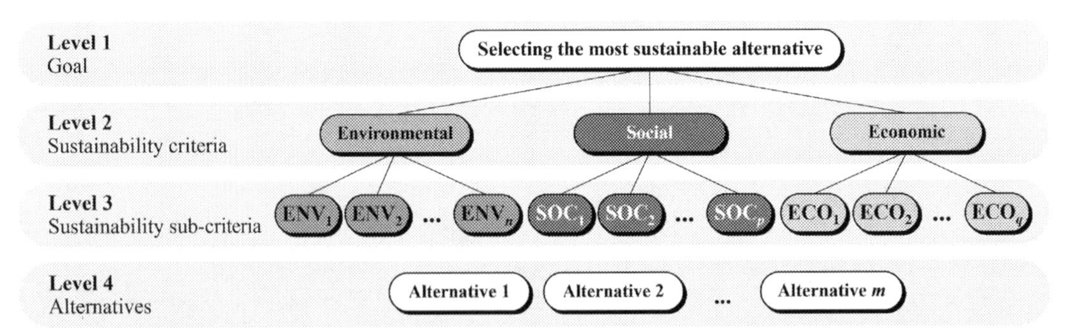

FIGURE 2.4 Illustration of the hierarchy model of AHP for selecting the most sustainable alternative. *Credit: Elaborated by the authors.*

linguistic variables, a range of values (S_M) within which impacts magnitude can be included should be defined. For instance, when evaluating the local job creation subcriterion, a *medium* rating might cover a range of values ($S_{M_{medium}}$) between 50 and 100 jobs.year. Consulting with current regulations and conducting in-depth interviews with stakeholders and experts in sustainability and relevant fields to the LCSA may be helpful in completing this task.

Term sets usually comprise three to seven fuzzy sets that are well defined and commonly overlap by 50% in the values they encompass (Shepard, 2005). Note that the number of fuzzy sets must align with the quality of the input data and their importance in the assessment model. Moreover, when comparing alternatives (t) against a specific subcriterion (m), their performances (x_{tm}) can be used to define U_m. However, if only two alternatives are being considered in LCSA, this procedure loses its meaning because each of the alternatives will assume an extreme position (maximum and minimum) in U_M.

Establishing membership functions ($\mu_M(x)$) bears significant importance in determining the degree of association (μ_M) between an element (x) and a fuzzy set. This pivotal step forms an integral part of the process of transforming numerical values into linguistic terms. To put it differently, a membership function ($\mu_{M_{low}}(x_{tm})$) can describe the degree of membership of a specific project's environmental impact within the *low* fuzzy set (Fig. 2.5).

Next, a rule base that relates the fuzzy sets and linguistic variables must be delineated to evaluate the alternatives' environmental, social, and economic performances according to their results obtained in sustainability subcriteria. Rules are typically in the IF-THEN form. For example, "If the terrestrial ecotoxicity impact is *low* and water consumption is *high*, then the environmental rating is *good*." Once again, the insights of experts on the matter are highly valuable. After establishing all the rule bases, input values (x_{tm}), membership

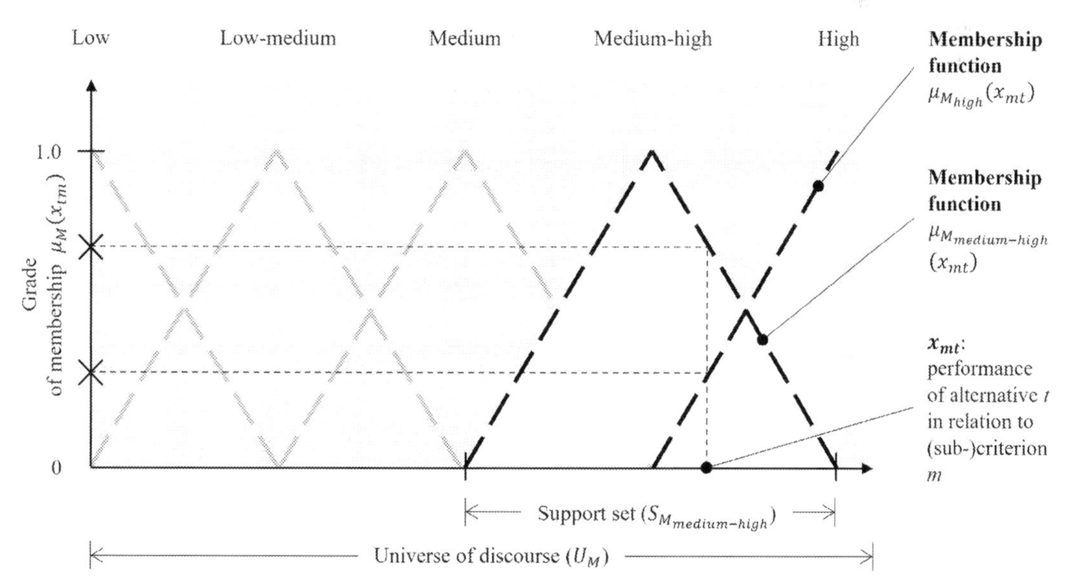

FIGURE 2.5 Illustration of triangular fuzzy sets for determining the significance of a specific project's impact on a sustainability criterion. *Credit: Elaborated by the authors.*

functions ($\mu_M(x)$), and rule bases can be combined to derive an output. Finally, it is necessary to convert this fuzzy output into a crisp value that can be easily interpreted.

To achieve an overall sustainability assessment outcome, following identical procedures for the environmental, social, and economic dimensions is necessary. This means that the same process must be used to convert the dimension criteria into the sustainability index. These outcomes can then be utilized to compare and rank various alternatives according to their sustainability performance in LCSA.

4. Conclusions

Considering that the absence of a standardized methodological framework for LCSA poses notable challenges to its practitioners, this chapter explored the integration of LCSA with MCDM methods, specifically focusing on the AHP method and Fuzzy Logic. The central idea was to demonstrate the usefulness of these two relevant multicriteria tools in conducting LCSA studies based on the existing literature.

In this sense, the main challenges faced by practitioners were initially mapped out as the subjective nature of certain sustainability-related issues, particularly those pertaining to the social dimension, and the integration of the results of LCA, SLCA, and LCC. Furthermore, a thorough discussion was conducted to gain a deeper understanding of these challenges and to identify potential solutions for overcoming them. On the one hand, the inherent subjectivity of sustainability and the numerous multidimensional results may not lead to conclusions and make decision making difficult. On the other hand, employing MCDM methods in combination with LCSA can effectively address these challenges and facilitate decision-making processes oriented toward sustainability.

Our findings reveal that the AHP method has often been used to define weights for aggregation criteria according to the opinions of experts and interested parties. On the other hand, fuzzy logic is compatible with standards of socioenvironmental quality already established. It is also more suitable for situations where sustainability issues cannot a priori be quantified.

While AHP and fuzzy logic are valuable tools in the realm of MCDM, they are merely two of many available methods, and their usage should be adapted to best fit the specific needs and context of LCSA. Furthermore, it is crucial to recognize that the responsibility to make the final decisions lies with individuals, and these methods do not absolve them of that responsibility. Lastly, it is necessary to explore other relevant tools to determine their potential to support LCSA studies.

References

Afrinaldi, F., Zhang, H.C., 2014. A fuzzy logic based aggregation method for life cycle impact assessment. Journal of Cleaner Production 67, 159–172. https://doi.org/10.1016/j.jclepro.2013.12.010.

Ahmad, S., Tahar, R.M., 2014. Selection of renewable energy sources for sustainable development of electricity generation system using analytic hierarchy process: a case of Malaysia. Renewable Energy 63, 458–466. https://doi.org/10.1016/j.renene.2013.10.001.

Amer, M., Daim, T.U., 2011. Selection of renewable energy technologies for a developing county: a case of Pakistan. Energy for Sustainable Development 15, 420–435. https://doi.org/10.1016/j.esd.2011.09.001.

Andriantiatsaholiniaina, L.A., Kouikoglou, V.S., Phillis, Y.A., 2004. Evaluating strategies for sustainable development: fuzzy logic reasoning and sensitivity analysis. Ecological Economics 48. https://doi.org/10.1016/j.ecolecon.2003.08.009.

Benoit-Norris, C., Cavan, D.A., Norris, G., 2012. Identifying social impacts in product supply chains: overview and application of the social hotspot database. Sustainability 4, 1946–1965. https://doi.org/10.3390/su4091946.

Bitter, J., Printz, S., Lahl, K., Vossen, R., Jeschke, S., 2017. Fuzzy Logic Approach for Sustainability Assessment Based on the Integrative Sustainability Triangle—An Application for a Wind Power Plant, pp. 50–61. https://doi.org/10.14621/ce.20170206 vol. 3.

Bjørn, A., Moltesen, A., Laurent, A., Owsianiak, M., Corona, A., Birkved, M., Hauschild, M.Z., 2018. Life cycle inventory analysis. In: Life Cycle Assessment. Springer International Publishing, Cham, pp. 117–165. https://doi.org/10.1007/978-3-319-56475-3_9.

CALCAS, 2009. D20 Blue Paper on Life Cycle Sustainability Analysis, Co-ordination Action for Innovation in Life-Cycle Analysis for Sustainability.

Ciroth, A., Eisfeldt, F., 2016. PSILCA—A Product Social Impact Life Cycle Assessment Database, Database Version 1.0.

Corona, B., 2016. Análisis de sostenibilidad del ciclo de vida de una configuración innovadora de tecnología termosolar. Universidad Politécnica de Madrid.

Demirtas, O., 2013. Evaluating the best renewable energy technology for sustainable energy planning. International Journal of Energy Economics and Policy 3, 23–33.

Elkington, J., 1998. Cannibals with Forks: The Triple Bottom Line of Sustainability. New Soc. Publ.

Ghazvinei, P., Mir, M., Darvishi, H., Ariffin, J., 2017. Solid Waste—Management Models. https://doi.org/10.1007/978-3-319-43228-1_3.

Goedkoop, M., Spriensma, R., 2000. The Eco-Indicator 99—A Damage Oriented Method for Life Cycle Assessment, Methodology Report. PRé Consultants.

Gomes, L.F.A.M., Gomes, C.F.S., 2019. Princípios e métodos para tomada de decisão: Enfoque multicritério, 6a Edição. ed. Atlas, São Paulo.

González, B., Adenso-Díaz, B., González-Torre, P.L., 2002. A fuzzy logic approach for the impact assessment in LCA. Resources, Conservation and Recycling 37. https://doi.org/10.1016/S0921-3449(02)00069-1.

Güereca, L.P., Agell, N., Gassó, S., Baldasano, J.M., 2007. Fuzzy approach to life cycle impact assessment: an application for biowaste management systems. International Journal of Life Cycle Assessment 12. https://doi.org/10.1065/lca2006.10.276.

Guinée, J.B., 2016. Life cycle sustainability assessment: what is it and what are its challenges? In: Clift, R., Druckman, A. (Eds.), Taking Stock of Industrial Ecology. Springer, Cham, pp. 45–68. https://doi.org/10.1007/978-3-319-20571-7_3.

Guinée, J.B., Heijungs, R., Huppes, G., Zamagni, A., Masoni, P., Buonamici, R., Ekvall, T., Rydberg, T., 2011. Life cycle assessment: past, present, and future. Environmental Science and Technology 45 (1), 90–96. https://doi.org/10.1021/es101316v.

Hemdi, A.R., Saman, M.Z.M., Sharif, S., 2013. Sustainability evaluation using fuzzy inference methods. International Journal of Sustainable Energy 32, 169–185. https://doi.org/10.1080/14786451.2011.605947.

Herva, M., Franco-Uría, A., Carrasco, E.F., Roca, E., 2012. Application of fuzzy logic for the integration of environmental criteria in ecodesign. Expert Systems with Applications 39, 4427–4431. https://doi.org/10.1016/j.eswa.2011.09.148.

Huijbregts, M., Steinmann, Z., Elshout, P., Stam, G., Verones, F., Vieira, M., Zijp, M., Hollander, A., van Zelm, R., 2017. ReCiPe 2016: a harmonised LCIA method at midpoint and endpoint level. International Journal of Life Cycle Assessment 22, 138–147. https://doi.org/10.1007/s11367-016-1246-y.

Hunkeler, D., Lichtenvort, K., Rebitzer, G., 2008. Environmental Life Cycle Costing, Environmental Life Cycle Costing. CRC Press, Boca Raton. https://doi.org/10.1201/9781420054736.

Hwang, C.-L., Yoon, K., 1981. Methods for Multiple Attribute Decision Making. Multiple Attribute Decision Making, 186. Springer, Berlin, Heidelberg, pp. 58–191.

ISO, 2006. Environmental management - Life Cycle Assessement - Requirements and Guidelines. International Organization for Standardization. ISO 14044.

ISO, 2006. Environmental Management - Life Cycle Assessment - Principles and Framework. International Organization for Standardization. ISO 14040.

Kalbar, P., Das, D., 2020. Advancing life cycle sustainability assessment using multiple criteria decision making. In: Life Cycle Sustainability Assessment for Decision-Making. https://doi.org/10.1016/b978-0-12-818355-7.00010-5.
Karger, C.R., Hennings, W., 2009. Sustainability evaluation of decentralized electricity generation. Renewable and Sustainable Energy Reviews 13 (3), 583–593. https://doi.org/10.1016/j.rser.2007.11.003.
Kazimieras Zavadskas, E., Antucheviciene, J., Kar, S., 2019. Multi-objective and multi-attribute optimization for sustainable development decision aiding. Sustainability 11, 3069. https://doi.org/10.3390/su11113069.
Kloepffer, W., 2008. Life cycle sustainability assessment of products (with Comments by Helias A. Udo de Haes, p. 95). In: International Journal of Life Cycle Assessment. https://doi.org/10.1065/lca2008.02.376.
Kouloumpis, V., Azapagic, A., 2018. Integrated life cycle sustainability assessment using fuzzy inference: a novel FELICITA model. Sustainable Production and Consumption 15, 25–34. https://doi.org/10.1016/j.spc.2018.03.002.
Kühnen, M., Hahn, R., 2017. Indicators in social life cycle assessment: a review of frameworks, theories, and empirical experience. Journal of Industrial Ecology 21 (6), 1547–1565. https://doi.org/10.1111/jiec.12663.
Lassio, J.G., Magrini, A., Castelo Branco, D., 2021. Life cycle-based sustainability indicators for electricity generation: a systematic review and a proposal for assessments in Brazil. Journal of Cleaner Production 311, 127568. https://doi.org/10.1016/j.jclepro.2021.127568.
Lindfors, A., 2021. Assessing sustainability with multi-criteria methods: a methodologically focused literature review. Environmental and Sustainability Indicators 12, 100149. https://doi.org/10.1016/j.indic.2021.100149.
Martín-Gamboa, M., Iribarren, D., García-Gusano, D., Dufour, J., 2017. A review of life-cycle approaches coupled with data envelopment analysis within multi-criteria decision analysis for sustainability assessment of energy systems. Journal of Cleaner Production 150, 164–174. https://doi.org/10.1016/j.jclepro.2017.03.017.
Moltesen, A., Bjørn, A., 2018. LCA and sustainability. In: Life Cycle Assessment. Springer International Publishing, Cham, pp. 43–55. https://doi.org/10.1007/978-3-319-56475-3_5.
Munda, G., 2005. "Measuring sustainability": a multi-criterion framework. Environment, Development and Sustainability 7, 117–134. https://doi.org/10.1007/s10668-003-4713-0.
Munda, G., 1995. Multicriteria evaluation in a fuzzy environment theory and applications in ecological economics. Contributions to Economics 34.
Novick, D., 1959. The Federal Budget as an Indicator of Government Intentions and the Implications of Intentions. RAND Corporation, Santa Monica, CA.
Ortega, N.R. de S., 2001. Aplicação da teoria de conjuntos Fuzzy a problemas da biomedicina. Universidade de São Paulo, São Paulo. https://doi.org/10.11606/T.43.2001.tde-04122013-133237.
Owsianiak, M., Bjørn, A., Laurent, A., Molin, C., Ryberg, M.W., 2018. LCA applications. In: Life Cycle Assessment. Springer International Publishing, Cham, pp. 31–41. https://doi.org/10.1007/978-3-319-56475-3_4.
Phillis, Y., Andriantiatsaholiniaina, L., 2001. Sustainability: an ill-defined concept and its assessment using fuzzy logic. Ecological Economics 37, 435–456. https://doi.org/10.1016/S0921-8009(00)00290-1.
Pizzol, M., Laurent, A., Sala, S., Weidema, B., Verones, F., Koffler, C., 2017. Normalisation and weighting in life cycle assessment: quo vadis? International Journal of Life Cycle Assessment 22, 853–866. https://doi.org/10.1007/s11367-016-1199-1.
Rödger, J.-M., Kjær, L.L., Pagoropoulos, A., 2018. Life cycle costing: an introduction. In: Hauschild, H., Rosenbaum, R., Olsen, S. (Eds.), Life Cycle Assessment. Springer International Publishing, Cham, pp. 373–399. https://doi.org/10.1007/978-3-319-56475-3_15.
Rosenbaum, R.K., Bachmann, T.M., Gold, L.S., Huijbregts, M.A.J., Jolliet, O., Juraske, R., Koehler, A., Larsen, H.F., MacLeod, M., Margni, M., McKone, T.E., Payet, J., Schuhmacher, M., van de Meent, D., Hauschild, M.Z., 2008. USEtox—the UNEP-SETAC toxicity model: recommended characterisation factors for human toxicity and freshwater ecotoxicity in life cycle impact assessment. International Journal of Life Cycle Assessment 13, 532–546. https://doi.org/10.1007/s11367-008-0038-4.
Saaty, T.L., 1990. How to make a decision: the analytic hierarchy process. European Journal of Operational Research 48 (1), 9–26. https://doi.org/10.1016/0377-2217(90)90057-I.
Sanyé-Mengual, E., Sala, S., 2022. Life cycle assessment support to environmental ambitions of EU policies and the sustainable development goals. Integrated Environmental Assessment and Management 18, 1221–1232. https://doi.org/10.1002/ieam.4586.
Shepard, R.B. (Ed.), 2005. Quantifying Environmental Impact Assessments Using Fuzzy Logic. Springer Series On Environmental Management. Springer, New York. https://doi.org/10.1007/0-387-28098-7.

Sonetti, G., Lombardi, P., 2020. Multi-criteria decision analysis of a building element integrating energy use, environmental, economic and aesthetic parameters in its life cycle. Green Energy and Technology. https://doi.org/10.1007/978-3-030-23786-8_26.

Toniolo, S., Tosato, R.C., Gambaro, F., Ren, J., 2020. Life cycle thinking tools: life cycle assessment, life cycle costing and social life cycle assessment. In: Life Cycle Sustainability Assessment for Decision-Making. Elsevier, pp. 39–56. https://doi.org/10.1016/B978-0-12-818355-7.00003-8.

UN, 2015. Transforming Our World: The 2030 Agenda for Sustainable Development United Nations United Nations Transforming Our World: the 2030 Agenda for Sustainable Development. A/RES/70/1. United Nations.

UNEP/SETAC Life Cycle Initiative, 2011. Towards Life Cycle Sustainability Assessment: Making Informed Choices on Products.

UNEP, 2021. The Methodological Sheets for Subcategories in Social Life Cycle Assessment (S-LCA). United Nations Environment Programme (UNEP).

UNEP/SETAC Life Cycle Initiative, 2009. Guidelines for Social Life Cycle Assessment of Products, Management.

UNEP, 2020. Guidelines for Social Life Cycle Assessment of Products and Organizations 2020.

Wang, J.-J., Jing, Y.-Y., Zhang, C.-F., Zhao, J.-H., 2009. Review on multi-criteria decision analysis aid in sustainable energy decision-making. Renewable and Sustainable Energy Reviews 13, 2263–2278. https://doi.org/10.1016/j.rser.2009.06.021.

Wernet, G., Bauer, C., Steubing, B., Reinhard, J., Moreno-Ruiz, E., Weidema, B., 2016. The ecoinvent database version 3 (part I): overview and methodology. International Journal of Life Cycle Assessment 21, 1218–1230. https://doi.org/10.1007/s11367-016-1087-8.

Wulf, C., Werker, J., Ball, C., Zapp, P., Kuckshinrichs, W., 2019. Review of sustainability assessment approaches based on life cycles. Sustainability 11, 5717. https://doi.org/10.3390/su11205717.

Zadeh, L.A., 1965. Fuzzy sets. Information and Control 8 (3), 338–353. https://doi.org/10.1016/S0019-9958(65)90241-X.

Zanghelini, G.M., Cherubini, E., Soares, S.R., 2018. How multi-criteria decision analysis (MCDA) is aiding life cycle assessment (LCA) in results interpretation. Journal of Cleaner Production 172, 609–622. https://doi.org/10.1016/j.jclepro.2017.10.230.

Zanni, S., Awere, E., Bonoli, A., 2020. Life cycle sustainability assessment: an ongoing journey. In: Life Cycle Sustainability Assessment for Decision-Making. https://doi.org/10.1016/b978-0-12-818355-7.00004-x.

CHAPTER

3

Life cycle sustainability assessment of concentrating solar power technologies: A comparison between solar tower power and thermal parabolic trough generations in Brazil

João Gabriel Lassio[1,2], Denise Matos[1,2], Cibelle Trama[1,2], Amaro Olímpio Pereira, Jr.[1] and David Castelo Branco[1]

[1]Energy Planning Program, Federal University of Rio de Janeiro, Rio de Janeiro, Brazil; [2]Cepel - Electrical Energy Research Center, Rio de Janeiro, Brazil

1. Introduction

As stated in the Intergovernmental Panel on Climate Change's (IPCC) Sixth Assessment Report (AR6) (IPCC, 2021), "Climate change is a threat to human well-being and planetary health"—an affirmative statement with a very high level of confidence. According to the same report, "Global greenhouse gas (GHG) emissions in 2030 implied by nationally determined contributions (NDCs) announced by October 2021 make it likely that warming will exceed 1.5°C during the 21st century and make it harder to limit warming below 2°C" (IPCC, 2021).

The rise in Earth's temperature causes more extreme weather events, such as heatwaves, heavy precipitation, droughts, and tropical cyclones. These events have negative impacts on food security, human health, and biodiversity. To remain between 1.5 and 2.0°C level

https://doi.org/10.1016/B978-0-443-21651-0.00008-5

of temperature rise, a profound, rapid, and sustained reduction in GHG emissions is needed. That is why it is being called nowadays a climatic emergency (IPCC, 2021).

According to the IPCC AR6 (IPCC, 2021), energy transition entails mitigation pathways encompassing a change from fossil fuels, without carbon capture and storage, to very low or zero carbon energy sources, demand-side measures, and efficiency improvement. Increased electrification on the demand side completes the scenario in which humanity limits global warming to 2°C or less since net zero GHG emissions are reached by 2050. In this context, generating electricity from renewable energy sources is strategically important. Since the diversity of specific situations demands a wide range of available routes, all generating technologies should be considered, taking into account parameters such as flexibility, dispatchability, and seasonality (IEA, 2022; IPCC, 2021).

The Brazilian electric system comprises a National Interconnected System (SIN in the Portuguese acronym), a large-scale system until recently characterized as hydrothermal, with multiannual regularization. Nevertheless, over the last 20 years, the system's regularization capacity has diminished, wind power expansion has become highly intense, and the system is now described as a hydro-thermal-wind system, with decreasing regularization while increasing the participation of intermittent sources in its electricity mix (Brazil, 2021, 2022).

The installed capacity of wind power has shown significant growth within the Brazilian electricity mix, increasing from 0.25% in 2006 to 11.4% in 2021, surpassing the 20 GW mark (Brazil, 2022). After 2010, the expansion of wind power gained momentum and continues to outpace the country's average rate of installed capacity expansion. Solar energy using solar photovoltaic (PV) panels has also experienced a notable rise in the national electricity mix in the late 2010s. According to the Brazilian Statistical Yearbook of Electricity publications (Brazil, 2011, 2015, 2023), no centralized solar power plants were recorded in 2009 (Brazil, 2022). However, by 2021, solar energy accounted for 2.55% of the total installed capacity, presenting more than 4 GW.

As the penetration of renewable energy increases, challenges related to intermittency and variability emerge. During periods of high energy demand, thermal power plants play a crucial role in supporting the system. They provide stability and ensure a reliable power supply, compensating for the irregularities in renewable energy sources (Trama et al., 2021).

A renewable alternative that can provide dispatchability and flexibility to the system is concentrated solar power (CSP) generation. By concentrating solar energy onto a receiver to generate electricity, CSP technology provides cleaner energy than fossil fuels and offers the advantage of energy storage capabilities. As shown by Trama and colleagues (Trama et al., 2021), introducing CSP plants on an hourly price-setting model of the Brazilian hydro-thermal-wind system allowed a cost reduction in operating the system, replacement of the thermal generators, and complementarity with intermittent sources.

However, although there is an advantage in replacing thermal generation due to reduced GHG emissions, the sustainability concept is gathering momentum across various sectors of society, encouraging the consideration of socioenvironmental issues in its decision-making process. Regarding this matter, though CSP can be considered carbon-free in its operation stage, a broader evaluation of its impacts and benefits should consider its complete life cycle and the three dimensions of sustainability: environmental, social, and economic criteria.

Therefore, this chapter aims to conduct a life cycle sustainability assessment (LCSA) to analyze and compare the sustainability performance of two CSP technologies in the Brazilian context: (i) solar tower power (STP) with a thermal energy storage (TES) of 6 h; (ii) solar thermal parabolic trough (STPT) with a TES of 6 h. This analysis can provide valuable insights into the advantages and disadvantages of utilizing CSP as a viable alternative for electricity generation in Brazil, leading to more informed decisions in its energy sector.

2. Materials and methods

2.1 Concentrating solar power technologies

CSP generation, also referred to as concentrated solar thermal, is a form of harnessing solar energy using concentration technologies to generate electricity on a large scale. Various CSP generation systems exist, including STPT, STP, Fresnel, and parabolic dishes. While each system possesses unique characteristics, they all share a common approach of utilizing solar concentration to heat a fluid, which can be used for electricity generation. STPT and STP are particularly suitable for centralized power generation among the mentioned technologies. Currently, STPT presents the highest number of commercial plants in operation. These systems offer efficient and effective means of harnessing solar energy on a large scale, contributing to generating electricity from renewable sources (GIZ/MCTI, 2014a).

Several requirements must be considered when selecting a location for installing a CSP plant. These include abundant solar resources, with high direct solar irradiation levels exceeding 1800 kWh/m^2/year. Additionally, the availability of extensive land with suitable topographic conditions, low wind speeds, and access to water (primarily for cooling purposes in the case of wet cooling systems) are essential factors to consider. Moreover, the chosen location should have reasonable access infrastructure and be easily connected to the power grid (GIZ/MCTI, 2014a; Vieira et al., 2018).

One notable advantage of CSP technology is its capability to store thermal energy for later use. This feature allows excess heat generated during periods of high solar irradiation to be stored in a thermal medium, such as molten salts. Consequently, even when the sun is not present, such as at night or on cloudy days, the stored thermal energy can be utilized to continue generating electricity. This storage capability enhances the reliability and flexibility of CSP plants, enabling them to provide power beyond daylight hours (GIZ/MCTI, 2014b).

STPT technology consists of curved parabolic mirrors arranged in rows that track the sun's movement throughout the day. These mirrors reflect and concentrate sunlight onto a receiver at each parabola's focal point. The receiver is typically a metallic pipe called a trough containing a heat transfer fluid, such as thermal oil. The fluid is heated to high temperatures when sunlight reflects and concentrates on the parabolic trough. This heat is then transferred to a pressurized water circuit, generating steam. The steam powers a turbine connected to an electric generator, which produces electricity (Fig. 3.1) (GIZ/MCTI, 2014b).

STP technology features a tall central tower that concentrates solar radiation captured by a series of heliostat mirrors positioned around it. These mirrors are controlled to reflect and concentrate sunlight onto a single point at the top of the tower, known as the solar receiver. The solar receiver receives concentrated solar radiation from the heliostat mirrors. The receiver comprises a particle flow receiver (usually a thermal fluid) or an air receiver. The receiver's purpose is to absorb the heat from concentrated solar radiation (GIZ/MCTI, 2014c).

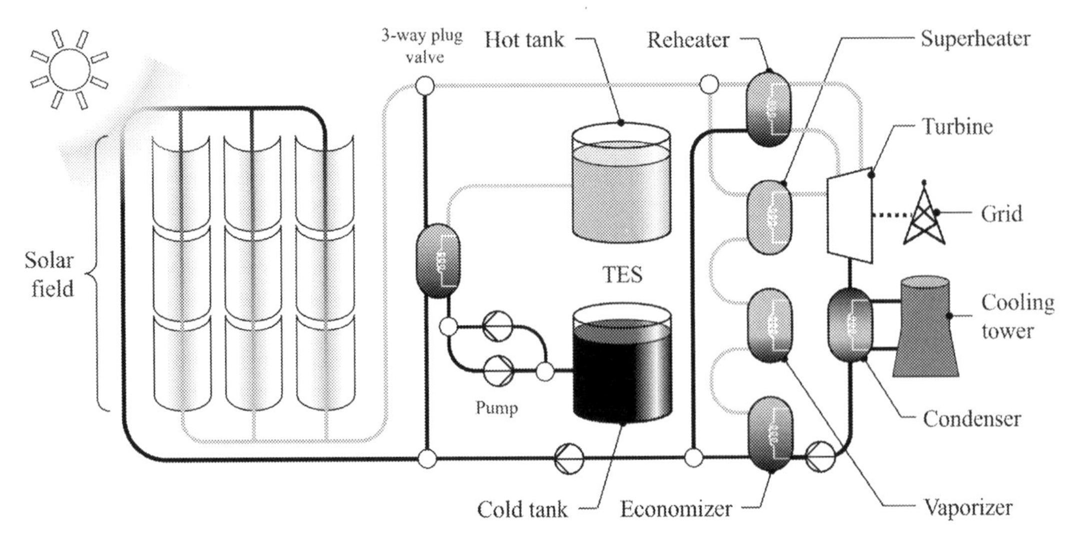

FIGURE 3.1 Scheme of an STPT power plant. *Credit: Elaborated by the authors, based on GIZ/MCTI, 2014a. Basic Mapping of General Pre-conditions for CSP Technologies in Brazil; GIZ/MCTI, 2014b. Parabolic Trough CSP Technology—State of the Art and Market Overview; GIZ/MCTI, 2014c. Tower CSP Technology—State of Art and Market Review.*

The heat absorbed by the solar receiver is transferred to a heat transfer medium such as thermal oil, molten salts, or heated air. This heat transfer medium circulates through a closed circuit and is responsible for storing and transporting the captured thermal energy, which is then conducted to a boiler or heat exchanger. The transferred heat converts water into steam (in the case of a steam cycle) or increases the temperature of the air (in the case of an air cycle). The high-pressure steam or air is directed toward a turbine, which generates electricity. The cooled steam or air is then condensed and recirculated to the boiler or heat exchanger (Fig. 3.2) (GIZ/MCTI, 2014c).

2.2 Life cycle sustainability assessment

LCSA has emerged as a response to the global discussions on sustainability, broadening the scope of life cycle assessment (LCA) to include social and economic aspects. This more holistic approach is suited for introducing environmental, social, and economic considerations into decision-making processes focused on sustainability. However, a standardized and widely accepted methodology is still needed to apply LCSA (UNEP/SETAC Life Cycle Initiative, 2011).

In this context, the research project "Coordination Action for Innovation in Life Cycle Analysis for Sustainability" (CALCAS) and the collaboration between the United Nations Environment Program (UNEP) and the Society of Environmental Toxicology and Chemistry (SETAC) provide valuable resources for implementing LCSA (Guinée, 2016). On the one hand, CALCAS offers an integrated framework and roadmap to guide the incorporation of environmental, social, and economic dimensions into LCSA (CALCAS, 2009). On the other hand, UNEP-SETAC produced a report titled "Towards a Life Cycle Sustainability

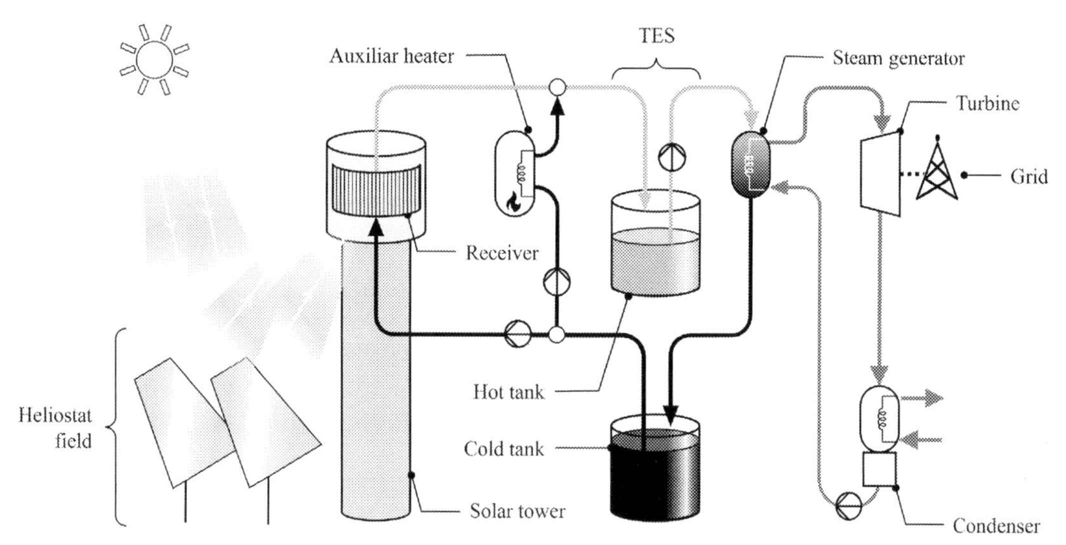

FIGURE 3.2 Scheme of an STP plant. *Credit: Elaborated by the authors, based on GIZ/MCTI, 2014b. Parabolic Trough CSP Technology—State of the Art and Market Overview.*

Assessment," which showcases the utilization and integration of LCA, social life cycle assessment (SLCA), and life cycle costing (LCC) as initial steps in applying LCSA (Kloepffer, 2008; UNEP/SETAC Life Cycle Initiative, 2011).

2.2.1 *Life cycle assessment*

LCA is an environmental evaluation tool that covers the complete life cycle of a product or service, encompassing activities from raw material extraction and processing to waste disposal and recycling. The International Organization for Standardization (ISO) has developed specific standards for LCA, namely ISO 14040 (ISO, 2006) and ISO 14044 (ISO, 2006), to ensure consistency and comparability in its application. These international standards outline the steps in conducting an LCA, including goal and scope definition, life cycle inventory (LCI), life cycle impact assessment (LCIA), and interpretation.

Ecoinvent (Wernet et al., 2016) is the most comprehensive and extensively used LCI database among the various existing options today to analyze elementary flows (Bjørn et al., 2018). Regarding assessing these elementary flows' environmental impact, notable LCIA methods are available and include eco-indicator (Goedkoop and Spriensma, 2000), ReCiPe 2016 (Huijbregts et al., 2017), and USEtox (Rosenbaum et al., 2008).

2.2.2 *Social life cycle assessment*

SLCA assesses the impacts of a product or service's life cycle on the stakeholders' well-being (UNEP, 2020). To assist this analysis, UNEP (UNEP, 2020) offers a framework for SLCA, outlining specific elements and providing detailed guidelines to guide its application process. Moreover, this working group proposes generic and specific social indicators, suggesting potential data sources for collecting them (UNEP, 2021).

Regarding more specifically secondary data sources, the Social Hotspot Database (SHDB) (Benoit-Norris et al., 2012) and the Product Social Impact Life Cycle Assessment (PSILCA) (Ciroth and Eisfeldt, 2016) stand out as the main social LCI databases. These secondary data sources are crucial in SLCA studies since they allow practitioners to collect relevant social data and focus on a country or sector-level social performance (Kühnen and Hahn, 2017; Toniolo et al., 2020).

2.2.3 *Life cycle costing*

LCC allows for evaluating a product's costs throughout its life cycle, taking into account the interactions within the economic dimension of sustainability (Rödger et al., 2018). Its origins date back to the 1950s when Novick (1959) evaluated investments in military weaponry using a life cycle-based approach. In general, this comprehensive assessment tool supports practitioners in identifying cost-saving opportunities and enhancing the decision-making process (Petrillo et al., 2016).

While LCC is already being utilized in different industry sectors, a standardized methodology that aligns with the LCSA remains necessary. The closest approach to achieving this alignment is the methodology presented by Hunkeler et al. (2008). These authors integrate LCC with the LCA framework, providing a robust analysis of both economic and environmental factors within life cycle-based assessments.

2.3 Applying LCSA to the CSP technologies: goal and scope definition and LCI

This section addresses the goal and scope definition and LCI steps of LCSA concerning the interaction of these CSP layouts within sustainability's environmental, social, and economic dimensions. In this sense, this chapter conducts a life cycle-based analysis following the ISO 14040 (ISO, 2006) standard on two different CSP plants to be installed in Bom Jesus da Lapa, northeast Brazil. One consists of an STPT, and the other corresponds to an STP—today's two most attractive commercial-scale CSP technologies. Both present 100 MW (MW), corresponding to the commercial power plants' installed capacity to be interconnected to the SIN. These technologies also present a useful life of 30 years, air cooling, and a 6-hour TES—typical energy storage capacity adopted by commercial CSP plants (Denholm et al., 2013; Hernández Moris et al., 2021; Jorgenson et al., 2013; Yagi et al., 2021). Based on these assumptions, the System Advisory Model (SAM) software (NREL, 2018) obtained the solar multiple for each CSP technology by conducting sensitivity analyses. These analyses identified the optimal layouts in terms of the lowest levelized cost of energy (LCOE) (Table 3.1). SAM is a computational program developed by the National Renewable Energy Laboratory (NREL) for financial modeling and performance simulation of renewable energy projects, such as photovoltaic systems, CSP plants, wind farms, fuel cells, solar water heating, biomass combustion, geothermal power generation, and wave energy. In the case of CSP, this software converts hourly solar irradiation and other meteorological conditions into thermal energy and, subsequently, electrical energy (NREL, 2018).

The system boundaries are defined from the extraction of raw materials to the decommissioning and final disposal of the CSP plants. However, this analysis does not include transport activities, connection to the distribution network, and transmission lines. It was assumed

TABLE 3.1 Main CSP layouts' technical parameters.

	CSP technologies	
Technical parameters	STPT	STP
Installed capacity	100 MW	100 MW
TES	6 h	6 h
Tank storage volume	1870.79 MW_{ht} (volume)	1616.5 MW_{ht} (volume)
Cooling system	Air cooling	Air cooling
Solar multiple	1.7	1.7
Collector field area	2.59×10^6 m^2	6.48×10^6 m^2
Tower height	N/A	195.85 m
Energy production	382,026.43 MWh/yr	447,712.03 MWh/yr
Useful lifetime	30 years	30 years
Country origin	Spain	USA

that STPT is from Spain and STP is from the United States (USA). The impact categories were selected based on a previous study by Lassio et al. (2021), which proposes 28 life cycle-based sustainability indicators for evaluating electricity generation from different energy sources in Brazil, considering the environmental, social, and economic dimensions.

In the following items, it is possible to verify more details regarding the scope of the LCSA of the CSP technologies considered and its LCI according to sustainability's environmental, social, and economic dimensions.

2.3.1 *Environmental dimension*

The environmental assessment of the LCSA utilized SimaPro 9, Ecoinvent database 3.8 (Wernet et al., 2016) ReCiPe 2016 Midpoint (Huijbregts et al., 2017)—taking into account 17 of its impact categories: (i) global warming (EN01), (ii) stratospheric ozone depletion (EN02), (iii) ionizing radiation (EN03), (iv) ozone formation (EN04), (v) fine particulate matter formation (EN05), (vi) terrestrial acidification (EN06), (vii) freshwater eutrophication (EN07), (viii) marine eutrophication (EN08), (ix) terrestrial ecotoxicity (EN09), (x) freshwater ecotoxicity (EN10), (xi) marine ecotoxicity (EN11), (xii) human carcinogenic toxicity (EN12), (xiii) human noncarcinogenic toxicity (EN13), (xiv) land use (EN14), (xv) mineral resource scarcity (EN15), (xvi) fossil resource scarcity (EN16), and (xvii) water consumption (EN17). In addition, an 18th impact category related to the effects of CSP projects on environmentally protected areas (EN18) was also considered.

A functional unit of 1.0 kWh of electricity generated was chosen as the reference unit for the two considered CSP technologies' input and output streams. To ensure the relevance of these renewable energy technologies and align with the Brazilian context, we adapted data from the attributional approach of the Ecoinvent, scaling up a 50 MW STPT to 100 MW

TABLE 3.2 Assumptions adopted in calculating the scale-up factors used in the LCA.

	CSP technologies	
Main characteristics	**STPT**	**STP**
Installed capacity	50–100 MW	20–100 MW
Building structures and processes	(×) 1.6	(×) 2.56
Collector field area	Ratio between collector field areas (1.72)	Ratio between collector field areas (5.48)
Power block	Ratio between installed capacity of the CSP plants	Ratio between installed capacity of the CSP plants
Receiver system	N/A	Ratio between tower heights

and a 20 MW STP to 100 MW. The key adaptations are summarized in Table 3.2. Note that for the data related to building structures and activities obtained from Ecoinvent, scale-up factors of 1.6 for STPT and 2.56 for STP were applied based on the assumptions adopted by Corona (2016). Additionally, this analysis considered information from the literature and the SAM software (NREL, 2018).

With regard more specifically to the impact category of water consumption, the SAM software provided the water requirements associated with cleaning the mirrors and the power block. Such consumption corresponds to 89,209 m^3/year for STPT and 81,567 m^3/year for STP.

2.3.2 Social dimension

In analyzing the social dimension, four social indicators are considered: (i) local employment (SO01); (ii) health and safety (of workers) (SO02); (iii) social acceptance (SO03); and (iv) traditional communities (SO04). Social LCI data generally come from secondary sources, mainly the available literature and reports from sectors and international organizations. The local employment indicator was based on the research conducted by Ram et al. (2020) on creating new jobs associated with the current energy transition. Concerning the indicator of health and safety of workers, the required information was obtained from a previous study carried out by Hämäläinen et al. (2009) on the global trend of work-related injuries and illnesses at regional and national levels. In both social indicators, the performance of the countries of origin of CSP technologies in the production of materials and equipment stage is considered. On the other hand, Brazil's performance in the other stages of the life cycle of CSP technologies is considered (Table 3.3).

Given that there are currently no CSP projects in Brazil, obtaining primary data on the community's perception of this technology is challenging. Thus, two different previous studies were considered to incorporate the social acceptance indicator in this SLCA. The first corresponds to a survey by Sena et al. (2016), which investigates renewable energy's impacts and social acceptance in northeastern Brazil. Its findings highlight that the population generally favors photovoltaic solar technology. In parallel, Hanger et al. (2016) addressed the community acceptance of large-scale solar energy installations in developing countries,

TABLE 3.3 Estimates of job creation and occupational accidents related to CSP technologies.

	Job creation (jobs/MW) Ram et al. (2020)		Occupational accidents (%) Hämäläinen et al. (2009)	
Life cycle stage	STPT	STP	STPT	STP
Production	4.2	4.0	3.93 (Spain)	4.68 (USA)
Construction	50.24	50.24	15.17 (Brazil)	15.17 (Brazil)
O&M	1.88	1.88	15.17 (Brazil)	15.17 (Brazil)
Decommissioning	4.18	4.18	15.17 (Brazil)	15.17 (Brazil)

specifically focusing on CSP. The authors showed a favorable situation for this energy source. Therefore, the same high level of acceptance was considered for the different CSP technologies. Additionally, the local Communities indicator corresponds to civil liberties assessments derived from reports published by Freedom House (2021). Thus, this SLCA incorporates the performance evaluations of both the origin countries of the CSP technologies (Spain: 53/60; USA: 51/60) and Brazil itself (43/60).

2.3.3 *Economic dimension*

In this LCSA, LCC focuses on the internal revenues and costs associated with the complete energy generation chain from the CSP technologies considered from the producer's perspective. Such revenues and costs are categorized according to relevant stages of their value chain. They are also aggregated into a single final cost and correlated to a functional unit of 1.0 kWh. Thus, six economic indicators are considered: (i) research and development (R&D) costs (EC01); (ii) investment costs (EC02); (iii) operation and maintenance (O&M) costs (EC03); (iv) disposal costs (EC04); (v) levelized cost of energy (LCOE) (EC05); and (vi) payback period (EC06). Economic-financial data for most of these indicators were obtained with the SAM software (NREL, 2018) and run considering a discount rate of 8% (Trama et al., 2021). R&D costs and disposal costs indicators were calculated based on IRENA (2012) and San Miguel and Corona (2018), respectively (see Table 3.4).

TABLE 3.4 Economic-financial parameters considered in LCC analysis.

	CSP technologies	
Economic-financial parameter	STPT	STP
R&D costs	2.9% of total costs	
Investment costs	USD 5957.91/kW	USD 6004.63/kW
Fixed O&M costs	USD 66/kW-yr	USD 66/kW-yr
Variable O&M costs	USD 4/MWh	USD 3.5/MWh
Disposal costs	USD 104.6/kW	

2.3.4 *Integrating LCA, SLCA, and LCC results*

For integrating LCA, SLCA, and LCC results and obtaining a single sustainability score, we suggest weighting and aggregating procedures through the AHP method. Implementing this method demands not only the use of dimensionless input parameters but also the involvement of stakeholders in weight criteria. On the one hand, we employed internal normalization techniques to convert LCA, SLCA, and LCC results into a dimensionless scale (Pizzol et al., 2017). On the other hand, we consulted eight researchers within the energy sector to support defining the order of priority of sustainability indicators and dimensions.

Hence, based on the sustainability indicators and dimensions taken into account in this analysis, the hierarchical structure of the AHP method was organized into four levels. The first level represents the overarching objective of identifying the most sustainable CSP technology. The second level encompasses the sustainability dimensions, namely environmental, social, and economic. Moving further down the hierarchy structure, the third level decomposes the sustainability dimensions into their respective indicators. Lastly, the fourth level concerns the two evaluated CSP technologies.

3. Results and discussion

3.1 Life cycle sustainability assessment

3.1.1 *Environmental dimension*

Fig. 3.3 illustrates the superior environmental performance of STP technology across 11 out of the 17 indicators measured by ReCiPe 2016 (Huijbregts et al., 2017), including stratospheric

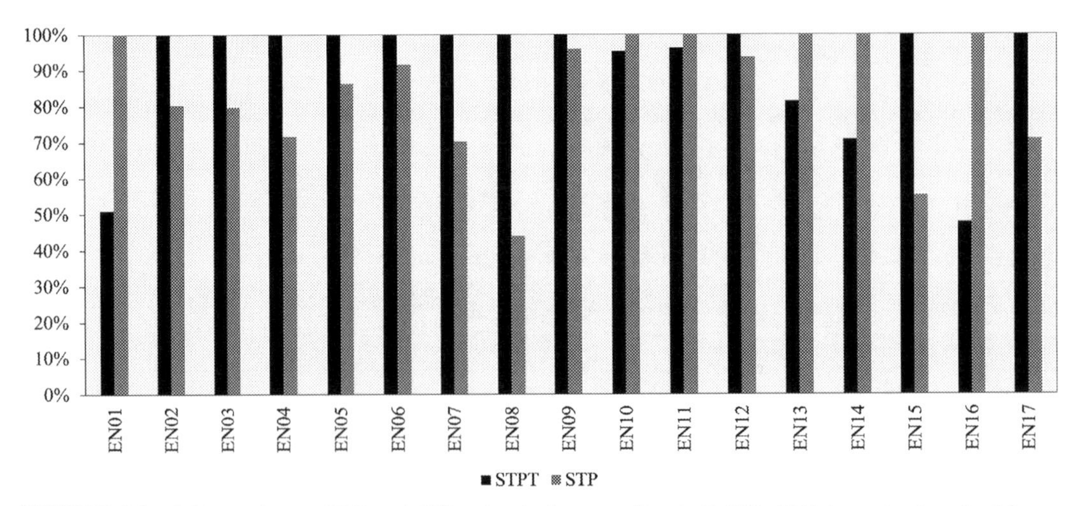

FIGURE 3.3 LCA results on STPT and STP technologies according to ReCiPe 2016 impact categories (characterization; ReCiPe 2016 midpoint (H) V1.07/world (2010) H). *Credit: Elaborated by the authors.*

ozone depletion (EN02), ozone formation (EN04), and water consumption (EN17). The impact categories where STPT technology exhibits better results than STP include global warming (EN01) and land use (EN14). On the other hand, STPT technology presents the best performance concerning the protected areas indicator (EN18) and obtained a score of 0.25, while STP technology acquired 0.22.

3.1.2 Social dimension

Table 3.5 lists the SLCA results for the two CSP technologies. The STPT technology performs best in local employment (SO01) and traditional communities (SO04) indicators. On the other hand, STP technology presents the best performance in the health and safety (of workers) indicator (SO02). Note that both CSP technologies achieved the same level of social acceptance (SO03).

3.1.3 Economic dimension

Concerning the economic dimension, Table 3.6 shows that the STP technology performs better than the STPT technology in all considered economic indicators: R&D costs (EC01), investment costs (EC02), O&M costs (EC03), disposal costs (EC04), LCOE (EC05), and payback period (EC06). It is worth mentioning that this result is due mainly to the fact that STP generates more energy annually than STPT.

3.2 Interpretation

For integrating LCA, SLCA, and LCC results and obtaining the sustainability dimension indicators, each environmental, social, and economic indicator was weighted according to experts' judgment. In the environmental dimension, the three most relevant impact categories are global warming (EN01), water consumption (EN17), and land use (EN14). Concerning social indicators, they have been ranked in the following order of importance to the social dimension: local employment (SO01), social acceptance (SO03), traditional communities

TABLE 3.5 SLCA results on STPT and STP technologies.

			CSP technologies	
Social indicator	**Code**	**Unit**	**STPT**	**STP**
Local employment	SO01	Jobs/MW	60.5	60.3
Health and safety	SO02	Accidents/kWh	7.59×10^{-8}	6.50×10^{-8}
Social acceptance	SO03	Dimensionless	High	High
Traditional communities	SO04	Dimensionless	0.48	0.47

TABLE 3.6 LCC results on STPT and STP technologies.

Economic indicator	Code	Unit	CSP technologies	
			STPT	STP
R&D costs	EC01	USD/kWh	1.67×10^{-3}	1.44×10^{-3}
Investment costs	EC02	USD/kWh	5.20×10^{-2}	4.47×10^{-2}
O&M costs	EC03	USD/kWh	4.02×10^{-3}	3.51×10^{-3}
Disposal costs	EC04	USD/kWh	9.13×10^{-7}	7.70×10^{-7}
LCOE	EC05	USD/kWh	0.19	0.16
Payback period	EC06	years	7.86	7.64

TABLE 3.7 LCSA results on STPT and STP technologies.

Indicator	CSP technologies	
	STPT	STP
Environmental	0.533	0.467
Social	0.497	0.503
Economic	0.469	0.531
Sustainability	0.512	0.488

(SO04), and health and safety (of workers) (SO02). In the case of the economic dimension, this list is as follows: investment costs (EC02), payback time (EC06), LCOE (EC05), O&M costs (EC03), R&D costs (EC01), and disposal costs (EC04). As a result, Table 3.7 shows that STPT performs better in the environmental dimension, while STP performs better in the social and economic dimensions.

In order to obtain a single sustainability score, the sustainability dimensions were assigned different weights. This procedure chose the environmental dimension as the most important for sustainability, followed by the social and economic dimensions. Table 3.7 also presents the sustainability index for the two CSP technologies analyzed and indicates that STPT proves to be a priori the preferred option in terms of sustainability.

Considering that this step is vital to effectively and comprehensively present, analyze, and compare LCSA results, a graphical representation of the results is also used, displaying not only the single score but also the alternatives' performances regarding each sustainability dimension and indicator (see Fig. 3.4). Note that this approach simplifies the interpretation process, making it more straightforward, comprehensible, and transparent for decision makers and stakeholders.

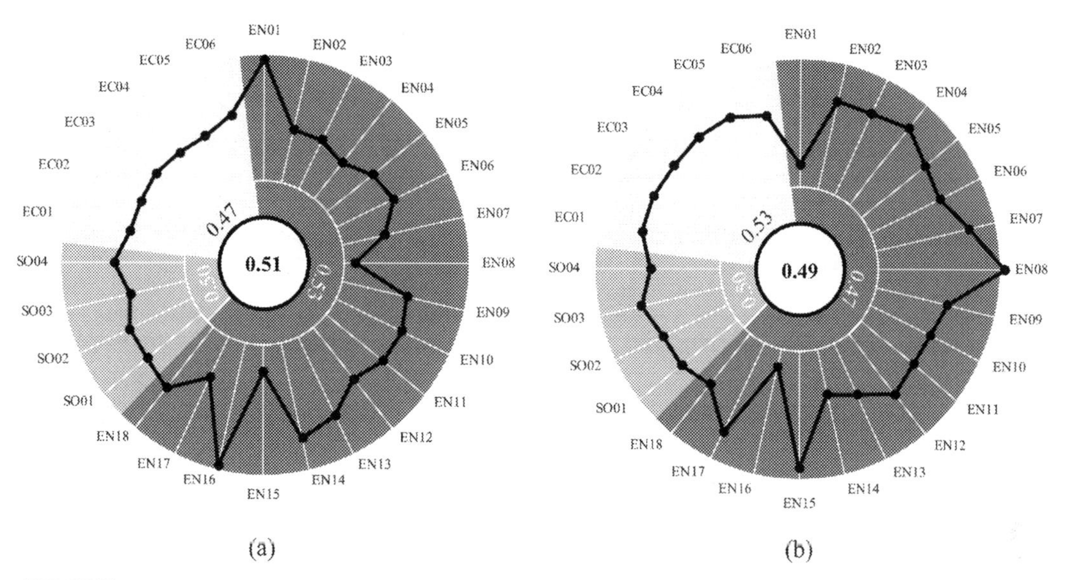

FIGURE 3.4 Graphical LCSA results on STPT (A) and STP (B) technologies. *Credit: Elaborated by the authors.*

4. Conclusion

This chapter offered a theoretical understanding of LCSA and addressed the scientific community's demand for its application on various products and services. Consequently, the effectiveness of this methodology has been exemplified through a case study evaluating the sustainability of two more suitable CSP technologies for electricity generation in Brazil: STPT and STP. This demonstration holds significance as the global economy progressively embraces a sustainable and low-carbon pathway by adopting renewable energy sources.

Our results showed that the STPT technology performed better than the STP regarding sustainability. On the one hand, STPT obtained the highest index for global sustainability and performed better in the environmental dimension. On the other hand, STP obtained the highest indicators for social and economic dimensions. Since the present study was based on specific assumptions, system boundaries, and LCI adaptations, these results may not be universally applicable. However, it contributes to the LCSA and energy fields by comparing leading CSP technologies in Brazil more comprehensively and accurately, which in turn can enhance decision-making quality during the power planning process.

LCSA is pivotal in supporting the transition toward a sustainable future by identifying the most sustainable options available. Comprehensively presenting its results addresses communication challenges and effectively highlights the trade-offs associated with various alternatives. Furthermore, LCSA proves to be an interesting sustainability tool across different industry sectors. This is because it can facilitate the environmental, social, and economic improvements of technologies and products. Finally, LCSA is also able to assist policymakers in designing scenarios for the future guided by the principles of sustainable development.

Future research endeavors could enhance the existing work by incorporating data from primary sources, particularly in the social dimension. Researchers can obtain a more comprehensive and accurate understanding of the social implications associated with CSP technologies by utilizing data collected directly from relevant stakeholders and communities.

References

Benoit-Norris, C., Cavan, D.A., Norris, G., 2012. Identifying social impacts in product supply chains: overview and application of the social hotspot database. Sustainability 4, 1946–1965. https://doi.org/10.3390/su4091946.

Bjørn, A., Moltesen, A., Laurent, A., Owsianiak, M., Corona, A., Birkved, M., Hauschild, M.Z., 2018. Life cycle inventory analysis. In: Life Cycle Assessment. Springer International Publishing, Cham, pp. 117–165. https://doi.org/10.1007/978-3-319-56475-3_9.

Brazil, 2023. 2022 Statistical Yearbook of Electricity.

Brazil, 2022. Brazilian Energy Balance 2021. Brasília.

Brazil, 2021. The Ten-Year Energy Expansion Plan 2030.

Brazil, 2015. 2014 Statistical Yearbook of Electricity.

Brazil, 2011. 2010 Statistical Yearbook of Electricity.

CALCAS, 2009. D20 Blue Paper on Life Cycle Sustainability Analysis, Coordination Action for Innovation in Life-Cycle Analysis for Sustainability.

Ciroth, A., Eisfeldt, F., 2016. PSILCA—A Product Social Impact Life Cycle Assessment Database, Database Version 1.0.

Corona, B., 2016. Análisis de sostenibilidad del ciclo de vida de una configuración innovadora de tecnología termosolar. Universidad Politécnica de Madrid.

Denholm, P., Wan, Y.Y.H., Hummon, M., Mehos, M., 2013. An Analysis of Concentrating Solar Power With Thermal Energy Storage in a California 33 % Renewable Scenario (Technical Report NREL/TP-6A20-58186). Tech. Rep. NREL/TP-6A20-58186.

Freedom, H., 2021. Freedom in the World 2021.

GIZ/MCTI, 2014a. Basic Mapping of General Pre-conditions for CSP Technologies in Brazil.

GIZ/MCTI, 2014b. Parabolic Trough CSP Technology—State of the Art and Market Overview.

GIZ/MCTI, 2014c. Tower CSP Technology—State of Art and Market Review.

Goedkoop, M., Spriensma, R., 2000. The Eco-Indicator 99—A Damage Oriented Method for Life Cycle Assessment, Methodology Report. PRé Consultants.

Guinée, J.B., 2016. Life cycle sustainability assessment: what is it and what are its challenges? In: Clift, R., Druckman, A. (Eds.), Taking Stock of Industrial Ecology. Springer, Cham, pp. 45–68. https://doi.org/10.1007/978-3-319-20571-7_3.

Hämäläinen, P., Leena Saarela, K., Takala, J., 2009. Global trend according to estimated number of occupational accidents and fatal work-related diseases at region and country level. Journal of Safety Research 40, 125–139. https://doi.org/10.1016/j.jsr.2008.12.010.

Hanger, S., Komendantova, N., Schinke, B., Zejli, D., Ihlal, A., Patt, A., 2016. Community acceptance of large-scale solar energy installations in developing countries: evidence from Morocco. Energy Research and Social Science 14, 80–89. https://doi.org/10.1016/j.erss.2016.01.010.

Hernández Moris, C., Cerda Guevara, M.T., Salmon, A., Lorca, A., 2021. Comparison between concentrated solar power and gas-based generation in terms of economic and flexibility-related aspects in Chile. Energies 14, 1063. https://doi.org/10.3390/en14041063.

Huijbregts, M., Steinmann, Z., Elshout, P., Stam, G., Verones, F., Vieira, M., Zijp, M., Hollander, A., van Zelm, R., 2017. ReCiPe 2016: a harmonised LCIA method at midpoint and endpoint level. International Journal of Life Cycle Assessment 22, 138–147. https://doi.org/10.1007/s11367-016-1246-y.

Hunkeler, D., Lichtenvort, K., Rebitzer, G., 2008. Environmental Life Cycle Costing, Environmental Life Cycle Costing. CRC Press, Boca Raton. https://doi.org/10.1201/9781420054736.

IEA, 2022. World Energy Outlook 2022.

IPCC, 2021. Climate Change 2021: The Physical Science Basis. Contribution of Working Group I to the Sixth Assessment Report of the Intergovernmental Panel on Climate Change. Cambridge University Press 6. https://doi.org/10.1017/9781009157896.

IRENA, 2012. Renewable energy cost analysis—concentrating solar power. IRENA Working Papers 1.

ISO, 2006. Environmental management - Life Cycle Assessement - Requirements and Guidelines. International Organization for Standardization. ISO 14044.

ISO, 2006. Environmental Management - Life Cycle Assessment - Principles and Framework. International Organization for Standardization. ISO 14040.

Jorgenson, J., Denholm, P., Mehos, M., Turchi, C., 2013. Estimating the performance and economic value of multiple concentrating solar power technologies in a production cost model. A Report by the National Renewable Energy Laboratory, United States. National Renewable Energy Laboratory. Technical report NREL/TP-6A20-58645.

Kloepffer, W., 2008. Life cycle sustainability assessment of products. International Journal of Life Cycle Assessment 13, 89–95. https://doi.org/10.1065/lca2008.02.376.

Kühnen, M., Hahn, R., 2017. Indicators in social life cycle assessment: a review of frameworks, theories, and empirical experience. Journal of Industrial Ecology 21 (6), 1547–1565. https://doi.org/10.1111/jiec.12663.

Lassio, J.G., Magrini, A., Castelo Branco, D., 2021. Life cycle-based sustainability indicators for electricity generation: a systematic review and a proposal for assessments in Brazil. Journal of Cleaner Production 311, 127568. https://doi.org/10.1016/j.jclepro.2021.127568.

Novick, D., 1959. The Federal Budget as an Indicator of Government Intentions and the Implications of Intentions. RAND Corporation, Santa Monica, CA.

NREL, 2018. System Advisory Model (SAM).

Petrillo, A., De Felice, F., Jannelli, E., Autorino, C., Minutillo, M., Lavadera, A.L., 2016. Life cycle assessment (LCA) and life cycle cost (LCC) analysis model for a stand-alone hybrid renewable energy system. Renewable Energy 95, 337–355. https://doi.org/10.1016/j.renene.2016.04.027.

Pizzol, M., Laurent, A., Sala, S., Weidema, B., Verones, F., Koffler, C., 2017. Normalisation and weighting in life cycle assessment: quo vadis? International Journal of Life Cycle Assessment 22, 853–866. https://doi.org/10.1007/s11367-016-1199-1.

Ram, M., Aghahosseini, A., Breyer, C., 2020. Job creation during the global energy transition towards 100% renewable power system by 2050. Technological Forecasting and Social Change 151, 119682. https://doi.org/10.1016/j.techfore.2019.06.008.

Rödger, J.-M., Kjær, L.L., Pagoropoulos, A., 2018. Life cycle costing: an introduction. In: Hauschild, H., Rosenbaum, R., Olsen, S. (Eds.), Life Cycle Assessment. Springer International Publishing, Cham, pp. 373–399. https://doi.org/10.1007/978-3-319-56475-3_15.

Rosenbaum, R.K., Bachmann, T.M., Gold, L.S., Huijbregts, M.A.J., Jolliet, O., Juraske, R., Koehler, A., Larsen, H.F., MacLeod, M., Margni, M., McKone, T.E., Payet, J., Schuhmacher, M., van de Meent, D., Hauschild, M.Z., 2008. USEtox—the UNEP-SETAC toxicity model: recommended characterisation factors for human toxicity and freshwater ecotoxicity in life cycle impact assessment. International Journal of Life Cycle Assessment 13, 532–546. https://doi.org/10.1007/s11367-008-0038-4.

San Miguel, G., Corona, B., 2018. Economic viability of concentrated solar power under different regulatory frameworks in Spain. Renewable and Sustainable Energy Reviews 91, 205–218. https://doi.org/10.1016/j.rser.2018.03.017.

Sena, L.A. de, Ferreira, P., Braga, A.C., 2016. Social acceptance of wind and solar power in the Brazilian electricity system. Environment, Development and Sustainability 18, 1457–1476. https://doi.org/10.1007/s10668-016-9772-0.

Toniolo, S., Tosato, R.C., Gambaro, F., Ren, J., 2020. Life cycle thinking tools: life cycle assessment, life cycle costing and social life cycle assessment. In: Life Cycle Sustainability Assessment for Decision-Making. Elsevier, pp. 39–56. https://doi.org/10.1016/B978-0-12-818355-7.00003-8.

Trama, C.P., Júnior, A.O.P., Guimarães, A.P.C., Diniz, A.L., Vieira, L.D.S.R., 2021. Cost–benefit analysis of solar thermal plants with storage in a hydrothermal system. Energies 14. https://doi.org/10.3390/en14185662.

UNEP/SETAC Life Cycle Initiative, 2011. Towards Life Cycle Sustainability Assessment: Making Informed Choices on Products.

UNEP, 2021. The Methodological Sheets for Subcategories in Social Life Cycle Assessment (S-LCA). United Nations Environment Programme (UNEP).

UNEP, 2020. Guidelines for Social Life Cycle Assessment of Products and Organizations 2020.
Vieira, L.D.S.R., Guimarães, A.P.C., Lisboa, P., 2018. Concentrating solar power. In: Renewables. Atitude Editorial, pp. 44–51.
Wernet, G., Bauer, C., Steubing, B., Reinhard, J., Moreno-Ruiz, E., Weidema, B., 2016. The ecoinvent database version 3 (part I): overview and methodology. International Journal of Life Cycle Assessment 21, 1218–1230. https://doi.org/10.1007/s11367-016-1087-8.
Yagi, K., Sioshansi, R., Denholm, P., 2021. Using concentrating-solar-power plants as economic carbon-free capacity resources. Energy Conversion and Management X 12, 100112. https://doi.org/10.1016/j.ecmx.2021.100112.

CHAPTER

4

Exploring sustainable workforce management: Trends, solution approaches, and practices

César Augusto Henao[1], *Andrés Felipe Porto*[2,3] *and Virginia I. González*[1]

[1]Universidad del Norte, Barranquilla, Colombia; [2]Department of Industrial Engineering, Institución Universitaria Americana, Barranquilla, Colombia; [3]Universitat Politècnica de Catalunya, Barcelona, Spain

1. Introduction

In an era characterized by global challenges such as climate change, resource depletion, and social inequality, the need for sustainability has become paramount. In addition, the global economic recovery after the COVID-19 pandemic is hampered by rising inflation, policy uncertainties, supply-chain disruptions, and labor market challenges (Ibn-Mohammed et al., 2021). As organizations and decision-makers strive to integrate sustainable practices into their operations, it is crucial to assess the impact of various trends and approaches within the sustainability framework. Consequently, this chapter aims to explore different emerging trends in sustainable workforce management, solution approaches, and sustainable practices for improving the decision-making processes and informing policy development. Thus, this chapter will discuss three interesting topics related to SDG 8, which aims to promote sustained, inclusive, and sustainable economic growth, full and productive employment, and decent work for all.

The first topic will delve into the emerging trends in sustainable workforce management. This includes the adoption of work flexibility arrangements such as flexible work schedules, remote work, job sharing, and the promotion of diversity and inclusion in the workforce planning processes, among others. Each trend will be analyzed in terms of its potential impact on the company's profitability, employee well-being, and overall productivity. By understanding these trends, organizations can identify opportunities to align their workforce

Evolution and Trends of Sustainable Approaches
https://doi.org/10.1016/B978-0-443-21651-0.00012-7

management practices with sustainable principles, particularly Target 8.5, an effort to achieve full and productive employment and decent work for all women and men.

The second topic will focus on discussing the solution approaches commonly employed to solve complex workforce management problems. Specifically, it will explore the pivotal role played by operations research techniques in supporting sustainable workforce management. This includes the utilization of data analytics, optimization models, heuristics, and metaheuristics. By leveraging these solution techniques, organizations can assess the impact of work flexibility arrangements on sustainable workforce management. Consequently, organizations can design a more efficient and sustainable workforce management system, enabling them to optimize workforce allocation, reduce labor costs, and enhance overall productivity and employee satisfaction. This will contribute to Target 8.2 intending to achieve higher levels of economic productivity.

The third topic will emphasize the significance of incorporating sustainable workforce management practices into decision-making and policy processes. It will highlight the importance of collaboration among employers, employees, and state regulatory agencies responsible for labor and workplace standards. Thus, the topic will underscore the need for regulatory frameworks that incentivize sustainable practices and therefore provide guidelines that align employer and employee development goals. By integrating sustainability into decision-making and policy processes, a more resilient and future-oriented workforce management system can be achieved, with the intention of supporting Target 8.3 by promoting development-oriented policies that support productive activities and decent job creation.

2. Emerging trends in sustainable workforce management

Employers are now needed to adopt flexible work arrangements that allow them to proactively find and retain talent and, in turn, respond to evolving workforce demands. Thus, given that employers strive to integrate sustainable workforce management practices into their operations, it becomes crucial to evaluate the advantages and disadvantages of various emerging trends within the sustainability framework. This chapter aims to delve into the following seven emerging trends.

2.1 Job sharing

Job sharing involves two or more employees sharing the responsibilities of a single full-time position. Typically, each employee works a part-time schedule, but they work on contrasting shifts or alternate days, depending on the preferred level of flexibility. This approach promotes a sustainable balance between work and personal life, allowing employees to split their responsibilities, and together they cover the duties and workload associated with the position. Thus, job sharing serves as a potential solution to the dilemmas encountered by employees in managing both their professional and personal-family spheres (Arkin and Dobrofsky, 2022).

TABLE 4.1 Advantages and disadvantages of job sharing.

Advantages	Disadvantages
Work-life balance: Job sharing allows employees to achieve a better work-life balance by reducing their individual work hours, which can lead to reduced stress and improved overall well-being.	*Communication challenges:* Coordinating job-sharing arrangements may require extra effort to ensure effective communication between employees sharing the position.
Diverse perspectives: With multiple employees involved, job sharing brings a variety of skills, experiences, and perspectives to the position, potentially enhancing creativity and problem-solving.	*Potential for conflicts:* Differences in work styles, goals, or approaches among employees sharing the position can lead to conflicts or misunderstandings.
Knowledge transfer: Job sharing can facilitate the transfer of knowledge and skills among employees, ensuring that critical expertise is not concentrated in a single individual.	*Training and onboarding:* Getting new employees up to speed in a job-sharing arrangement may take longer and require more resources compared to a single full-time employee.
Coverage: It provides extended coverage, minimizing the risk of work disruptions due to absences or unexpected events, as there are multiple individuals familiar with the position.	*Administrative complexities:* Managing schedules, benefits, and compensation for multiple employees in a job-sharing arrangement can be administratively complex.

According to Crampton et al. (2003), engaging in job sharing can encompass two main distinct approaches to dividing responsibilities: shared responsibility and divided responsibility. On one side, shared responsibility occurs when two employees equally participate in all the duties associated with a single full-time position. In this approach, there is no formally defined division of tasks. The position partners are interchangeable, capable of seamlessly taking over from one another. On the other hand, divided responsibility comes into play when two employees jointly hold a full-time position, but tasks are allocated by project or client group. In this arrangement, job sharers carry out distinct tasks and offer support to one another as a backup.

Table 4.1 lists some of the main advantages and disadvantages associated with job sharing.

2.2 Remote work

Remote work involves a work arrangement in which employees perform their job duties and tasks from a location outside the traditional workplace, such as their home, a coworking space, or any other location. When remote work requires the use of personal electronic devices or an internet connection, it is known as telecommuting or telework (Groen et al., 2018; Vartiainen, 2021). Moreover, when employees commit to achieving specific goals and have the flexibility to work at times and in locations of their choosing, outside the conventional company settings, it is termed Smart Working. The experience gained during the COVID-19 pandemic and advancements in technology accelerated the adoption and/or

TABLE 4.2 Advantages and disadvantages of remote work.

Advantages	Disadvantages
Work-life balance: Remote work allows employees to avoid long commutes, spend more time with family, and have greater control over their daily routines.	*Supervision challenges:* Managing remote employees can be tougher, requiring trust, specific tracking tools, and ensuring their safety and health in remote locations.
Flexible schedules: Remote employees can adapt their work schedules to meet their personal needs while effectively accomplishing job duties.	*Communication hurdles:* Effective communication can be more complex in a remote environment due to reliance on technology, potentially leading to misunderstandings and lack of clarity.
Time and cost savings: Eliminating daily commutes reduces transport externalities (e.g., pollution, noise, congestion, traffic accidents) and saves employees time and money on transportation, meals, and work attire.	*Lack of in-person interaction:* Face-to-face communication and in-person collaboration may be more challenging in a remote setting, potentially impacting team cohesion and creativity.
Remote talent acquisition: Companies are expanding their talent acquisition efforts globally, allowing them to reduce geographic limitations and find the best-fit candidates for their positions.	*Feelings of isolation:* Some employees may experience feelings of isolation or loneliness when working remotely, which can affect their emotional well-being and engagement.

adaption of remote work, and this trend continues to shape the workforce landscape (Spieske et al., 2023).

With the normalization of remote work, many organizations are adopting hybrid remote work models, combining in-person and remote work (Wontorczyk and Roznowsk, 2022). This approach allows employees to split their time between working from home and coming into the traditional workplace (Pianese et al., 2023). Many companies adopt remote work policies to attract and retain talent, reduce costs, and provide employees with more flexible working conditions (Chatterjee et al., 2022). In addition, it provides flexibility while maintaining some level of in-person collaboration and team building. Note that, remote work can be temporary or a permanent arrangement, depending on the policies and preferences of both the employee and the employer.

Table 4.2 lists some of the main advantages and disadvantages associated with remote work.

2.3 Flexible work schedules

One of the prominent trends in sustainable workforce management is the adoption of flexible work schedules. This workforce management practice serves dual purposes. First, due to seasonality and variability in workforce demand, as well as unplanned employee absenteeism, a mismatch occurs between staffing and workforce demand (Henao, 2015; Henao et al., 2022). Consequently, this results in increased costs associated with overstaffing and understaffing (Porto et al., 2020). Therefore, employers can implement flexible work schedules that enable a better alignment between workforce supply and demand, allowing for the minimization of overstaffing and understaffing costs. Second, employers are increasingly recognizing the value of allowing employees to have greater control over their work hours. As a result, a workforce management approach that considers employees' preferences in their work schedules can enhance employee satisfaction and well-being, elevate retention rates, and increase employee productivity. Below, we briefly describe some of the main strategies for creating flexible work schedules.

1. *Flexible contracts*: Traditional full-time employee hiring is allowed, but part-time employee hiring is also permitted. Part-time contracts typically offer different shift lengths and different sequences of work and rest days to better accommodate workforce demand (Taskiran and Zhang, 2017; Porto et al., 2019).
2. *Flextime*: Employee work shifts can have varying start and end times, providing two main advantages: (1) greater flexibility in aligning shift supply with the seasonality of workforce demand, and (2) employees can select their preferred start and end times within specific guidelines established by their employer (Henao et al., 2015; Cuevas et al., 2016).
3. *Compressed workweeks*: Employees who are working a full week's hours in fewer than five traditional workdays. This arrangement allows employees to have longer days at work but fewer workdays, enhancing work-life balance while retaining full-time status (Munyon et al., 2023).
4. *Annualized hours*: Contracts with agreed-upon annual work hours, but then these hours are distributed irregularly across the weeks of the year (Lusa et al., 2008). That is, during weeks of high demand or busy seasons, employees may work more hours, and during weeks of low demand, they may work fewer hours, all while ensuring that their annual work hour commitment is met. Thus, the key feature of annualized hours is the flexibility it offers in managing seasonal demands or variable workload patterns (Porto et al., 2023a, 2023b, 2024).
5. *On-call workers/Temporary workers*: Employees who are available as needed, either as in-house on-call workers or external temporary hires. They are typically used on days or during times when the workforce demand exceeds forecasts or when there is employee absenteeism (Milner and Pinker, 2001; Mac-Vicar et al., 2017).
6. *Overtime*: The use of extra hours to extend the duration of a work shift beyond regular hours (Porto et al., 2022). However, while overtime minimizes understaffing, its excessive use raises overstaffing costs and reduces employee well-being.
7. *Multiple breaks*: Flexibility in the quantity, location, and duration of breaks within a work shift, according to company policies. The purpose of this approach is to insert a break (or multiple breaks) during a staff surplus, reducing the number of employees during the break but extending the shift and consequently increasing postbreak staffing (Rekik et al., 2010; Álvarez et al., 2020; Kiermaier et al., 2020). A *split shift* is a particular case of an extended break in the middle of a shift.

Table 4.3 lists some of the main advantages and disadvantages associated with flexible work schedules.

TABLE 4.3 Advantages and disadvantages of flexible work schedules.

Advantages	Disadvantages
Maximizing demand coverage and minimizing understaffing/overstaffing: Businesses can easily adjust staffing levels to meet fluctuations in demand due to its seasonality and variability.	***Regulatory compliance:*** Ensuring compliance with labor laws and regulations in different regions can be complex. Furthermore, as labor laws become stricter, less flexibility in work schedules can be implemented.
Enhanced employee retention: Companies with flexible schedules have higher employee retention rates when considering employee preferences in scheduling.	***Challenges in coordination:*** Coordinating employee schedules can be complex, especially in larger organizations.
Enhanced productivity: Adopting flexible work schedules empowers employers to boost workforce productivity, ensuring that labor hours are utilized and allocated in the most efficient manner.	***Implementation costs:*** Employees assigned to undesirable work schedules to cut over/understaffing costs may require monetary compensation or other benefits from employers.

2.4 Multiskilling initiatives

In response to the variability in workforce demand and unplanned absenteeism, and as a complementary approach to flexible work schedules, employers are investing in programs that enable the development of a multiskilled workforce. Multiskilled employees are those who are trained to effectively perform a range of tasks, allowing employers to transfer available staff from overstaffed tasks to those that are understaffed (Mercado and Henao, 2021; Battaïa and Dolgui, 2022). Thus, this form of workforce flexibility enables companies to reduce costs associated with overstaffing and understaffing (Henao et al., 2019). Nonetheless, attaining such a multiskilled workforce involves both time and financial investments, prompting employers to determine how much, where, and how to add multiskilling (Henao et al., 2022, 2024).

The literature on multiskilling shows that there are three main methods for designing training plans that ensure a sustainable multiskilled workforce. These are the following:

1. *Base multiskilling*: While a portion of the workforce remains single-skilled, another portion will be multiskilled, thus possessing the ability to work on two or more types of tasks (Henao et al., 2015; Restrepo et al., 2017).
2. *Multiskilling with 2-chaining*: It stands out as the predominant and highly supported method for creating multiskilling structures (Simchi-Levi and Wei, 2012). In this 2-chaining approach, training and allocation decisions are managed through closed chains of different lengths, but multiskilled employees can be trained in a maximum of two task types (Fontalvo Echavez et al., 2021).
3. *Multiskilling with k-chaining and $k \geq 2$*: This method closely resembles the previous one, with the distinction being the inclusion of multiskilled employees trained in two, three, or even more types of tasks. While a 2-chaining approach generally yields significant

TABLE 4.4 Advantages and disadvantages of a multiskilled workforce.

Advantages	Disadvantages
Labor flexibility: Multiskilled employees can easily adapt to changing job demands, helping to address fluctuating workloads and staffing needs.	*Training costs:* Initial training to develop multiple skills can be time consuming and costly. Thus, full multiskilling is not only costly but also unnecessary, so partial multiskilling schemes are enough.
Cost savings: Multiskilling reduces the need for hiring single-skilled staff, overtime, and part-time staff, leading to cost savings in recruitment and salaries.	*Skill maintenance:* It may be challenging to keep all acquired skills at a high level over time, leading to skill degradation resulting from the forgetting phenomenon.
Improved productivity: Multiskilled employees can switch between tasks, optimizing resource allocation and enhancing overall productivity.	*Resistance to change:* Some employees may resist or struggle with acquiring new skills, impacting their morale and job satisfaction.
Enhanced problem-solving: Multiskilled workers tend to have a more comprehensive understanding of processes, making them better at troubleshooting and finding innovative solutions.	*Workload variation and burnout risk:* Multiskilled employees may experience workload imbalances and risk burnout as they are often called upon to perform various tasks.

savings in over/understaffing costs, the k-chaining approach with $k \geq 2$ proves to be more responsive in highly variable demand scenarios (Wang and Zhang, 2015; Simchi-Levi and Wei, 2015; Abello et al., 2021; Mercado et al., 2022).

These three methods for adding multiskilling can be applied under two approaches: homogeneous workforce and heterogeneous workforce. A homogeneous workforce is characterized by the fact that the productivity of all multiskilled employees is at its peak in all the tasks they are trained for (Muñoz et al., 2024). Conversely, a heterogeneous workforce is characterized in that the productivity of multiskilled employees can vary across the tasks in which they are trained, according to the frequency with which they were assigned to work on each task type. Particularly, when there is little disparity among trained tasks, a homogeneous workforce aligns with the 2-chaining method (Henao et al., 2016). However, when tasks vary significantly in complexity and a k-chaining approach with $k \geq 2$ is used, opting for a heterogeneous workforce is more suitable. This is because multiskilled employees often go through the phenomenon known as learning, forgetting, and relearning (Vergara et al., 2021; Henao et al., 2023).

Finally, Table 4.4 lists some of the main advantages and disadvantages of a multiskilled workforce.

2.5 Employee retention benefits packages

In addition to flexible work arrangements, companies are offering their employees a set of perks, incentives, and benefits. The goal of these retention benefits packages is to make the company an attractive place to work and enhance job satisfaction (Balkin and Werner, 2023; Memili et al., 2023). Additionally, it fosters a positive and motivated workforce that is more likely to contribute to the company's success (Coetzee and Stoltz, 2015). Some common components of employee retention benefits packages are for example:

1. *Competitive compensation*: Offering competitive salaries and bonuses to ensure that employees are fairly compensated for their work. This can include performance-based raises and bonuses.
2. *Healthcare benefits and wellness programs*: Providing comprehensive health insurance coverage for employees, including medical, dental, and vision plans. In turn, promoting employee well-being through wellness initiatives such as gym memberships, mental health support, and stress management programs. This helps employees feel secure about their health and well-being.
3. *Retirement plans*: Offering retirement savings plans with additional employer contributions or matching to help employees save for their future.
4. *Flexible work arrangements*: Allowing employees to have more control over their work schedules, which can include options like remote work, flextime, or compressed workweeks.
5. *Career development*: Offering opportunities for skill development, training, and advancement within the company. This can include tuition reimbursement, mentoring programs, and career paths.
6. *Employee recognition and rewards*: Recognizing and rewarding employees for their hard work and achievements, which can include performance-based bonuses, awards, and public recognition.
7. *Family-friendly benefits*: Providing benefits that support employees with families, such as childcare assistance, parental leave, or flexible scheduling for caregivers.

Table 4.5 lists some of the main advantages and disadvantages associated with benefits packages.

TABLE 4.5 Advantages and disadvantages of benefits packages.

Advantages	Disadvantages
Reduced employee turnover: By offering attractive benefits, companies can retain talented and experienced employees, reducing turnover and the associated costs of hiring and training new workers.	***Rising expectations:*** Once certain benefits are offered, employees may expect them to be maintained or even improved over time, placing pressure on the company to sustain or increase benefit costs.
Improved employee morale and satisfaction: Employee retention packages enhance employee satisfaction and well-being, which can boost morale and productivity in the workplace.	***Equity and envy:*** Feelings of envy or inequity among employees may arise if not everyone receives the same benefits or if some feel that others are receiving preferential treatment.
Competitiveness in the job market: Offering appealing benefits can help the company attract high-quality candidates in a competitive job market.	***Financial cost:*** Providing additional benefits can be costly for the company, which could affect short-term profitability.

2.6 Promotion of diversity and inclusion

Companies are actively working to create inclusive environments that value differences, such as race, ethnicity, gender, age, sexual orientation, disability, and socioeconomic status (Katiraee et al., 2019, 2021; Calzavara et al., 2020; Henao et al., 2023). Thus, sustainable workforce management places a strong emphasis on promoting diversity, equity, and inclusion in the workforce. This approach encompasses recruitment, training, leadership commitment, and ongoing monitoring to ensure that diversity and inclusion principles are integrated into the fabric of the organization, ultimately driving innovation and enhancing social responsibility. Here is a breakdown of what this approach entails:

1. *Diversity*: This aspect involves actively recruiting and retaining employees from a variety of backgrounds. The goal is to ensure that the workforce reflects a wide range of perspectives and experiences.
2. *Inclusion*: Inclusion goes beyond simply having a diverse workforce; it focuses on creating a workplace culture where every employee feels valued, respected, and empowered to contribute their unique insights and abilities. It involves promoting an environment where individuals from diverse backgrounds can thrive, collaborate, and participate fully in all aspects of the organization.
3. *Equity*: This aspect seeks to address and rectify historical disparities and biases that may exist within the workplace. It involves implementing policies and practices that ensure fairness and equal opportunities for all employees, regardless of their background or characteristics.

Table 4.6 lists some of the main advantages and disadvantages associated with promoting diversity and inclusion practices.

TABLE 4.6 Advantages and disadvantages of a diverse and inclusive workforce.

Advantages	Disadvantages
Enhanced innovation: Diverse teams tend to generate a variety of perspectives and approaches, leading to innovative ideas and creative solutions.	***Resistance to change:*** Some employees may resist changes in culture and organizational structure that come with diversity and inclusion.
Increased talent attraction: Companies that value diversity and inclusion are often more attractive to diverse talents, expanding the organization's skill set and expertise.	***Communication challenges:*** Managing adequate communication and training to promote diversity and inclusion may require additional time and resources.
Adaptation to global markets: In an increasingly diverse world, companies with diverse teams are better equipped to understand and serve global markets.	***Requires ongoing time and effort:*** Promoting diversity and inclusion is not a one-time goal but an ongoing process that requires dedication and continuous effort.
Conflict reduction and increased productivity: An inclusive environment tends to reduce conflicts and improve morale, which, in turn, can boost productivity and employee retention.	***Complex measurement and evaluation:*** Determining the impact of diversity and inclusion on company performance can be challenging and may require careful metrics and assessments.

2.7 Workforce management 4.0

Workforce management should consider the novel skill sets demanded by digital transformation and Industry 4.0 (Galaske et al., 2018). Thus, the integration of artificial intelligence and automation in the workforce is an emerging trend with the potential to revolutionize various industries (Khang et al., 2023). Organizations are exploring ways to implement these advanced digital technologies and data-driven processes to enhance the efficiency, productivity, and adaptability of their workforce. Below are some key aspects of workforce management in the context of Industry 4.0.

1. *Skills and training*: Organizations focus on developing and upskilling their employees to ensure they have the necessary skills to work with emerging technologies such as automation, artificial intelligence, and data analytics.
2. *Digital work environments*: Workplaces are transformed into digital ecosystems where employees use connected devices and platforms for communication, collaboration, and task management.
3. *Data-driven decision-making*: Workforce management leverages data analytics to make informed decisions about resource allocation, performance evaluation, and talent acquisition.
4. *Human-robot collaboration*: Employees work alongside robots and automated systems, with a focus on tasks that require creativity, problem-solving, and emotional intelligence, while machines handle repetitive and routine work.
5. *Continuous learning and adaptation*: Employees are encouraged to engage in lifelong learning and adapt to evolving technologies and processes, ensuring they remain relevant and valuable to the organization.

Table 4.7 lists some of the main advantages and disadvantages of workforce management 4.0.

TABLE 4.7 Advantages and disadvantages of workforce management 4.0.

Advantages	Disadvantages
Improved efficiency: Workforce management 4.0 can optimize resource allocation and scheduling, leading to increased productivity and reduced operational costs.	***Implementation costs:*** Adopting new technologies can be expensive due to the need for new software, hardware, and employee training.
Enhanced data-driven decision-making: Access to real-time data and analytics aids informed talent decisions, workforce recruitment and allocation, skills improvement, and performance evaluation for employers.	***Resistance to change:*** Employees may resist changes associated with automation and digitization, leading to potential productivity disruptions.
Agility and adaptability: An agile workforce responds swiftly to market changes, maintaining competitiveness and capitalizing on opportunities.	***Skill gaps:*** Transitioning to workforce management 4.0 may reveal skill gaps in the existing workforce, requiring significant investment in training and upskilling.
Remote work and flexibility: The flexibility offered by digital tools allows for remote work options, which can improve work-life balance and attract a wider talent pool.	***Data security concerns:*** Managing vast amounts of data can pose security risks if not handled properly, potentially exposing sensitive information.

Finally, each of these seven trends brings its own set of benefits and challenges to the table. To fully understand their potential impact, organizations must assess their unique contexts and consider factors such as industry type, workforce demographics, and organizational culture. By aligning workforce management practices with sustainable principles, particularly in pursuit of Target 8.5 of Sustainable Development Goal 8, it is possible to achieve full and productive employment and decent work for all. So, organizations can position themselves for long-term success and contribute to a more sustainable future.

3. Solution approaches for solving complex workforce scheduling problems

This section focuses on exploring and discussing the most popular solution approaches for solving complex workforce scheduling problems (WSPs) that integrate flexible work arrangements. Henao et al. (2015) and Porto et al. (2022) explain that WSPs involve a three-level hierarchical process: *planning*, *scheduling*, and *assignment*. Such that, decisions at lower levels are influenced by those at higher levels. Let is briefly describe the decisions made at each of these levels:

1. *Planning*: Strategic decisions for an extended horizon (e.g., seasons, 1 year), addressing staffing (employee numbers by task type and contract type) and training for single-skilled and multiskilled employees.
2. *Scheduling*: Tactical decisions within weekly or monthly horizons, including shift scheduling, days-off scheduling, and tour scheduling (combining shifts and days off).
3. *Assignment*: Operational decisions on employee task-type assignments without specifying shifts or days off.

Now, operations research (OR) is a field of study that uses mathematical models, optimization techniques, and analytical methods to solve complex decision-making problems in various industries and domains. When dealing with complex WSPs, OR practitioners typically employ a range of techniques and methodologies to find optimal or near-optimal solutions. The choice of technique depends on the specific characteristics of the WSP. In fact, it is often necessary to use a combination of techniques or adapt existing methods to meet the unique requirements of a WSP. Here are some key techniques and approaches used for solving WSPs:

1. *Linear programming* (LP): LP is a mathematical optimization technique used to solve WSPs where a linear objective function is to be maximized or minimized, subject to linear constraints.
2. *Integer programming* (IP): IP extends linear programming to handle discrete decision variables. It is useful when decision variables must take integer values.
3. *Mixed-integer programming* (MIP): MIP combines both continuous and integer decision variables, making it suitable for a wider range of WSPs.
4. *Nonlinear programming* (NLP): NLP deals with WSPs where the objective function or constraints are nonlinear. It is applied to problems with complex relationships, like in the case of WSPs, where the learning-forgetting phenomenon is factored in.

5. *Dynamic programming* (DP): DP is used to solve WSPs that can be broken down into smaller subproblems, allowing for efficient recursive solutions.
6. *Constraint programming* (CP): CP is used to solve WSPs where variables have constraints on their possible values.
7. *Simulation* (SM): SM involves creating a computer model of a system or process to observe its behavior and analyze different scenarios. It is used when the underlying system is too complex or uncertain to be solved analytically.
8. *Queuing theory* (QT): QT studies the behavior of queues or waiting lines. It is used to optimize service processes, such as call center staffing, traffic management, and manufacturing lines.
9. *Heuristics/Metaheuristics* (H/M): H/M are problem-solving techniques that use rules of thumb or approximation algorithms to find good solutions quickly, especially when exact methods are computationally expensive.
10. *Multiobjective optimization* (MOO): In some situations, there are multiple conflicting objectives to be optimized simultaneously. MOO techniques aim to find a set of solutions that represent trade-offs between these objectives.
11. *Goal programming* (GP): GP is used when there are multiple, often conflicting, objectives to be optimized. It aims to find a solution that minimizes the deviations from specified target values for each objective.
12. *Markov decision processes* (MDPs): MDPs are used in situations where decisions are made over time and have probabilistic outcomes, such as in reinforcement learning and optimization of sequential processes.
13. *Decomposition algorithms* (DAs): DA such as Bender's decomposition, are optimization techniques used to solve large-scale, complex problems by breaking them down into smaller, more manageable subproblems.
14. *Combining data science and mathematical optimization* (DS-MO): DS-MO, an emerging trend, integrates data science tools with optimization models to enhance decision-making in WSPs. By incorporating real-world data and historical patterns, this method leads to more effective, data-driven scheduling decisions.

Furthermore, WSPs can be addressed through deterministic or robust approaches. The choice between these two paradigms hinges on the nature of the problem and the modeling assumptions employed. On one side, deterministic methods assume that the parameters used are fixed and known; therefore, any perturbation around the preestablished parameter values could render the initially found solution suboptimal or even infeasible (Henao et al., 2023). On the other hand, robust methods explicitly incorporate uncertainty in various parameters of the WSP within the mathematical formulation (Mercado et al., 2022). The choice of the most suitable approach depends on factors such as the complexity of the problem, the availability of data, and the desired level of robustness when dealing with real-world scheduling scenarios (Henao et al., 2022). Here are some commonly used robust methods for handling WSPs with uncertain parameters:

1. *Closed-form* (CF): CF is an equation that provides an exact, algebraic solution for a WSP without the need for iterative or numerical techniques. Partial knowledge of the stochastic parameter is used to formulate a CF equation.

2. *Robust optimization* (RO): RO focuses on finding solutions that perform well under a range of uncertain scenarios. It seeks to minimize the worst-case outcome, considering variations in parameter values while maintaining feasibility. This approach does not require knowledge of the probability distribution associated with each uncertain parameter.
3. *Two-stage stochastic optimization* (TSSO): In TSSO, decisions are made in two stages. Initially, a decision is made based on available information, and then, after uncertainty is resolved, adjustments are made to optimize the objective further. Unlike RO, TSSO requires full knowledge of the probability distribution.
4. *Distributionally robust optimization* (DRO): DRO is an approach in which optimal solutions are evaluated by considering the worst-case scenario among a set of probability distributions for uncertain variables. This method relies on partial distributional information, which differentiates it from the RO and TSSO methods.
5. *Fuzzy optimization* (FO): FO handles uncertainty using fuzzy logic and fuzzy sets. It allows for the representation of imprecise information and vagueness in decision-making, making it suitable for problems with subjective or uncertain data.

Finally, by selecting and using the most appropriate solution techniques for each WSP, employers can more effectively evaluate the effects of flexible work arrangements on sustainable workforce management. As a result, businesses can create a more streamlined and enduring workforce management system, allowing for improved resource allocation, cost reduction, heightened overall productivity, and increased employee satisfaction. This aligns with the objectives of Target 8.2 from SDG 8, which seeks to attain elevated levels of economic productivity.

4. Sustainable workforce management practices

This section discusses the importance of establishing regulatory frameworks that encourage and promote sustainable workforce management practices. These frameworks serve as pivotal guidelines that align the developmental goals of both employers and employees. Thus, these frameworks should primarily originate from state regulatory agencies responsible for labor and workplace standards. Furthermore, as employers adhere to these regulatory frameworks, they also have the capacity to institute their own regulations that prioritize the well-being of both their employees and their businesses.

This discussion is pertinent because, in prior sections, we explored the potential of adopting flexible work arrangements and utilizing operations research techniques to enhance business productivity and enhance employee satisfaction. However, it is crucial to acknowledge that, without due consideration for the well-being and preferences of employees in the formulation of workforce management practices, improvements in business productivity may inadvertently undermine the creation and retention of quality jobs. Thus, below, we briefly offer a set of recommendations or beginnings on how regulatory frameworks could be crafted to successfully implement each of the seven emerging trends outlined in subsection 4.2. It is understood that successful implementation is one in which, by aligning employer and employee objectives effectively, it is possible to maximize both business productivity and employee well-being.

4.1 Job sharing

Regulatory frameworks should include several key aspects to support successful job-sharing arrangements. First and foremost, they should offer flexibility, allowing employers to establish job-sharing agreements while defining responsibilities and providing guidelines for seamless coordination. Equitable compensation should also be a cornerstone, ensuring that compensation and benefits are distributed fairly among job sharers, taking into account their shared responsibilities. Moreover, these frameworks should establish communication standards to facilitate effective interaction between job sharers and their employers, ultimately promoting smooth collaboration.

4.2 Remote work

To support effective remote work practices, regulatory frameworks should involve several essential components. Firstly, they should include clear telecommuting regulations, defining expectations for work hours and goals, equipment, data security, and employee well-being. Secondly, these frameworks should encourage employers to establish remote work policies that prioritize work-life balance and flexibility for their employees. Lastly, performance evaluation criteria should be established to ensure fair treatment of remote employees, thereby fostering a conducive remote work environment.

4.3 Flexible work schedules

Regulatory frameworks should include several crucial components to adequately tackle the challenges linked to the implementation of flexible work schedules. First, ensure strict regulatory compliance by conducting thorough research on labor laws and adapting policies accordingly. This proactive approach will help navigate complexities as labor laws evolve. Second, tackle coordination challenges in larger organizations by implementing scheduling software and providing training to supervisors or managers responsible for scheduling. Third, address implementation costs by prioritizing fair compensation and benefits for employees with less desirable schedules, alongside transparent and equitable compensation structures. Additionally, consider offering nonmonetary incentives, such as extra time off or flexible benefits, to mitigate challenges associated with these schedules.

4.4 Multiskilling initiatives

To effectively implement and manage multiskilling initiatives, regulatory frameworks should encompass several crucial elements. Firstly, companies should implement efficient training programs, emphasizing essential skills rather than full multiskilling, thus minimizing both required training times and financial investments. Secondly, these frameworks should ensure that employees receive comprehensive and ongoing training to acquire and sustain a diverse skill set, thereby preventing declines in learning and productivity due to the forgetting phenomenon. Thirdly, optimization techniques should be implemented to efficiently allocate tasks and work shifts to multiskilled employees, optimizing the utilization of this workforce. Fourthly, to promote a culture of continuous learning and skill development to effectively address resistance to change. Offering incentives and advancement opportunities can further motivate employees to become multiskilled. Lastly, develop workload allocation strategies that distribute tasks fairly among multiskilled employees, while implementing monitoring systems to prevent burnout and maintain a balanced workload.

4.5 Employee retention benefits packages

To enhance employee retention and satisfaction through benefits packages, employers can establish regulatory frameworks that incorporate several essential elements. Firstly, they should define clear guidelines for competitive compensation, including considerations for bonuses and performance-based raises. Secondly, these frameworks should prioritize equity, ensuring that each employee receives the same benefits if they achieve the same level of productivity, thus avoiding preferential treatment. Thirdly, they should encourage the adoption of flexible work arrangements and create avenues for career development within the organization. By addressing these aspects, regulatory frameworks can contribute to creating a positive and motivating work environment. Finally, state regulatory agencies could establish regulatory frameworks where tax advantages or other benefits are provided to companies that invest in employee benefit packages.

4.6 Promotion of diversity and inclusion

Regulatory frameworks should incorporate various essential elements to advance diversity and inclusion effectively in the workplace. Firstly, they should define practices that encourage the recruitment and retention of a diverse workforce. Secondly, continuous training programs should be implemented to nurture an inclusive culture that values and respects diverse perspectives, effectively manages potential resistance to change among certain current employees, and provides tools for effective communication among employees from different backgrounds. Thirdly, these frameworks should address workplace disparities and ensure equal opportunities for all employees. By addressing these aspects, regulatory frameworks can contribute to a more equitable and harmonious work environment.

4.7 Workforce management 4.0

To foster a thriving workforce in the digital age, it is imperative to consider several key elements within regulatory frameworks. First and foremost, there ought to be a strong emphasis on skills development, encouraging ongoing training and upskilling initiatives to prepare employees for emerging technologies. Secondly, stringent data management guidelines must be established, defining data security and privacy regulations to safeguard sensitive information effectively. Thirdly, it is essential to promote change management strategies within these frameworks to adeptly address any resistance that may arise when implementing automation and digitization processes. By encompassing these elements, regulatory frameworks can facilitate a seamless transition into the digital landscape while ensuring the well-being and readiness of the workforce.

Finally, it can be affirmed that the effective formulation of regulatory frameworks paves the way for the adoption of sustainable workforce management practices that mutually benefit both employers and employees. This aligns seamlessly with Target 8.3, which advocates for the development of policies that bolster productive endeavors, facilitate the creation of quality jobs, foster entrepreneurship, spur creativity and innovation, and promote the formalization and expansion of enterprises of various scales.

5. Overall conclusion

This chapter has explored three interconnected topics that, when combined, offer a comprehensive approach to tackling contemporary challenges in workforce management and achieving SDG 8's targets. The initial topic introduced seven emerging trends, including flexible work arrangements, highlighting the importance of adapting to evolving labor dynamics. The subsequent topic extended this discussion by presenting a comprehensive overview of solution approaches for complex workforce scheduling problems, demonstrating the applicability of various operations research techniques and methodologies. Lastly, the third topic underscored the critical role of regulatory frameworks in promoting sustainable workforce management practices, emphasizing the need to balance business productivity with employee well-being through guidelines and policies.

Together, these topics offer a cohesive framework for decision-makers in workforce management. They not only identify the evolving landscape of labor practices but also provide tools, methods, and regulatory insights to effectively navigate this terrain. The discussions presented on these three interrelated topics enable decision-makers to harness the potential of flexible work arrangements, optimize scheduling, and promote sustainability while fostering a work environment that values diversity, inclusion, and employee satisfaction. In essence, these topics collectively empower decision-makers with a comprehensive toolkit to address the multifaceted challenges and opportunities in contemporary workforce management.

6. Declaration of generative AI and AI-assisted technologies in the writing process author-disclosure

During the preparation of this work, the authors used ChatGPT to improve the clarity, coherence, and conciseness of some paragraphs because the authors are not English native speakers. After using this tool/service, the authors reviewed and edited the content as needed and took full responsibility for the content of the publication.

References

Abello, M.A., Ospina, N.M., De la Ossa, J.M., Henao, C.A., González, V.I., 2021. Using the k-chaining approach to solve a stochastic days-off-scheduling problem in a retail store. In: Rossit, D.A., Tohmé, F., Mejía, G. (Eds.), Production Research. ICPR-Americas 2020, Communications in Computer and Information Science, vol 1407. Springer, Cham. https://doi.org/10.1007/978-3-030-76307-7_12.

Álvarez, E., Ferrer, J.C., Muñoz, J.C., Henao, C.A., 2020. Efficient shift scheduling with multiple breaks for fulltime employees: a retail industry case. Computers & Industrial Engineering 150, 106884. https://doi.org/10.1016/j.cie.2020.106884.

Arkin, W., Dobrofsky, L.R., 2022. Job sharing. In: Working Couples. Routledge, pp. 122–137. https://doi.org/10.4324/9781003276159.

Balkin, D.B., Werner, S., 2023. Theorizing the relationship between discretionary employee benefits and individual performance. Human Resource Management Review 33 (1), 100901. https://doi.org/10.1016/j.hrmr.2022.100901.

Battaïa, O., Dolgui, A., 2022. Hybridizations in line balancing problems: a comprehensive review on new trends and formulations. International Journal of Production Economics 108673. https://doi.org/10.1016/j.ijpe.2022.108673.

Calzavara, M., Battini, D., Bogataj, D., Sgarbossa, F., Zennaro, I., 2020. Ageing workforce management in manufacturing systems: state of the art and future research agenda. International Journal of Production Research 58 (3), 729–747. https://doi.org/10.1080/00207543.2019.1600759.

Chatterjee, S., Chaudhuri, R., Vrontis, D., 2022. Does remote work flexibility enhance organization performance? Moderating role of organization policy and top management support. Journal of Business Research 139, 1501–1512. https://doi.org/10.1016/j.jbusres.2021.10.069.

Coetzee, M., Stoltz, E., 2015. Employees' satisfaction with retention factors: exploring the role of career adaptability. Journal of Vocational Behavior 89, 83–91. https://doi.org/10.1016/j.jvb.2015.04.012.

Crampton, S., Douglas, C., Hodge, J., Mishra, J., 2003. Job sharing: challenges and opportunities. Seidman Business Review 9 (1), 11.

Cuevas, R., Ferrer, J.C., Klapp, M., Muñoz, J.C., 2016. A mixed integer programming approach to multi-skilled workforce scheduling. Journal of Scheduling 19 (1), 91–106. https://doi.org/10.1007/s10951-015-0450-0.

Fontalvo Echavez, O., Fuentes Quintero, L., Henao, C.A., González, V.I., 2021. Two-stage stochastic optimization model for personnel days-off scheduling using closed-chained multiskilling structures. In: Rossit, D.A., Tohmé, F., Mejía, G. (Eds.), Production Research. ICPR-Americas 2020, Communications in Computer and Information Science, vol 1407. Springer, Cham. https://doi.org/10.1007/978-3-030-76307-7_2.

Galaske, N., Arndt, A., Friedrich, H., Bettenhausen, K.D., Anderl, R., 2018. Workforce management 4.0 - assessment of human factors readiness towards digital manufacturing. In: Trzcielinski, S. (Ed.), Advances in Ergonomics of Manufacturing: Managing the Enterprise of the Future. AHFE 2017, Advances in Intelligent Systems and Computing, vol 606. Springer, Cham. https://doi.org/10.1007/978-3-319-60474-9_10.

Groen, B.A., Van Triest, S.P., Coers, M., Wtenweerde, N., 2018. Managing flexible work arrangements: teleworking and output controls. European Management Journal 36 (6), 727–735. https://doi.org/10.1016/j.emj.2018.01.007.

Henao, C.A., 2015. Diseño de una Fuerza Laboral Polifuncional Para el Sector Servicios: Caso Aplicado a la Industria del Retail (Tesis Doctoral, Pontificia Universidad Católica de Chile, Santiago, Chile). Available: https://repositorio.uc.cl/handle/11534/11764.

Henao, C.A., Muñoz, J.C., Ferrer, J.C., 2015. The impact of multi-skilling on personnel scheduling in the service sector: a retail industry case. Journal of the Operational Research Society 66 (12), 1949–1959. https://doi.org/10.1057/jors.2015.9.

Henao, C.A., Ferrer, J.C., Muñoz, J.C., Vera, J., 2016. Multiskilling with closed chains in a service industry: a robust optimization approach. International Journal of Production Economics 179, 166–178. https://doi.org/10.1016/j.ijpe.2016.06.013.

Henao, C.A., Muñoz, J.C., Ferrer, J.C., 2019. Multiskilled workforce management by utilizing closed chains under uncertain demand: a retail industry case. Computers & Industrial Engineering 127, 74–88. https://doi.org/10.1016/j.cie.2018.11.061.

Henao, C.A., Batista, A., Porto, A.F., González, V.I., 2022. Multiskilled personnel assignment problem under uncertain demand: a benchmarking analysis. Mathematical Biosciences and Engineering 19 (5), 4946–4975. https://doi.org/10.3934/mbe.2022232.

Henao, C.A., Mercado, Y.A., González, V.I., Lüer-Villagra, A., 2023. Multiskilled personnel assignment with k-chaining considering the learning-forgetting phenomena. International Journal of Production Economics 265, 109018. https://doi.org/10.1016/j.ijpe.2023.109018.

Henao, C.A., Porto, A.F., González, V.I., 2024. A benchmark dataset for the retail multiskilled personnel planning under uncertain demand. Data Science. In press.

Ibn-Mohammed, T., Mustapha, K.B., Godsell, J., Adamu, Z., Babatunde, K.A., Akintade, D.D., Koh, S.C.L., 2021. A critical analysis of the impacts of COVID-19 on the global economy and ecosystems and opportunities for circular economy strategies. Resources, Conservation and Recycling 164, 105169. https://doi.org/10.1016/j.resconrec.2020.105169.

Katiraee, N., Battini, D., Battaïa, O., Calzavara, M., 2019. Human diversity factors in production system modelling and design: state of the art and future researches. IFAC-PapersOnLine 52 (13), 2544–2549. https://doi.org/10.1016/j.ifacol.2019.11.589.

Katiraee, N., Calzavara, M., Finco, S., Battini, D., Battaïa, O., 2021. Consideration of workers' differences in production systems modelling and design: state of the art and directions for future research. International Journal of Production Research 59 (11), 3237–3268. https://doi.org/10.1080/00207543.2021.1884766.

Khang, A., Rani, S., Gujrati, R., Uygun, H., Gupta, S.K. (Eds.), 2023. Designing Workforce Management Systems for Industry 4.0: Data-Centric and AI-Enabled Approaches. CRC Press. https://doi.org/10.1201/9781003357070.

Kiermaier, F., Frey, M., Bard, J.F., 2020. The flexible break assignment problem for large tour scheduling problems with an application to airport ground handlers. Journal of Scheduling 23, 177–209. https://doi.org/10.1007/s10951-019-00635-5.

Lusa, A., Corominas, A., Munoz, N., 2008. A multistage scenario optimisation procedure to plan annualised working hours under demand uncertainty. International Journal of Production Economics 113 (2), 957–968. https://doi.org/10.1016/j.ijpe.2007.11.009.

Mac-Vicar, M., Ferrer, J.C., Muñoz, J.C., Henao, C.A., 2017. Real-time recovering strategies on personnel scheduling in the retail industry. Computers & Industrial Engineering 113, 589–601. https://doi.org/10.1016/j.cie.2017.09.045.

Memili, E., Patel, P.C., Holt, D.T., Swab, R.G., 2023. Family-friendly work practices in family firms: a study investigating job satisfaction. Journal of Business Research 164, 114023. https://doi.org/10.1016/j.jbusres.2023.114023.

Mercado, Y.A., Henao, C.A., 2021. Benefits of multiskilling in the retail industry: k-chaining approach with uncertain demand. In: Rossit, D.A., Tohmé, F., Mejía, G. (Eds.), Production Research. ICPR-Americas 2020, Communications in Computer and Information Science, vol 1407. Springer, Cham. https://doi.org/10.1007/978-3-030-76307-7_10.

Mercado, Y.A., Henao, C.A., González, V.I., 2022. A two-stage stochastic optimization model for the retail multiskilled personnel scheduling problem: a k-chaining policy with $k\geq2$. Mathematical Biosciences and Engineering 19 (1), 892–917. https://doi.org/10.3934/mbe.2022041.

Milner, J., Pinker, E., 2001. Contingent labor contracting under demand and supply uncertainty. Management Science 47 (8), 1046–1062. https://doi.org/10.1287/mnsc.47.8.1046.10233.

Muñoz, R., Muñoz, J.C., Ferrer, J.C, González, V.I., Henao, C.A., 2024. When should shelf stocking be done at night? A workforce management optimization approach for retailers. Computers & Industrial Engineering 190, 110025. https://doi.org/10.1016/j.cie.2024.110025.

Munyon, T.P., LeClaire, C., Pace, L., Boldin, T., 2023. What makes a compressed workweek successful? Organizational Dynamics 52 (2), 100982. https://doi.org/10.1016/j.orgdyn.2023.100982.

Pianese, T., Errichiello, L., da Cunha, J.V., 2023. Organizational control in the context of remote working: a synthesis of empirical findings and a research agenda. European Management Review 20 (2), 326–345. https://doi.org/10.1111/emre.12515.

Porto, A.F., Henao, C.A., López-Ospina, H., González, E.R., 2019. Hybrid flexibility strategy on personnel scheduling: retail case study. Computers & Industrial Engineering 133, 220–230. https://doi.org/10.1016/j.cie.2019.04.049.

Porto, A.F., Henao, C.A., López-Ospina, H., González, E.R., González, V.I., 2020. Dataset for solving a hybrid flexibility strategy on personnel scheduling problem in the retail industry. Data in Brief 106066. https://doi.org/10.1016/j.dib.2020.106066.

Porto, A.F., Henao, C.A., Lusa, A., Polo Mejía, O., Porto Solano, R., 2022. Solving a staffing problem with annualized hours, multiskilling with 2-chaining, and overtime: a retail industry case. Computers & Industrial Engineering 167, 107999. https://doi.org/10.1016/j.cie.2022.107999.

Porto, A.F., Henao, C.A., Lusa, A., Porto-Barceló, R.I., 2024. Robust optimization model solving an annual multiskilled staffing problem for retail industry. In: Bautista-Valhondo, J., Mateo-Doll, M., Lusa, A., Pastor-Moreno, R. (Eds.), Proceedings of the 17th International Conference on Industrial Engineering and Industrial Management (ICIEIM) – XXVII Congreso de Ingeniería de Organización (CIO2023), Lecture Notes on Data Engineering and Communications Technologies, vol 206. Springer, Cham.

Porto, A.F., Lusa, A., Henao, C.A., Porto, R., 2023a. Planning annualized hours with flexible contracts. In: García Márquez, F.P., Segovia Ramírez, I., Bernalte Sánchez, P.J., Muñoz del Río, A. (Eds.), IoT and Data Science in Engineering Management. CIO 2022, Lecture Notes on Data Engineering and Communications Technologies, vol 160. Springer, Cham, pp. 374–378. https://doi.org/10.1007/978-3-031-27915-7_66.

Porto, A.F., Lusa, A., Henao, C.A., Porto Solano, R., 2023b. Annualized hours, multiskilling, and overtime on annual staffing problem: a two-stage stochastic approach. In: Izquierdo, L.R., Santos, J.I., Lavios, J.J., Ahedo, V. (Eds.), Industry 4.0: The Power of Data. CIO 2021, Lecture Notes in Management and Industrial Engineering. Springer, Cham. https://doi.org/10.1007/978-3-031-29382-5_12.

Rekik, M., Cordeau, J.F., Soumis, F., 2010. Implicit shift scheduling with multiple breaks and work stretch duration restrictions. Journal of Scheduling 13 (1), 49–75. https://doi.org/10.1007/s10951-009-0114-z.

Restrepo, M.I., Gendron, B., Rousseau, L.M., 2017. A two-stage stochastic programming approach for multi-activity tour scheduling. European Journal of Operational Research 262 (2), 620–635. https://doi.org/10.1016/j.ejor.2017.04.055.

Simchi-Levi, D., Wei, Y., 2012. Understanding the performance of the long chain and sparse designs in process flexibility. Operations Research 60 (5), 1125–1141. https://doi.org/10.1287/opre.1120.1081.

Simchi-Levi, D., Wei, Y., 2015. Worst-case analysis of process flexibility designs. Journal Operations Research 63 (1), 166–185. https://doi.org/10.1287/opre.2014.1334.

Spieske, A., Gebhardt, M., Kopyto, M., Birkel, H., Hartmann, E., 2023. The future of industry 4.0 and supply chain resilience after the COVID-19 pandemic: empirical evidence from a Delphi study. Computers & Industrial Engineering 181, 109344. https://doi.org/10.1016/j.cie.2023.109344.

Taskiran Gamze, K., Zhang, X., 2017. Mathematical models and solution approach for cross-training staff scheduling at call centers. Computers & Operations Research 87, 258–269. https://doi.org/10.1016/j.cor.2016.07.001.

Vartiainen, M., 2021. Telework and remote work. In: Oxford Research Encyclopedia of Psychology.

Vergara, S., Del Villar, J., Masson, J., Pérez, N., Henao, C.A., González, V.I., 2021. Impact of labor productivity and multiskilling on staff management: a retail industry case. In: Rossit, D.A., Tohmé, F., Mejía, G. (Eds.), Production Research. ICPR-Americas 2020, Communications in Computer and Information Science, vol 1408. Springer, Cham. https://doi.org/10.1007/978-3-030-76310-7_18.

Wang, X., Zhang, J., 2015. Process flexibility: a distribution-free bound on the performance of k-chain. Operations Research 63 (3), 555–571. https://doi.org/10.1287/opre.2015.1370.

Wontorczyk, A., Rożnowski, B., 2022. Remote, hybrid, and on-site work during the SARS-CoV-2 pandemic and the consequences for stress and work engagement. International Journal of Environmental Research and Public Health 19 (4), 2400. https://doi.org/10.3390/ijerph19042400.

CHAPTER

5

The challenges of undertaking and innovating with sustainable socio-environmental impact businesses: Analysis of narratives from socio-environmental entrepreneurs in southern Bahia

Aline Patrícia Mano[1], *Rafael Bertoldo dos Santos*[2] *and Katianny Gomes Santana Estival*[3]

[1]Department of Engineering and Computing, Universidade Estadual de Santa Cruz, Ilhéus, Brazil; [2]Broto Incubadora Universitária, Universidade Estadual de Santa Cruz, Ilhéus, Brazil; [3]Department of Administrative and Accounting Sciences, Universidade Estadual de Santa Cruz, Ilhéus, Brazil

1. Introduction

In the 21st century, social businesses (originated nomenclature) emerge as an alternative for combating poverty and providing solutions for generating work and income with a focus on positive and relevant socio-environmental outcomes. The term "social business" was coined by Muhammad Yunus, founder of the Grameen Bank in Bangladesh, a social microcredit bank for poor entrepreneurial mothers, in 1976. Yunus (2010) and Gonçalves-Dias and Nakagawa (2019).

A study by AUPA highlighted the likely emergence of social businesses in Brazil. The document initially mentions Catholic philanthropy and some pioneers of initiatives in the field of social entrepreneurship, accelerators, incubators, and the like, such as: Ashoka Brasil since

Evolution and Trends of Sustainable Approaches
https://doi.org/10.1016/B978-0-443-21651-0.00011-5

1986; AVINA Foundation since 1994; Banco Palmas/Fortaleza since 1998; Potencia Ventures/Artemisia since 2004; and others.

The activities of social businesses are closely related to, if not almost all, of the Sustainable Development Goals (SDGs). In this research, we will focus on the SDGs that have a convergence with sustainability, such as SDG 6—Clean Water and Sanitation; SDG 7—Affordable and Clean Energy; SDG 11—Sustainable Cities and Communities; SDG 12—Responsible Consumption and Production; and SDG 13—Climate Action. Businesses operating in these sectors are referred to as socio-environmental businesses.

The social vulnerability in Brazil is evident. It is recorded that over 33 million Brazilians experienced severe food insecurity, according to data from the second National Survey on Food Insecurity in the Context of the COVID-19 Pandemic in Brazil, conducted by the Brazilian Network for Research on Food Sovereignty and Security in 2022. In Bahia, one of the poorest states in the country, according to data from the PENSSAN Network, hunger affected 1.9 million Bahians, representing 12.9% of the state's population. Over half of the population (62.6%) experienced some level of food insecurity, meaning six out of every 10 households in Bahia.

The state of Bahia also faces various challenges related to environmental issues, climate, flora, and sustainability. For example, there are data on deforestation in the state that highlight an alarming circumstance. Between 2020 and 2021, Bahia was the second Brazilian state with the highest loss of native vegetation in the Atlantic Forest, losing 5267 ha of forest, according to the Atlantic Forest Atlas (SOS Mata Atlântica Foundation and INPE).

In 2021, water and supply crizes were also recorded in the state of Bahia. The state also gained national attention when it was hit by floods at the beginning of the same year, revealing environmental, climatic, and human-induced imbalances. These scenarios find alternatives in socio-environmental businesses that mitigate environmental impacts, carbon sequestration, promote clean energy, sanitation, resource reuse, reverse logistics, etc.

Therefore, considering the relevance of socio-environmental businesses, we emphasize the importance of this work as an alternative for improving the performance and consolidation of socio-environmental businesses in the southern region of Bahia. In this way, our objective is to analyze the challenges faced by socio-environmental entrepreneurs in the southern region of Bahia, based on their narratives. To achieve this, this chapter begins with the characterization of the profile of socio-environmental entrepreneurs operating in the southern region of Bahia, followed by an analysis of the difficulties faced by this group and the proposal of a set of alternatives to improve the identified challenges.

2. Theoretical reference

2.1 NIS: Concepts and nomenclatures

We will address the various nomenclatures and concepts related to social impact businesses (SIBs). The recent connotation in literary production is highlighted, along with some similarly significant expressions, such as inclusive businesses, sustainable businesses, and social enterprises, among others.

According to Yunus (2010), SIBs must adhere to certain principles to be considered as such: overcoming poverty as the enterprise's goal, financial-socio-environmental sustainability, no distribution of dividends to investors, and profits remaining within the enterprise for its expansion in line with the proposed socio-environmental action principles.

A study titled "What are Impact Businesses," conducted by the Alliance for Impact Investments and Businesses and Pipe Social (what is PIPE SOCIAL? Remember that this chapter will be published in an international book) in 2019, collected over 280 perceptions regarding the theme of social impact. This included concepts and determinants required for an enterprise to be considered an impact business. Concepts were evaluated across 13 countries and 49 specialized organizations, among others. Internationally, the most significant aspect is the intentionality of impact and the impact generated by the enterprise.

The study defined the following concept for social impact (Ice Pipe School, 2019a, p. 30): "enterprises that have a clear intention to address a socio-environmental problem through their main activity … operate according to market logic, with a business model seeking financial returns, and commit to measuring the impact they generate." The same study also delved into characterizing a minimum set of criteria that define a social impact business: (1) intentional resolution of socio-environmental problems; (2) the main activity is impact-focused; (3) pursuit of financial return; and (4) monitoring of the generated impact.

According to SEBRAE (Brazilian Service for Micro and Small Enterprises Support), there has been a growth in innovative social impact initiatives in Brazil. These enterprises emerge from profitable initiatives and offer solutions to socio-environmental problems, while also following market logic; hence, profits can be shared with partners or reinvested (Cruz, 2019).

Divergences in the literature also exist concerning some guidelines for SIBs. Yunus (2010) considers SIBs as enterprises with a mission to solve social problems, financially self-sustainable, and not distributing dividends. On the other hand, Chu (2007) and other authors advocate for the distribution of dividends and profits to SIB investors as a way to attract more investment. This action can enable and ensure the creation of new SIBs, in addition to motivating resource acquisition.

It becomes essential to highlight certain international perspectives. For example, according to Higashi et al. (2017), the European initiative with ties to the social economy centralizes SIB activities around civil society organizations with a nature of public functions focused on the collective. The North American concept considers private companies and organizations with a market logic dedicated to socio-environmental solutions as SIBs. Another European view is that of "social enterprises," directed toward including vulnerable populations in access to services, products, and opportunities. The European Commission (CE) and the Emergence of Social Enterprise in Europe (EMES), internationally recognized bodies, converge in the significance of social enterprises as organizations with a social aspect that aims to provide goods and services to the disadvantaged, with limited profit generation, often independent, and regulated in various European countries. The emergence of social enterprises was inspired by the offer of affordable public services and the creation of opportunities for vulnerable populations (Carnauba, 2018).

The North American perspective creates various segments and possibilities for actions and social impact entrepreneurship, such as corporate philanthropy that directs resources to social programs as a competitive advantage; social-purpose companies operating in the base of the pyramid market with a social mission; and hybrid entities aiming for shareholder profit generation while partially investing in social projects; among others (Carnauba, 2018; Santos et al., 2016).

2.2 SIBs from the socio-environmental entrepreneurship perspective

We begin this section by characterizing entrepreneurship as a culture of scientific and technological knowledge, entrepreneurial spirit, innovation capability, vision for new business opportunities and activities, leadership, and project management. The entrepreneur is an individual with an idea who becomes influenced by the environmental context in which they are embedded, generating goods and services subject to market judgment; the one who makes things happen and forecasts facts to anticipate the organization's future (Dornelas, 2008).

Entrepreneurship was consolidated in Brazil in the 1990s, influenced by the United States in the global dissemination of entrepreneurship. This was due to the culturally spread capitalist "prosperity" promoted by Americans, fueled by globalization and hegemonic imperialism. The emergence of small enterprises in Brazil and their need to remain in the market spurred entrepreneurship. Trends such as the need for microbusiness formalization, combined with unemployment pressure, led the Brazilian Federal Government to create the Entrepreneurial Brazil Program in 1999. This program trained millions of Brazilians and generated various lines of credit and financing for entrepreneurs (Dornelas, 2008).

According to Hisrich et al. (2014), entrepreneurship is vital for creating businesses, business growth, and consequently, a nation's prosperity, with innovation being a key factor. The authors highlight two circumstances related to entrepreneurs and their performance: first, entrepreneurial opportunities equivalent to situations where new services and raw materials can be marketed at costs higher than their production costs, thus generating profit; second, entrepreneurial action corresponding to the creation of new products, processes, and services.

Sustainable entrepreneurship is a recent research area that seeks to combine the concepts of entrepreneurship and sustainability. However, the nature and interaction between these two dimensions create uncertainties. Sustainable entrepreneurship involves the engagement of both social and environmental segments in enterprise development, going beyond basic requirements to demonstrate efficiency in pursuing sustainability. To be considered sustainability-oriented, the enterprise must combine opportunities and intentions to create economic, social, and ecological value simultaneously. The main reasons for sustainable entrepreneurship include the potential of sustainable practices to generate financial gains and the intention of entrepreneurs to contribute directly to social and environmental well-being.

Ebert (2017), CEO of TerpenOil Green Chemistry, highlighted in an article for GVExecutivo Magazine from Fundação Getúlio Vargas, the excerpt about the sustainable entrepreneur and their challenges:

The sustainable entrepreneur is someone who considers environmental, economic, and social aspects in their core business, delivers innovative solutions to how goods and services are consumed, and proposes business formats that contribute to economic sustainability. Their primary goal is, therefore, to maximize the environmental value created through their ideas, based on viable and effective entrepreneurial models. Finding business solutions for environmental challenges requires sustainable entrepreneurs to seek innovative models that simultaneously provide value to the environment and are viewed as feasible by consumer markets. To achieve this, nature itself can serve as inspiration.

Furthermore, it is worth a critique of profound relevance; within entrepreneurship lies the capitalist foundation. According to Ferraz (2022), there exists a portrayal of the entrepreneurial spirit as a panacea for social issues. The author invites reflection on the character of social development through capitalist foundations, which is not equivalent to the path of social transformation. Social entrepreneurship is not the ideal tool to fill the gap left by the State, as it individualizes the resolution of problems that are fundamentally social, attributing them to the entrepreneur's persona. Social, gender, and human empowerment issues stand in stark contrast to capitalism and its essence inherent in entrepreneurship, whether private or social.

3. Methodological procedures

In order to understand the challenges faced by socio-environmental entrepreneurs operating in the southern region of Bahia, Brazil, two case studies were conducted, involving the application of guided interview scripts. The research adopts a qualitative approach based on the analysis of participants' narratives.

3.1 Survey participants

To identify active SIBs in the study region, the southern territory of Bahia, Brazil, initially, the third Impact Business Map of 2021 from PIPE SOCIAL was examined. This map is a national reference for identifying, cataloging, and characterizing the profiles of active SIBs across different regions of Brazil. It contains data on the segment's performance, gender, racial, and educational stratification of the social entrepreneurs responsible for the SIBs, as well as other characteristics related to the social enterprises, such as formalization and legal nature. However, after searching the aforementioned map, no registered SIBs were found in the southern region of Bahia.

Due to this absence, alternatives were sought to identify SIBs operating in the southern region of Bahia, which included searching on social networks and seeking recommendations from researchers working in the area.

We note some uncertainty regarding the number of SIBs operating in the studied territory; nevertheless, contacts for 14 SIBs were identified and obtained, of which 02 (two) had a notable socio-environmental focus.

3.2 Interview application

The primary data source for this research was information collected through interviews, which, according to Gomes (2010), is the method with the highest potential for a single or multiple case studies. However, attention should be paid to instances of imprecize articulation in verbal accounts. Consequently, the information extracted from interviews was cross-referenced with systematically recorded data from the SIBs, such as minutes, protocols, meeting minutes, records of actions taken, notes, recordings of SIB activities, videos, and various documents. This step is vital for validating the facts reported in the interviews, along with other sources of evidence used to enhance the study's reliability.

4. Results and analysis

The Table 5.1 provides a summary of the description of each participating NIS.

It is important to conduct an analysis of the SIBs regarding their social mission and the minimum compliance requirements for classification as SIBs, according to Pipe Ice; Pipe Social (2019a,b). These criteria characterize a social impact business: (1) intentionality in solving socio-environmental problems; (2) the main activity is impact-focused; (3) pursuit of financial return; and (4) monitoring of the generated impact. Table 5.2 shows the classification of the studied SIBs according to these four criteria.

It was found that out of the studied SIBs, only SIB (B) satisfactorily fulfilled the full compliance of the minimum filters proposed by Pipe Social. None of the SIBs have permanent employees, as the majority of them are categorized as MEIs (microentrepreneurs).

Only one (01) SIB initiated its activities with a business plan, SIB (B), as an initiative emerged within the SEBRAE program to encourage and stimulate the emergence of impact businesses. A business plan is an essential document for any entrepreneur seeking success in their ventures. As stated by Professor William Sahlman from Harvard Business School:

> A well-crafted business plan is essential to achieve success in a business. It's the map that guides the entrepreneur on the path to established goals and helps avoid deviations that could compromise the success of the venture.

According to Dave Lavinsky, founder of Growthink and author of the book "Start at the End: How Companies Can Grow Bigger and Faster by Reversing Their Business Plan" in 2012, "a good business plan should be something you actually use on a day-to-day basis to manage your business, not just a document you create once and forget." This quote emphasizes the importance of a business plan not just as a static document but as a dynamic,

TABLE 5.1 Description of NIS research participants.

Identification	Description
NIS (A) Area: Ecotourism and cycle tourism PJ: MEI Phase: Booming business	SIB founded in 2019, a business originating from the 38-year-old social entrepreneur's thesis project, a master's degree holder in ecology. During the pandemic, actions had to be paused, and the entrepreneur dedicated time to skill development. Postpandemic, there was a surge in the cycling tourism sector that allowed the SIB to gain traction. Financial management, marketing, and advertising were handled by the social entrepreneur himself, who is not a professional in those fields.
NIS (B) Area: Project management, sustainable solutions PJ: MEI Phase: Expanding	SIB founded in 2019, with a business plan, under the management of a 29-year-old social entrepreneur and production engineer. A well-established innovation startup operating in the ESG sector; marketing, financial and accounting management, and advertising are entrusted to responsible professionals. Despite the pandemic, customer engagement, partnerships, and advertising were maintained.

Interviews with research participants.

TABLE 5.2 Scale of classification as SIBs (Ice; Pipe Social, 2019a,b).

ID	Item 1	Item 2	Item 3	Item 4
NIS A	Average	High	High	Average
NIS B	High	High	High	High

(A) Intentionality in solving socio-environmental problems; Main activity is impact-focused; Pursuit of financial return; Monitoring of the generated impact. (B) Capacity to meet the items; Nulo (Null): Does not meet the item; Baixa (Low): Low capacity to meet the item; Média (Medium): Medium capacity to meet the item, reasonable; Alta (High): High capacity to meet the item, satisfactory.
Based on survey data carried out by the authors in 2022.

flexible, and innovative management tool that can be updated and adapted based on market changes and business objectives.

Regarding social, financial, or environmental indicators, SIB (A) utilizes a measurement method through indicators. The entrepreneur reported that they are in the testing, consolidation, and enhancement phase of their environmental impact indicators. Their SIB operates in the ecotourism sector.

According to the Socioenvironmental Indicators Guide, Ethos (2018)

> Socioenvironmental and financial indicators are fundamental to understanding the long-term viability of an enterprise. It's not enough to assess only the financial return, but it's also necessary to evaluate how the enterprise impacts society and the environment. After all, an enterprise that generates short-term profit but harms society and the environment in the long run cannot be considered sustainable. Therefore, it's important to evaluate the socioenvironmental and financial indicators of an enterprise, as it provides a comprehensive view of its real contribution to sustainable development.

We also highlight the lack of consolidated methods for measuring the impacts generated by the SIBs, a limitation for most SIBs according to the specialized literature. Among the studied SIBs, none have consolidated indicators to measure the impact generated, which makes it difficult to assess the SIB's mission according to the precepts that determine guidelines for impact businesses. We observe that the studied SIBs are mostly in the phase of consolidating or expanding their businesses.

Another observed characteristic was the performance of multiple functions and tasks in the management of the SIBs by the entrepreneurs. They are responsible for managing media, advertising, finance, seeking new partners, managing teams, communicating with clients, improving new services or products, and more. This is due to the lack of financial resources for delegating strategic, technical, and operational functions. On the other hand, managing a social business is a complex and challenging task, making it difficult for a single person to perform all the necessary functions to ensure the success of the SIB.

Regarding advertising and promotion, we noticed that the main avenue used is social media. The social entrepreneurs themselves manage the SIBs' social media, including photo and video editing, engagement, traffic, and more. It is expected that the increasing reach of social media will allow SIBs to connect with their target audience, build loyal relationships, attract collaborators, investors, and partners.

According to Rogers (2016), "the internet is not just a new technology or channel, it's a revolution that is transforming businesses and society as a whole." Indeed, the Internet is a vital tool for businesses today, offering numerous opportunities for companies to connect with their customers, expand their digital presence, and increase profits. With the widespread access to the Internet worldwide, businesses can now reach global audiences and compete in markets that were previously inaccessible. The Internet is an indispensable tool for businesses in the digital age, providing a multitude of opportunities for businesses of all sizes and sectors.

Here, there is a reflection on the importance of social networks and the Internet with data that demonstrate a glaring limitation regarding the potential loss of social media for businesses. According to IBGE (2020), 29.5% of the population in the state of Bahia has no access to the Internet. These data reveal a structural challenge not only for SIBs but for social and digital inclusion, reflecting on education, citizenship, and other aspects that require inclusive public policies to be addressed.

These data combine a set of elements from a vulnerable social context in a state that sustains critical levels of social, digital, educational exclusion, and related issues. Fundamental issues that challenge social entrepreneurs, the actions of their SIBs, consolidation, expansion, and mission of these businesses with clear intentions to generate positive socioenvironmental impact, with a global vision and local action. However, these actions are deeply compromised due to the social indices presented in the state of Bahia, particularly concerning internet access among its population.

Regarding the impacts and challenges for SIBs during the acute period of the pandemic, some fragments from the transcripts of interviews with social entrepreneurs stand out:

"We had to experiment and diversify new relationship channels to keep up with trends."

"It was necessary to expand our digital presence (…)."

"The pandemic brought an obvious difficulty; we had 8 months with no revenue."

"(…) when things started to close, businesses began to shut down, we felt it."

"I didn't stop because I'm very resilient and honestly, even after 8 months at home with everything in the showroom in one of the rooms, I still managed to sell everything."

"And we had to change the business model in this period (…) regarding all economic, social, and environmental areas."

The reports about diversifying new tools and channels, a digital presence, and developing skills to manage pages, profiles, virtual platforms, etc., were the most recurring challenges reported for maintaining their SIBs during the pandemic period. Investments were reported in technical and operational support and equipment for safer and more stable transmissions to the target audience, actions on social media, live streams, virtual meetings, and others.

Training in digital platforms, digital media, and social media is essential for the success of entrepreneurs, in general, in an increasingly connected world. As Baer and Lemin (2018), a digital marketing specialist, states, "The future of business is digital, and digital training is essential for entrepreneurs' success." With the increasing use of digital technologies for marketing, commerce, and sales, a lack of skills in this area can put a social entrepreneur at a disadvantage. In this regard, the pandemic process prompted the development of this competency by the entrepreneur.

Furthermore, training in digital platforms and social media can help social entrepreneurs reach a larger and more engaged audience. According to Evans (2010), "social networks are a powerful tool for increasing the reach and connection with a business's target audience." Through the strategic use of digital platforms and social networks, social entrepreneurs can improve their online presence, build relationships with potential customers, and increase their chances of success in the digital market.

The pandemic required the reinvention of social entrepreneurs and entrepreneurs in general. Although the period was marked by difficulties and uncertainties, the world was experiencing the same situation, and there was an emphasis on digital commercial relationships and the promotion of business profiles, enterprises, and initiatives on the internet. This generated new skills, markets, and challenges, according to reports from social entrepreneurs.

The resilient quality of social entrepreneurs is noteworthy, as they navigated a period of global crisis with visible local impacts, unemployment, reduced family purchasing power, closures, and business bankruptcies. The interviewees highlighted in the following:

"Things started to improve in the digital world, the biggest contracts we got were through the web."

"Expansion of possibilities for serving external clients."

"We have six episodes of a podcast ready to go, we already have a website all set up, already designed and produced (…)."

"Ecotourism had a boom post-pandemic. In fact, it's enhanced this type of tourism, as many people who used to travel abroad seeking ecotourism experiences have now opted for experiences in Brazil."

"The pandemic kind of benefited us, those acceleration programs, one of them was in-person, but when it changed, we were able to access it online: opportunity created through the virtual medium."

"Sales, on the other hand, grew online, in the virtual world."

"We increased engagement on social media, we were always providing content on environmental education to people through the NIS, our profiles on social media, platforms, etc."

Numerous solutions were adopted by social entrepreneurs during the peak period of the pandemic to maintain their SIBs, transitioning their actions from physical to virtual. Some entrepreneurs reported expanding their client portfolios and contracts through online processes. Another notable point was the boost in e-commerce sales; the shift to the virtual realm required entrepreneurs to adopt practices and skills to adapt their businesses.

According to the "COVID-19 Commerce Insight" report by Salesforce,

> E-commerce grew rapidly as a direct result of the global pandemic. E-commerce has been one of the few sectors of the economy to thrive during the crisis, as consumers shift their purchasing to digital channels to avoid personal and in-store interactions ***Salesforce* (2021).**

A virtual presence is crucial for entrepreneurs in general; according to Statista's research, in 2020, the number of Internet users worldwide surpassed 4.9 billion, representing over 59% of the global population. Moreover, the pandemic accelerated the adoption of online shopping, with a significant 27.6% increase in global e-commerce sales in 2020. Having an online presence is not just about having a website or a social media profile but also about creating an effective digital marketing strategy. According to Salesforce's research (2021), "81% of consumers begin the purchasing process with online research." This highlights the importance of entrepreneurs investing in search engine optimization (SEO), online marketing, advertising, and other digital marketing strategies to ensure that their services and products are easily found by potential customers.

5. Conclusions

This chapter analyzed the challenges of undertaking socio-environmental impact businesses in the southern region of Bahia. The promotion and consolidation of socio-environmental impact businesses are alternatives to mitigate deforestation rates, for example, considering the innovative perspective of these businesses and the socio-environmental solutions they are capable of offering according to the reviewed literature. Our framework was based on analyses of impact businesses, as per Yunus (2010); Ice; Pipe Social (2019a,b); Barki et al. (2013); and Higashi et al. (2017). We also presented counterarguments from authors like Chu (2007), who advocates for distributing dividends and profits to investors of impact businesses as a means to attract capital to the ventures.

From the narratives of social entrepreneurs, it was observed that the vast majority started their businesses out of necessity, without strategic business planning, prior market research, feasibility assessment, guiding instruction, incubation, acceleration, or systemic technical support. The vital absence of these aforementioned elements can be justified by the need for generating work and income upon starting the business. There was also a latency in terms of qualification and specific training for business management, financial management, and marketing management. The analyzed businesses are categorized as microentrepreneurs (MEIs), and this type of entrepreneur tends to be multitasking, lacking space or resources to hire professionals to handle aspects like financial and accounting management, and market analysis, among others.

However, these social entrepreneurs have become dedicated scholars of their businesses, investing in training, qualification, and development of their knowledge, skills, and competencies. It is noteworthy that many deepened their knowledge during the pandemic.

Based on the narratives, we list some of the common aspects identified as challenging for social entrepreneurs.

(1) Elaboration and consolidation of a strategic business plan to initiate, maintain, and expand the impact business. This plan is also responsible for market diagnosis and feasibility analysis of products and services

(2) Training for social entrepreneurs in management: financial-accounting, marketing, digital traffic, as well as acquiring knowledge and proficiency in using management assistance tools (Udemy, 2021).

(3) Development of skills for pricing intangible products and services.

(4) Access to the internet for the target audience (customers). According to IBGE (2020), 29.5% of the population in the state of Bahia has no Internet access.

(5) Developing expertise in proposal formulation for submission and access to investments, resources, and funding from funding agencies.

We also emphasize that entrepreneurship, even socio-environmental, is not a magic solution to generate positive socio-environmental impacts, nor is it a solution to the capitalist system that exploits workers, accumulates wealth, and exacerbates poverty, devastating flora, fauna, polluting rivers, etc. We highlight that impact businesses and social entrepreneurs are strategic and vital actors aligned with the global commitment of the 17 SDGs, of which Brazil is a signatory. Therefore, they are essential components that require encouragement, incentive, and support to promote their objectives.

References

Baer, J., Lemin, D., 2018. Talk Triggers: The Complete Guide to Creating Customers through Word of Mouth. São Paulo: Alta Books.

Barki, E., et al., 2013. Social Impact Businesses in Brazil. São Paulo: Peirópolis.

Carnauba, I.W., 2018. Impact Businesses and Development: Different Perspectives Around the World and the Brazilian Ecosystem. 2018. Undergraduate Thesis in International Relations. University of Brasília, Brasília, p. 54.

Chu, M., 2007. Commercial returns at the base of the pyramid. Innovation: Technology, Governance, Globalization 2 (1–2), 115–146. MIT Press.

Cruz, C., 2019. Impact businesses. In: ABCR Festival 2019.

Dornelas, J.C.A., 2008. Entrepreneurship: Transforming Ideas into Businesses, third ed. Elsevier, Rio de Janeiro.

Ebert, M., Sustainable Entrepreneurship, 2017. Fundação Getúlio Vargas Magazine, V. 16, No. 05, Special Sustainability Section, pp. 37–41. ISSN 1806-8979; eISSN 2965-0747.

Evans, D., 2010. Social Media Marketing: The Next Generation of Business Engagement. John Wiley & Sons, Hoboken, NJ.

Ferraz, J. de M., 2022. Identity trap and critique of social entrepreneurship: exploration of oppression. Revista Katálysis 25 (2), 252–261. Available at: https://doi.org/10.1590/1982-0259.2022.e84255.

Gomes, A.A., 2010. Case study - planning and methods. Nuance: Studies on Education 15 (16), 215–221.

Gonçalves-Dias, S.L.F., Nakagawa, M.H., 2019. Title of the Paper or Book [Place of publication: Publisher]. (Note: If the place of publication or publisher is unknown, use [s.l.: s.n.]).

Higashi, L., Comini, G., D'Amario, E., 2017. Competitiveness and Challenges of Social Businesses: A Case Study from Brazil and Bangladesh. XX SEMEAD - Seminars in Administration. ISSN 2177-3866.

Hisrich, R., Peters, M.P., Shepherd, D.A., 2014. Entrepreneurship, ninth ed. AMGH Editora Ltda, São Paulo, p. 472.

Ice; Pipe Social, 2019a. What Are Impact Businesses: Characteristics Defining Enterprises as Impact Businesses. Instituto de Cidadania Empresarial - ICE.

Ice; Pipe Social, 2019b. What Are Impact Businesses: Characteristics that Define Ventures as Impact Businesses. Instituto de Cidadania Empresarial - ICE, p. 56.

Instituto Brasileiro de Geografia e Estatística (IBGE), 2020. Estimativas da população 2020. IBGE, Rio de Janeiro.

Instituto Ethos, 2018. Title of the Report. [Contributors: List Authors/editors as Last Name, Initials. Include Organizations or Companies Involved]. Fundação Dom Cabral & Pact São Paulo.

Rogers, D.L., 2016. The Digital Transformation Playbook: Rethink Your Business for the Digital Age. Columbia University Press, New York.

Salesforce, 2021. State of the Connected Customer. Salesforce, San Francisco, CA.

Santos, E.C.D.S., Gonçalves, M.C.D.S., Zaganelli, M.V., 2016. Gestão Financeira No Terceiro Setor: Estudo De Caso No Abrigo Sant'Ana – João Pinheiro (Mg). Revista de Estudos e Pesquisas Avançadas Do Terceiro Setor 3 (1), 128. https://doi.org/10.31501/repats.v3i1.7157.

Udemy, 2021. The Skills Gap and the Future of Work 2021. Udemy, San Francisco. Retrieved from: https://research.udemy.com/wp-content/uploads/2021/05/Udemy-Skills-Gap-Report-2021.pdf.

Yunus, M., 2010. Creating a World without Poverty: Social Business and the Future of Capitalism. PublicAffairs.

PART 2

Sustainable operations management

CHAPTER

6

Maintenance in operational condition and obsolescence management in a sustainable operations management context

Pascal Vrignat[1], *Mariem Besbes*[2], *Marc Zolghadri*[2] *and Frédéric Kratz*[3]

[1]PRISME Laboratory, Orléans University, Orléans, France; [2]Quartz Laboratory, ISAE-Supméca, Saint-Ouen, France; [3]PRISME Laboratory, INSA Centre-Val de Loire, Bourges, France

1. Introduction

Since the large-scale industrialization of our planet in the mid-19th century, the mission of maintenance policies has become increasingly crucial over time. Whatever the type of industry, system, application, service, and so on, maintenance in operational condition (MOC) has migrated from initially simple maintenance activities to tasks requiring upstream proactive and robust decisions by experts. These current maintenance policies are now unavoidable, even if several companies still consider maintenance to be an expense rather than an investment. The international community working on this subject under the aegis of "Prognostic and Health Management" has come together to propose various methods and tools that provide strategies for data processing (measurements), anomaly detection, diagnosis, and prognosis, all in line with the challenges associated with risk and cost reduction. These maintenance policies must also consider the issues associated with component obsolescence and shortages, as well as software obsolescence. The armed forces and the defense industry were the first to be dramatically affected by the problems of obsolescence and shortage. This is why the first obsolescence and DMSMS management methods and tools were developed by the US Department of Defense (DoD). The aim of these methods and tools is to reduce the unavailability of equipment while offering solutions at reasonable cost and with

Evolution and Trends of Sustainable Approaches
https://doi.org/10.1016/B978-0-443-21651-0.00004-8

acceptable risks. This chapter is positioned within a relatively broad framework studying the components of the MOC and their links with the management of obsolescence and DMSMS. It is divided into three sections. The first section is devoted to obsolescence and DMSMS. Section 2 presents the issues associated with the MOC theme. Section 3 explains the links between the main MOC activities and the management of obsolescence and DMSMS. We end with a conclusion.

2. Obsolescence and DMSMS: Twin disruptors of industry and society

Today, obsolescence affects many sectors (aerospace, automotive, energy, arms, telecommunications, etc.) (Mellal, 2020; Bartels et al., 2012). The shortage of raw materials and components and the reduction in manufacturing sources, namely diminishing manufacturing sources and material shortages, DMSMS, highly amplified since the first containment due to SARS-CoV-2 and successively aggravated by the blocking of the Suez Canal and the war in Ukraine (triggered in February 2022) and, more generally, international crises and instabilities, are paralyzing many supply chains. Obsolescence refers to being outdated, useless, or unused, while DMSMS refers to the process of rarefaction of raw materials or physical components. These two disruptive phenomena have been shown to have many similarities although they differ in some respects (Sparrow, 2015) for instance, there are similar techniques and methods for dealing with them (DSP 2022; Mandelbaum et al., 2019; Zolghadri et al., 2021; Salas Cordero et al., 2020; Trabelsi et al., 2021). For this reason, and for the sake of simplicity, in this chapter, we will use the term "obsolescence" to group the two phenomena together. Obsolescence has existed since the existence of the first techniques and tools created by man. However, the phenomenon has been attracting public attention, particularly since the early 2000s, with the emergence of so-called "programmed" obsolescence in the media. The latter is characterized by the intention of manufacturers to reduce the operational life of products to encourage customers and consumers to reequip. This article does not look at this type of obsolescence, but rather at obsolescence described as "natural" or "non-programmed," due to changes in knowledge, needs, regulations, techniques, or technology, for example. In everyday language, obsolescence is understood as "Evolution tending to render (something) obsolete" or "Decrease in the use value of a production good due not to material wear and tear, but to technical progress or the appearance of new products." The standard AFNOR (AFNOR, 2006) uses the dictionary definition "A good is obsolete when it is no longer used, or when it is obsolete, which does not necessarily mean that it is no longer available." The international standard IEC 62402 (IEC, 2019) defines obsolescence as "the transition from the state of availability to the state of unavailability of an entity from its manufacturer following the original specification." Other standards and benchmarks should be considered as references (SD-22, 2022; S3000L, 2021). Obsolescence is therefore an effect, not a cause. Quite often, the root causes of obsolescence are to be found in technological overtaking, the disappearance of the original supplier, lack of demand, changes in national or international regulations, and so on. It is now generally accepted that the main reason for obsolescence is linked to the difference between a long life cycle (a few decades for ships, submarines, nuclear power stations, etc.) and the short life cycles of their components (a few months for electronic or software composites) (Sandborn et al., 2007; Zolghadri et al., 2021). It is therefore reasonable to assume that highly complex systems contain components that are obsolete even before they are used for the first time. We define an obsolescent component as one that begins its commercial decline (i.e., at the end of its commercial maturity) before it is considered obsolete. There are several categories of obsolescence: physical or software (Merola, 2006), involuntary or programmed (Maycroft, 2009), direct or indirect, or total or partial (Desai and Mital, 2006; Mastrangelo et al., 2021). The causes of obsolescence are diverse and include, for example:

- the disappearance of original suppliers from the market for economic reasons, technical and technological advances,
- changes in standards and regulations,
- collapse in demand,
- changes in customer tastes, needs, or requirements,

Obsolescence does not only concern electronic products or components. Electrical components (industrial connectors), mechanical components (metal engine parts made from specific alloys), and chemical components (glue, paint) are also becoming obsolete. So, as the US Department of Defense's GuideBook points out, obsolescence is inevitable (DSP 2022).

It is also important to notice that the phenomenon of obsolescence has several consequences: technological, ecological, and even economic and social consequences (Dass and Lokhande, 2023; Grichi et al., 2018). In technological terms, obsolescence contributes to the introduction of technological improvements and innovations. But it can in turn have environmental consequences. For example, solving the problem of obsolescence can lead to overconsumption of materials, and therefore overexploitation of natural resources because more raw materials are needed. In economic terms, obsolescence can lead to a loss of money for consumers by forcing/incentivizing them to buy the same objects more often, with no gain in comfort. And finally, from a social point of view, obsolescence can be indirectly linked to poor working conditions, due to the increase in product manufacturing. Three management methods have been developed to deal with obsolescence (Sandborn, 2007). Reactive management consists of acting when obsolescence has already occurred. Proactive management is implemented for critical components at risk of obsolescence. Strategic management involves determining the optimal combination of mitigation and design update approaches using data from forecasts, logistics, and obsolescence dates.

3. Maintenance in operational condition

Maintenance is one of the main levers for improving the overall performance of industrial systems (Sénéchal and Léger, 2004). The issues at stake can be found in an organization (military or civilian) where several fleets of equipment coexist, often off-center from one another. In concrete terms, they are expressed at the level of the complex production execution system by the concept of MOC, which must guarantee the minimum contractual availability throughout the operational phase. As this phase can be long (e.g., 25 years for certain military equipment, 40 years in the rail industry, and much longer for certain urban infrastructures), we present MOC as a set of activities whose aim is to ensure that the system functions properly throughout the operating phase and to guarantee that it can be adapted to the necessary changes and supported at the end of its life. The most appropriate English translation of MOC is "In-service Support." There is general agreement that it can be broken down into two areas of activity: "technical support," with a strong focus on maintenance techniques and policies per se, and "logistical support," which guarantees the resources in terms of time and content required for the MOC process.

3.1 The maintenance paradigm

Throughout history, and to be more effective, maintenance policies have benefited from and followed the scientific and technological developments that were proposed to them. Fig. 6.1 describes this evolution, which has recently led manufacturers to follow the requirements and needs of "Industry 4.0." The production and/or service environment is made up of different types of equipment. All this equipment must be available and reliable to the highest possible level to guarantee continuity of service. The maintenance manager is responsible for maintaining the equipment in the condition in which it was originally designed and installed.

Several categories of maintenance policies are proposed in the literature (Erkoyuncu et al., 2017). Among these different categories and with the rise of Industry 4.0, condition-based maintenance (CBM) or predictive maintenance are significant examples. Sénéchal and

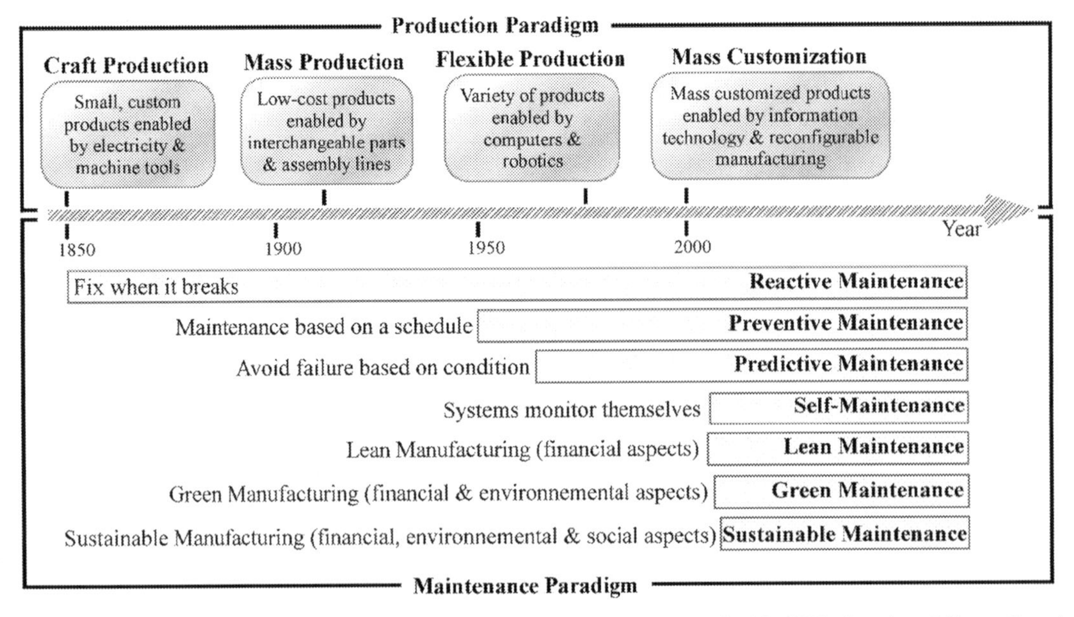

FIGURE 6.1 Maintenance history. *Adapted from Vogl, G.W., Weiss, B.A., Helu, M., 2019. A review of diagnostic and prognostic capabilities and best practices for manufacturing. Journal of Intelligent Manufacturing 30, 79–95 and Alves, A.C., Kahlen, F.-J., Flumerfelt, S., Siriban-Manalang, A. B., 2019. Lean Engineering for Global Development. Springer.*

Trentesaux Séné (2019), propose a new maintenance policy framework called sustainable condition-based maintenance (SCBM). SCBM requires the assessment of remaining life as a key indicator, an adaptation, and an extension of a generic prognostic process initially proposed for traditional condition-based maintenance. SCBM also includes an organization adapted to the prognosis of the environmental performance of systems. Zhang et al. (2021), study optimal opportunistic maintenance based on the health state of the system. In this framework, a state space partitioning approach that considers system degradation and spare parts inventory is proposed to analyze the requirements and actions for different types of maintenance under spare parts availability constraints (storage and ordering). Over the last few decades, maintenance theory has changed radically in line with new manufacturing paradigms (Fig. 6.1). Levitt (2009) defines "lean maintenance" as the provision of maintenance services to customers with as little waste as possible. The desired outcomes of this maintenance policy should minimize labor, spare parts, energy resources, and management effort. The intrinsic characteristic of Lean, associated with maintenance, must be combined with other well-known methods such as 5S, Kanban, and Six Sigma. Green maintenance involves managing maintenance operations in an environmentally friendly way. It includes all maintenance processes and strategies, that is, reactive, preventive, and proactive. This maintenance policy also incorporates the selection of raw materials and components needed to maintain equipment, the purchase, storage, and management of used materials, and the operation of all fluids and lubricants (Franciosi et al., 2018). In this context, the circular economy finds its full meaning in the development of new industrial ecosystems. Finally, we can note that the concept of "sustainable" maintenance appeared a few years ago. This concept implies, among other things, the elimination of breakdowns and other sources of energy waste. This is what led Vrignat et al. (2022), to carry out a literature review to study the role of maintenance in sustainable manufacturing.

3.2 The prognostics and health management community at the service maintenance policies

Prognostics and health management (PHM) for systems is a computationally based paradigm. PHM relies on physical knowledge, information, and data relating to the operation and maintenance of structures, systems, and components (Zio, 2016). The results of a PHM approach make it possible to detect equipment and process anomalies, diagnose degradation states and faults, and predict the evolution of degradation toward failure to estimate the remaining useful life (Fig. 6.2). They are used to support the decisions of a condition-based predictive maintenance policy for the efficient, reliable, and safe operation of a system or component. The data collected from sensors or other measurement and logging tools feeds three major tasks for PHM: (1) fault detection (detecting anomalies), (2) diagnosis (estimating the level of degradation), and (3) prognosis (predicting the remaining useful life (RUL)). Successful deployment of the PHM function, therefore, provides an essential basis for optimal decision-making. The availability of systems or components is then more efficient.

Several classifications have been proposed in the literature (Vrignat et al., 2022). These classifications are influenced by the specific terminologies and contexts of each scientific community. The objective of the PHM is to be able to determine in advance the most appropriate

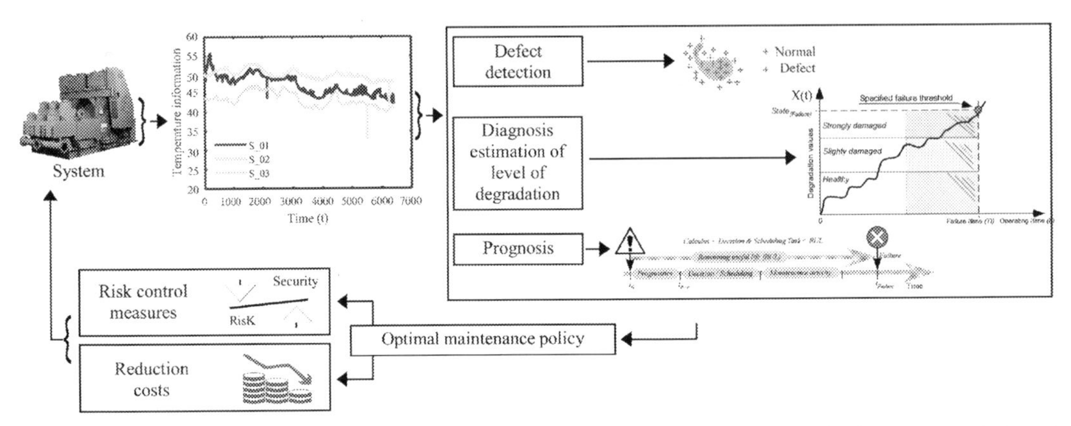

FIGURE 6.2 PHM tasks. *Adapted Zio, E., 2022. Prognostics and Health Management (PHM): where are we and where do we (need to) go in theory and practice. Reliability Engineering & System Safety 218, 108119.*

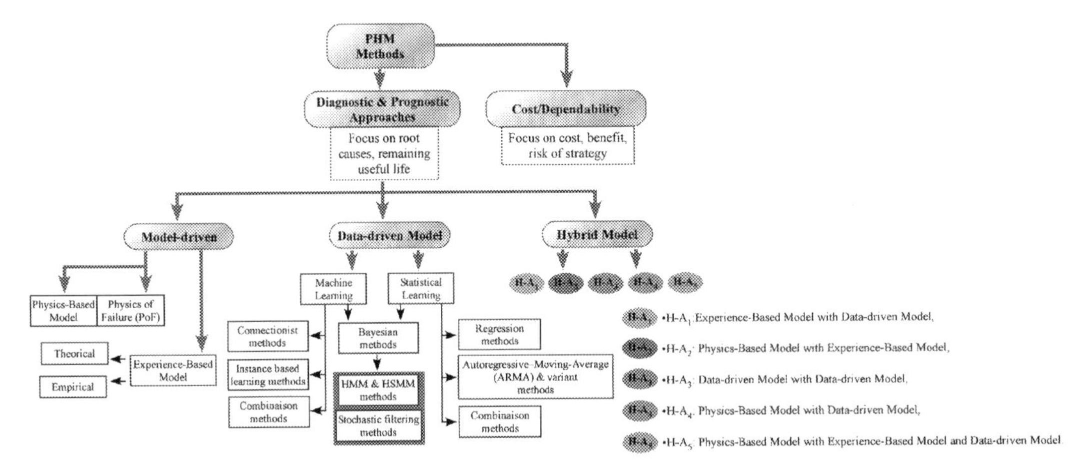

FIGURE 6.3 General description of the PHM methods.

methods for solving the problem of detection and diagnosis. A classification of diagnostic and prognostic methods into three main families is proposed in (Vrignat et al., 2022). Fig. 6.3 illustrates the three main approaches: (1) model-driven, (2) data-driven, and (3) hybrid.

4. Links between obsolescence management and MCO

4.1 Traditional MOC model

The missions entrusted to technical and logistical support are implicitly intertwined in a robust context of maintaining operational conditions and managing system obsolescence. In this context, we can state that "Obsolescence" and "Shortage" are locks that absolutely must be controlled if the maintained system and the associated service guarantee are not to be jeopardized. Fig. 6.4 shows an Actigram that models the interweaving of two key MOC activities from what might be described as a "Traditional" point of view. The availability of a system will be efficient if the manufacturer's specifications, the recommended conditions of use, the documentation, the maintenance policy, the appropriate tools, and the know-how of the people involved are respected and kept up to date. Cost constraints and the associated decisions complete this organizational strategy. Various indicators and direct or indirect measurements are collected over time to provide information for the decision-making process. When a function or a hardware or software component needs replacing, either the original supply or an authorized and qualified alternative supply must be processed. This operational activity for logistical support will be more effective if the supplier respects its contractual commitments and if the stock management policy in the factory, company, or holding company is optimal. It should be noted that this inventory management policy has a direct impact on the company's cost accounting in terms of the stock of components or functions capitalized.

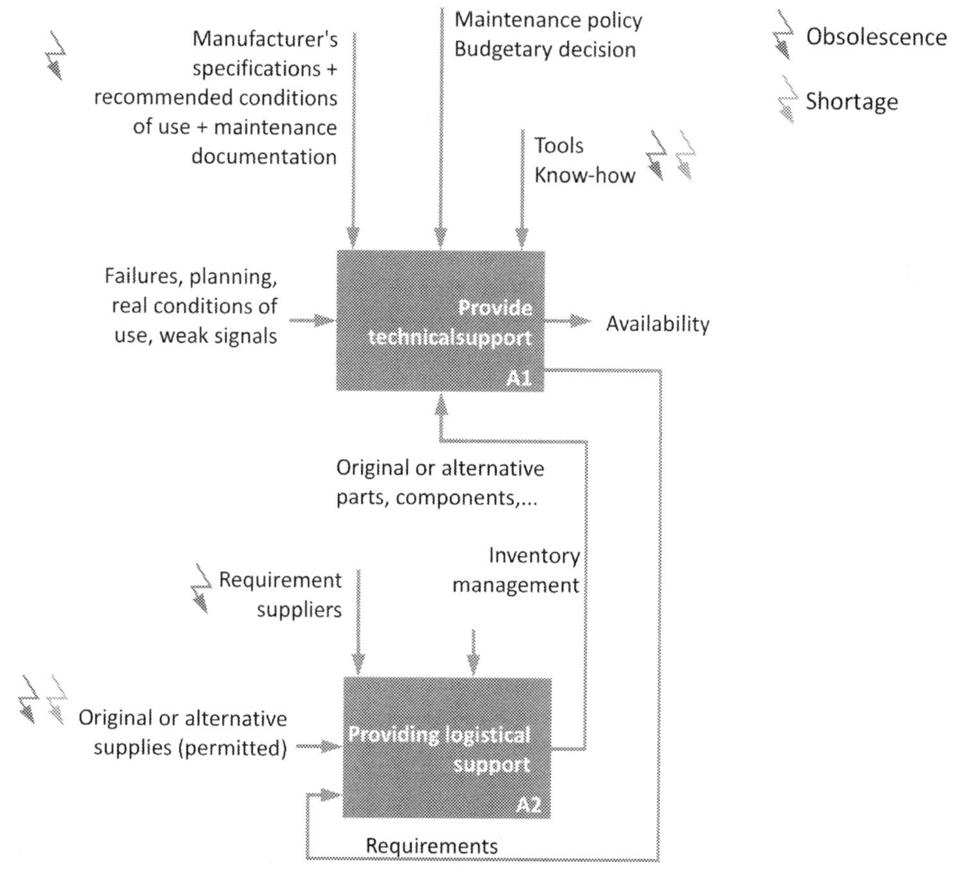

FIGURE 6.4 The MOC's two main activities: technical support and logistical support.

4.2 The MOC model enhanced by obsolescence and shortage management

The model in Fig. 6.4 must be completed by explicitly integrating the two functions "Obsolescence management" and "Shortage management." These two functions, labeled A3 and A4, must support the execution of the two MCO activities, that is, technical support and logistical support, labeled A1 and A2, respectively.

Some of the inputs and constraints of the two activities A1 and A2 may be subject to obsolescence and/or shortage. This is the case for specifications, tools, know-how, and so on. The broken arrows in red represent obsolescence in the sense of being exceeded. The green arrows represent a shortage of physical components, tools, know-how, or documentation. The role of activities A3 and A4 is to detect (by anticipation if possible) and deal with obsolescence and shortage. Cases of obsolescence (entities numbered 1, 2, 3, and 6 in Fig. 6.5) and cases of shortage (i.e., entities numbered 4 and 5) are managed by A3 and A4, respectively. Managing then means:

- determining the potential impact of these cases,
- determining their criticality,
- proposing remedial solutions,
- choosing the most effective and efficient one,
- implementing it and finally,
- monitor the effects.

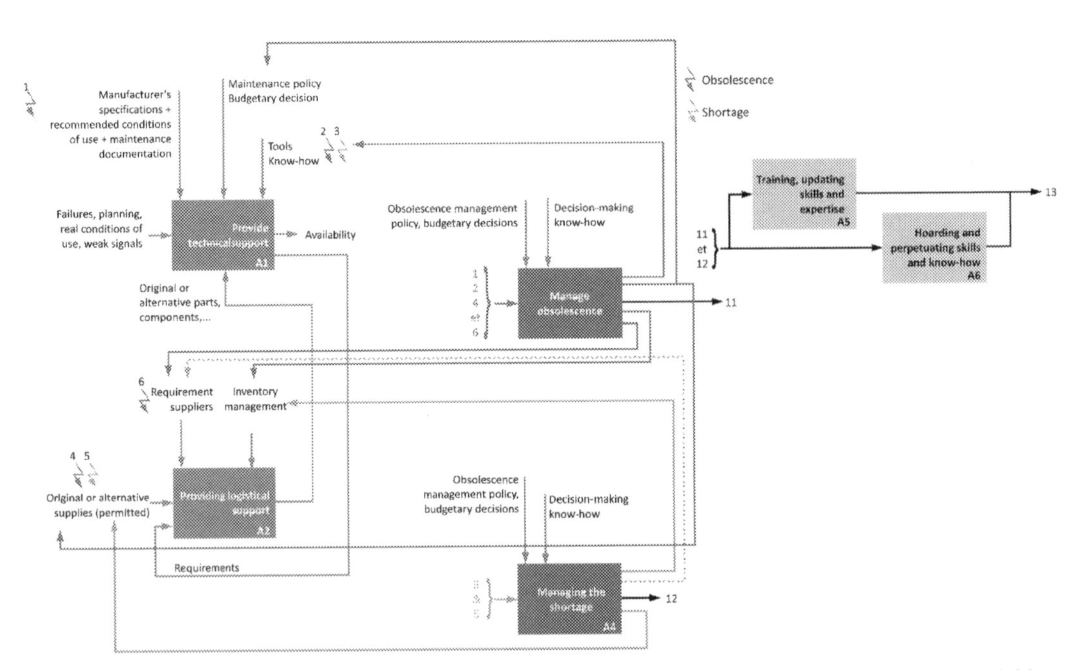

FIGURE 6.5 The MOC model augmented by obsolescence management and shortage management activities.

The authors in (IEC, 2019) or (DSP 2022) propose different remediation solutions (i.e., Last Buy (LBO), substitution, alternative, etc.). Activities A3 and A4 are driven by obsolescence management considerations while considering the efficiency of the various remediation solutions. They are conducted using decision support tools based on obsolescence and shortage management expertise. The feedback links between obsolescence and shortage management activities and technical and logistical support are explicitly represented in the model to formalize the impact that activities A3 and A4 should have on the progress of support activities.

4.3 A1-A3 and A2-A3 links

The "Manage Obsolescence, A3" activity has to monitor the obsolescence of manufacturer specifications (e.g., following a technological change), tooling, know-how, supplies (e.g., components or raw materials) and also requirements relating to suppliers. This presupposes a prior analysis of the entities most at risk of obsolescence. In practice, given the human resources available in companies, combined with the complexity of systems, it is impossible to monitor all the components of a system.

This activity will need to be able to deploy reactive but above all proactive obsolescence management resources. Once remediation solutions have been identified through obsolescence management, they need to be implemented. Companies therefore need to have the appropriate human and financial resources at their disposal. Pragmatically, this means:

- Updating maintenance policies,
- Defining action plans for tooling and know-how (see next subsection),
- Updating stock management policies (stock sizing, latest purchases, etc.)
- Possible changes to substitute or alternative components, and so on, for which all documentation must be maintained.

Shortage or DMSMS management focuses primarily on components. The aim is to protect logistical support against any shortages or even interruptions in supplies and component breakdowns. In certain situations, suppliers propose a time interval during which components remain available via so-called "qualified" channels (e.g., approved suppliers).

Detecting the availability of spare components is not an easy task in certain situations. In a stable component production market, rarefaction is very often linked to technological developments, the disappearance of suppliers, and so on. Shortages can therefore, to a certain extent, be modeled (mathematically) and predicted. But in other situations, such as war and earthquakes, it remains unpredictable. In these cases, you need an agile supply chain and close links with suppliers to identify and implement viable solutions. Once the solution for remedying the component shortage has been determined, the A4 activity must ensure that it is implemented for the benefit of logistical support.

This involves:

- Proposing, if necessary, updates to supplier requirements (i.e., new suppliers),
- Modifying stock management parameters for replacement components, alternatives, and so on,
- Updating supply specifications.

4.4 The particular case of obsolescence and the shortage of know-how and skills

Among the issues facing companies, obsolescence and the shortage of know-how and skills are among the most crucial to consider. Indeed, for systems with a (very) long service life (several decades, for example), it is understandable that technological developments only partially affect them. An aircraft or a train is designed and built using a certain technology at a given time. Once in operation, the technologies used only evolve as a result of operational necessity (e.g., the renewal of a standard) or following modernization planning (e.g., after 10 years or so of operation). As a result, the expertise, know-how, and skills needed to provide technical support for this equipment are becoming increasingly scarce as maintenance staff age and retire. It is, therefore, necessary to consider these maintenance obligations, the departure of maintenance workshop staff, and the documentation that is less and less easily accessible or available. A good example of this is the expertise required to adapt the Voyager 1 probe programs, 45 years after its launch in September 1977! Obsolescence and shortage management must play a key role in this aspect. This is why the model shows outputs 11 and 12 from activities A3 and A4, which are used for the two new activities, A5 and A6. The aim is to maintain the technical level of MCO personnel.

The obsolescence of know-how and skills refers, once again, to the fact that certain know-how or skills are outdated, useless, or even unusable. This may involve outdated technology. It is therefore necessary to have the means to capitalize on this know-how and skills within the company, particularly in the case of systems with a (very) long service life. Faced with the real risk of losing skills, knowledge, and know-how through the retirement of technical staff, but also the accessibility of technical documentation, test systems, maintenance equipment, and so on, the description of activity 6 "Hoarding and perpetuating know-how and skills" stresses the need to have a strategy for managing the company's intangible capital. The second activity modeled explicitly in Fig. 6.5, "Training and updating skills and know-how," is designed to address the shortage of know-how and skills and to ensure that technical staff are aware of technical, technological, and standards-related developments in the system, as well as the changes made to the system during the modernization phase. This involves building up a new body of knowledge, skills, and know-how. Companies have always faced this problem. However, the acceleration of various developments (technical, technological, regulatory, etc.) imposes serious constraints on the proactive management of these shortages.

5. Conclusion

The shortage of electronic components and, more broadly, of materials, and the obsolescence of software, technologies, regulations, and national or international standards are posing serious challenges for all players in the value chain. This can automatically lead to reduced availability of major civil and military infrastructures, assets (manufacturing plants, power stations, etc.), or equipment/systems (aircraft, trains, vehicles, etc.) because of the impossibility of maintaining systems. Repeated crises and various instabilities amplify the unavailability of systems and fleets of systems.

It is with this in mind that, today more than ever, the links that can exist between the management of shortages and obsolescence and the technical and logistical support activities of operational readiness maintenance are modeled and argued in this article.

Research work, enriched by industrial practices, is still needed to make this necessary symbiosis operational. Finally, it should be noted that the International Institute of Obsolescence Management (IIOM International) is a not-for-profit organization that exists to advance the science and practice of Obsolescence Management (https://www.theiiom.org/. International Institute of Obsolescence). IIOM exercises its powers through delegation agreements to independent legal entities in various countries and territories, designated as Chapters. IIOM International is dedicated to improving the knowledge and best practice of obsolescence practitioners through education, networking, and process development. IIOM is for professionals worldwide working in the field of obsolescence management and who want to further their knowledge and understanding of the obsolescence management discipline and who wish to network with other people working in the field.

References

AFNOR, 2006. Maintenance - Termes et définitions des éléments constitutifs des biens et de leur approvisionnement : NF X60-012.

Alves, A.C., Kahlen, F.-J., Flumerfelt, S., Siriban-Manalang, A.B., 2019. Lean Engineering for Global Development. Springer.

Bartels, B., Ermel, U., Sandborn, P., Pecht, M.G., 2012. Strategies to the Prediction, Mitigation and Management of Product Obsolescence. John Wiley & Sons.

Dass, A.K., Lokhande, S., 2023. Machine learning based prediction of obsolescence risk. International Journal of Intelligent Systems and Applications in Engineering 11 (4s), 293–301.

Desai, A., Mital, A., 2006. Design for maintenance: basic concepts and review of literature. International Journal of Product Development 3 (1), 77–121.

DSP, May 2022. Diminishing Manufacturing Sources and Material Shortages, A Guidebook of Best Practices for Implementing a Robust DMSMS Management Program. SD-22.

Erkoyuncu, J.A., Khan, S., Eiroa, A.L., Butler, N., Rushton, K., Brocklebank, S., 2017. Perspectives on trading cost and availability for corrective maintenance at the equipment type level. Reliability Engineering & System Safety 168, 53–69.

Franciosi, C., Iung, B., Miranda, S., Riemma, S., 2018. Maintenance for sustainability in the industry 4.0 context: a scoping literature review. IFAC-PapersOnLine 51 (11), 903–908.

Grichi, Y., Dao, T.-M., Beauregard, Y., 2018. A new approach for optimal obsolescence forecasting based on the random forest (RF) technique and meta-heuristic particle swarm optimization (PSO). In: Proceedings of the International Conference on Industrial Engineering and Operations Management, Paris, France, pp. 26–27.

IEC, 2019. IEC 62402: 2019 - Obolescence Management - Application Guide.

Levitt, J., 2009. The Handbook of Maintenance Management. Industrial Press Inc.

Mandelbaum, J., Patterson, C.M., Brown, R.S., 2019. Efficiencies through targeted interactions between the parts and DMSMS management communities. In: Institute for Defense Analyses Alexandria United States.

Mastrangelo, C.M., Olson, K.A., Summers, D.M., 2021. A risk-based approach to forecasting component obsolescence. Microelectronics Reliability 127, 114330.

Maycroft, N., 2009. Consumption, Planned Obsolescence and Waste. History of Art & Material Culture, University of Lincoln.

Mellal, M.A., 2020. Obsolescence—A review of the literature. Technology in Society 63, 101347.

Merola, L., 2006. The COTS software obsolescence threat. In: Fifth International Conference on Commercial-Off-The-Shelf (COTS)-Based Software Systems (ICCBSS'05). IEEE, p. 7.

S3000L, 2021. International procedure specification for logistics support analysis. In: AeroSpace, Security and Defense Industries Association of Europe (ASD). S-Series IPS specifications, pp. 1–737.

Salas Cordero, S., Zolghadri, M., Vingerhoeds, R., Baron, C., 2020. Identification and assessment of obsolescence in the early stages of system design. Journal of Integrated Design and Process Science 24 (3–4), 15–33.

Sandborn, P., 2007. Software obsolescence-Complicating the part and technology obsolescence management problem. IEEE Transactions on Components and Packaging Technologies 30 (4), 886–888.

Sandborn, P.A., Mauro, F., Knox, R., 2007. A data mining based approach to electronic part obsolescence forecasting. IEEE Transactions on Components and Packaging Technologies 30 (3), 397–401.

SD-22, 2022. A guidebook of best practices for implementing a robust DMSMS management program. In: Executive Agent for the Defense Standardization Program. Diminishing Manufacturing Sources and Material Shortages, pp. 1–335.

Sénéchal, O., Léger, J.-B., 2004. Tele-maintenance for improvement of performances in TPM and RCM. IFAC Proceedings Volumes 37 (5), 221–226.

Sénéchal, O., Trentesaux, D., 2019. A framework to help decision makers to be environmentally aware during the maintenance of cyber physical systems. Environmental Impact Assessment Review 77, 11–22.

Sparrow, R., 2015. Enhancement and obsolescence: avoiding an "enhanced rat race". Kennedy Institute of Ethics Journal 25 (3), 231–260.

Trabelsi, I., Zolghadri, M., Zeddini, B., Barkallah, M., Haddar, M., 2021. Prediction of obsolescence degree as a function of time: a mathematical formulation. Computers in Industry 129, 103470.

Vogl, G.W., Weiss, B.A., Helu, M., 2019. A review of diagnostic and prognostic capabilities and best practices for manufacturing. Journal of Intelligent Manufacturing 30, 79–95.

Vrignat, P., Kratz, F., Avila, M., 2022. Sustainable manufacturing, maintenance policies, prognostics and health management: a literature review. Reliability Engineering & System Safety 218, 108140.

Zhang, X., Liao, H., Zeng, J., Shi, G., Zhao, B., 2021. Optimal condition-based opportunistic maintenance and spare parts provisioning for a two-unit system using a state space partitioning approach. Reliability Engineering & System Safety 209, 107451.

Zio, E., 2016. Some challenges and opportunities in reliability engineering. IEEE Transactions on Reliability 65 (4), 1769–1782.

Zio, E., 2022. Prognostics and Health Management (PHM): where are we and where do we (need to) go in theory and practice. Reliability Engineering & System Safety 218, 108119.

Zolghadri, M., Addouche, S.-A., Baron, C., Soltan, A., Boissie, K., 2021. Obsolescence, rarefaction and their propagation. Research in Engineering Design 32 (4), 451–468.

CHAPTER

7

Circular economy strategies for improving furniture management in educational institutions

Pablo Maya-Duque, Brigitte Roldán-Ramírez, Betzy Cárdenas-Ibáñez and Luisa Jiménez-Ramírez

Departamento de Ingeniería Industrial, Facultad de Ingeniería, Grupo ALIADO, Universidad deAntioquia, Medellín, Colombia

1. Introduction

The current predominant economic model, known as the linear model, is mainly based on the use of large amounts of materials and low-cost, easily accessible energy. In this model, raw materials are transformed through processes that add value to the products, which are sold for use and eventually discarded at the end of their life cycle. However, this model reaches its limits in terms of how it obtains resources that are naturally finite and relies on an increasingly limited space to dispose of the generated waste. Considering this situation, the circular economy (CE) emerges as an alternative characterized by keeping products and materials within the economic system for as long as possible, thus reducing the pressure on resource extraction and waste disposal.

The concept of the circular economy paradigm has gained significant momentum worldwide. Various regions and countries have devised comprehensive plans and policies to promote the adoption of the circular economy. For instance, in 2020, Europe introduced the new Circular Economy Action Plan (CEAP), while China established its Development Plan for the Circular Economy (2021–25). Moreover, developing countries have actively embraced this transition, as evidenced by Colombia and Chile, both of which have already formulated their national circular economy strategies.

Initially, the implementation of the circular economy approach primarily concentrated on specific economic sectors, such as electrical and electronic equipment (WEEE),

Evolution and Trends of Sustainable Approaches
https://doi.org/10.1016/B978-0-443-21651-0.00009-7

automotive, construction, and packaging. However, there is a growing interest among researchers and practitioners in exploring its application in other sectors like textiles and furniture. The approaches vary from strategic and comprehensive studies that support the construction of public policies or sector and company strategies, to operational-level initiatives that aid in the implementation of specific circular economy strategies or business models.

The primary contribution of this chapter is the investigation of the furniture sector as a rapidly growing area within the circular economy paradigm, focusing specifically on universities and educational institutions as the target market segment. To achieve this objective, a literature review is conducted to identify potential circular economy initiatives applicable to this sector. These initiatives are then classified into nine main strategies. Subsequently, the identified strategies are characterized based on an evaluation of five key criteria, providing readers with a comprehensive understanding of their suitability for managing furniture in a circular manner at universities and educational institutions. Furthermore, this chapter delves into one of the identified strategies, the *product as a service* (PaaS), as a potentially disruptive approach within this context. The study of the PaaS strategy identifies essential factors that must be considered when configuring such a strategy and proposes a methodology to address these aspects while taking a holistic view of their interactions. To validate the proposed methodology, a case study is conducted at the Universidad de Antioquia in Colombia.

2. Conceptual framework

Two concepts that are valuable to define and delimit before delving into the description of circular economy strategies in furniture management are the very idea of what the circular economy is, and the context of the waste stream generated by furniture and bulky items.

2.1 Circular economy

As a sustainability paradigm that has gained traction in recent years, the concept of the circular economy does not have a unified definition and corpus of knowledge. However, a comprehensive analysis of 114 definitions conducted by Kirchherr et al. (2017) broadly defines the circular economy as an economic system that replaces the concept of end of life with the reduction, reuse, recycling, and recovery of materials in production, distribution, and consumption processes. It operates at multiple levels: the microlevel (products, companies, consumers), the meso level (eco-industrial parks), and the macro level (city, region, nation, and beyond), with the aim of achieving sustainable development while simultaneously creating environmental and economic quality. A similar definition is provided by Ellen MacArthur Foundation (2015). However, their definition also emphasizes the importance of decoupling economic growth from the consumption of finite resources. This is achieved by maximizing the utilization of products, components, and materials for as long as possible and, at the end of their service life, regenerating or reintegrating them into the production cycle, thereby minimizing waste and environmental impact.

The concept of the circular economy is typically defined by three principles:

- Eliminate and reduce waste and emissions at different stages of the product's life cycle, starting from its design phase, where the highest potential savings lie, up to its final disposal.
- Circulate products, components, and materials while maintaining their maximum value, promoting strategies such as reuse, remanufacturing, revaluation, and recycling.
- Regenerate the system so that elements and nutrients can return to their source and reinforce restoration efforts.

There are various circular economy strategies that enable the adoption of the principles. These strategies encompass a range of approaches, from public and environmental policies to specific business models, effectively covering the complete value chain and lifecycle of products. That is, they address everything from product design, through alternative circulation options to their final disposal or recovery.

2.2 The furniture sector and the bulky waste stream

The furniture sector is a significant contributor to the global economy, with production, consumption, and waste generation playing crucial roles in its dynamics. Over the past decade, furniture production has steadily increased due to rising demand fueled by population growth and urbanization. In the European Union, where around one-third of the furniture sold worldwide is produced, it represents an €84 billion market that employs approximately 1 million workers. Most of the producer companies in the sector are small or medium-sized enterprises, while consumption is mainly distributed among the domestic sector (82%) and business-to-business (B2B) consumption (18%).

However, the rapid growth in production and consumption has led to a concerning increase in furniture waste generation. In 2022, the world generated approximately 30 million tons of furniture waste, with much of it ending up in landfills or incineration sites. According to the European Federation of Furniture Manufacturers (UEA) statistics, furniture waste in the EU accounts for more than 4% of the total municipal solid waste (MSW) stream. In the EU, 80%–90% of it is incinerated or sent to landfill, with only around 10% being recycled. Similarly, the United States Environmental Protection Agency (EPA) reported that 80.2% of the furniture waste generated by Americans went to landfills, with just 0.3% being recovered for recycling. Reuse activity in the sector is also low. Where reuse does occur, it is mostly through commercial second-hand shops, social enterprise companies, or charities. Some furniture items are also exchanged via free and paid exchange platforms, such as eBay and Freecycle, though the number of items traded in this way is difficult to quantify. In the United Kingdom, data from The Furniture Reuse Network (FRN) indicates that its members delivered 120,000 tons of reuse, representing approximately 6% of the total furniture arising as waste.

This escalating waste generation poses environmental challenges, prompting governments, manufacturers, and consumers to explore more sustainable practices, such as recycling, upcycling, and circular economy initiatives. The market has seen an uptick in innovative and sustainable materials, driven by consumers' growing environmental consciousness and preference for eco-friendly options. However, as the furniture sector continues to evolve, there is a pressing need for stakeholders to collaborate and implement more sustainable strategies to reduce waste generation and its environmental impact. By embracing innovative design, recycling technologies, and responsible consumption, the furniture industry can move toward a more sustainable future, ensuring a balance between economic growth and environmental conservation.

3. Circular economy strategies for furniture management

A literature review was undertaken to identify documents that describe circular economy initiatives suitable for implementation in the furniture sector. A total of 31 documents were reviewed, including scientific papers (25) and technical reports (6), as illustrated in Fig. 7.1. Among these documents, a total of 36 initiatives were identified.

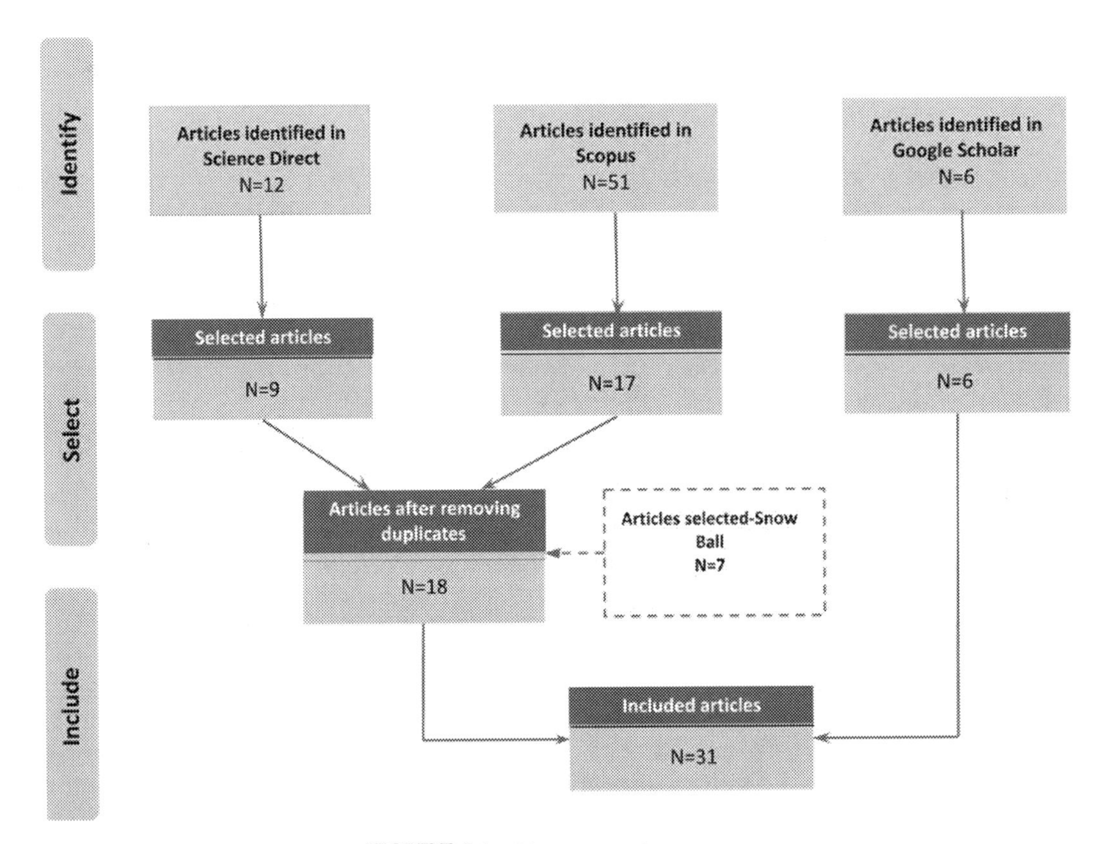

FIGURE 7.1 Literature review process.

3.1 Circular strategies for the furniture management at universities

The identified initiatives have been categorized into nine distinct and nonmutually exclusive strategies. The dendrogram in Fig. 7.2 visually represents the defined strategies and their corresponding associated initiatives.

We now present a succinct overview of each strategy.

- ***Circular design*** entails deliberately creating products, services, and systems in alignment with the principles of the circular economy. It involves considering the entire lifecycle of a product, emphasizing the development of durable, modular, easily disassembled, and sustainable materials.
- ***Awareness*** pertains to consumer behavior as a pivotal factor in enabling the mechanisms of the circular economy and establishing slower and closed loops within the supply chain of consumer goods and products. For instance, (Kirchherr et al., 2017) underscored the lack of consumer interest and awareness as the primary barrier for businesses and policymakers transitioning toward the circular economy.
- ***Circular procurement*** aims to procure or acquire products with circular characteristics or implement a flexible acquisition system that promotes economic and environmental sustainability (Van Oppen et al., 2018).
- ***Product as a service*** is a business model that allows consumers to access goods or services without ownership, emphasizing circularity (De Angelis et al., 2018), with a focus on extending product life cycles (Pieroni et al., 2020).
- ***Purchase with extended service*** enables the supplier not only to sell a product but also to provide necessary services during the product's usage phase. This approach facilitates product circularity through maintenance or repair systems (Alhola et al., 2019).
- ***Cascading use*** denotes the sequential utilization of resources for various purposes, typically through multiple material reuse phases before energy extraction or recovery operations (Campbell-Johnston, 2020). Its objective is to retain the value of materials and resources within the system for as long as possible by employing them sequentially in different applications.
- ***Take/buy-back schemes*** involve assigning responsibility to producers or retailers to repurchase products that were previously sold to customers. This is often achieved by establishing recycling and collection targets for a specific product or material. Another approach is to provide incentives for consumers to return used products to designated locations such as the point of sale.
- ***Extended producer responsibility (EPR)*** is an environmental policy approach in which a producer's responsibility for a product extends to its postconsumer stage in the life cycle. In practice, EPR involves producers taking responsibility for collecting end-of-life products and sorting them before their final treatment, ideally through recycling. EPR schemes enable producers to fulfill their responsibility either by providing the necessary financial resources and/or by assuming operational and organizational aspects of the process from municipalities. This can be done individually or collectively (OECD, 2016).
- ***Collaboration*** among consumers/clients, suppliers, or the government in designing and operating a green supply chain aims to reduce waste and environmental costs, increase customer satisfaction, and maximize return volumes.

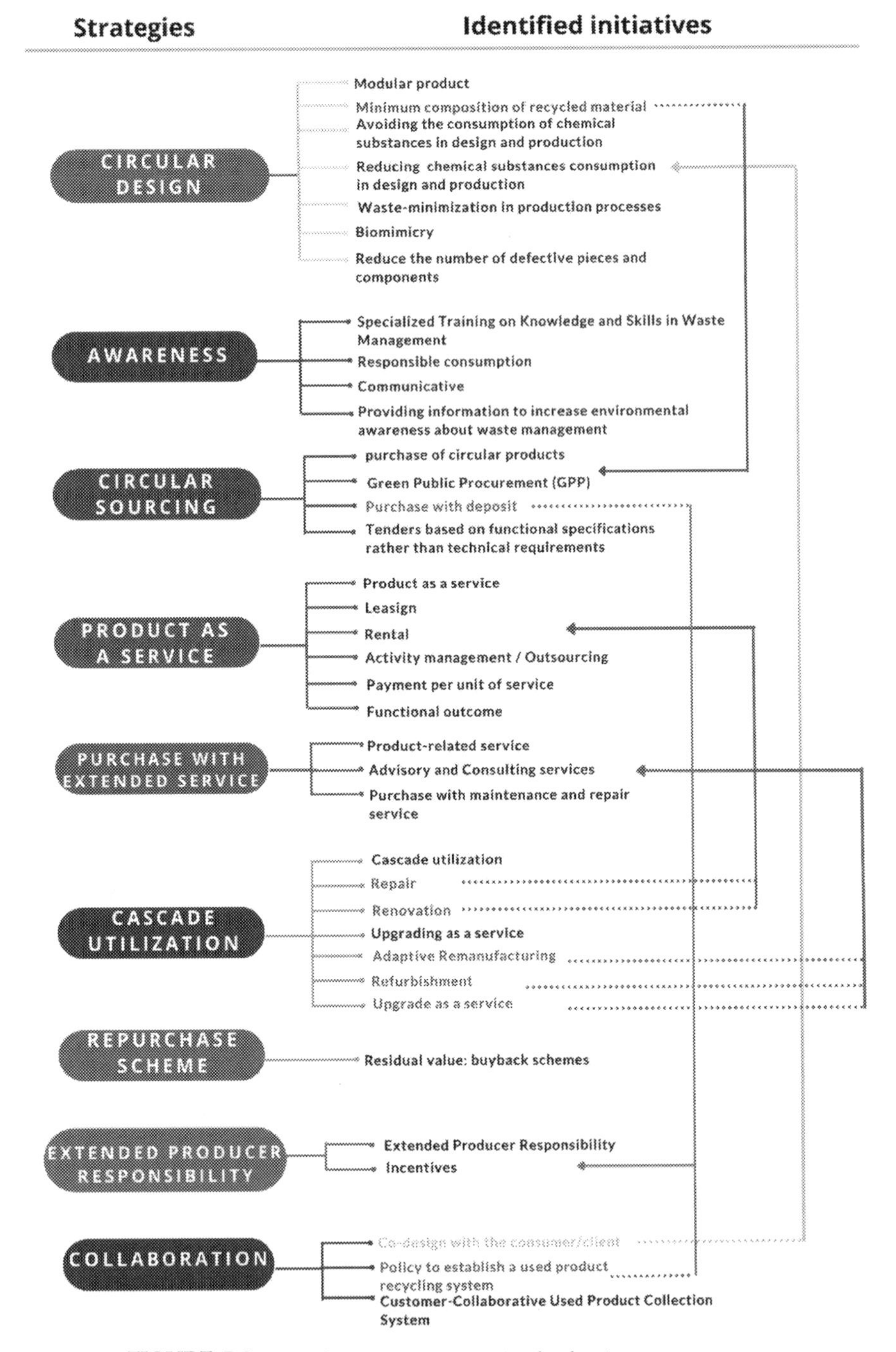

FIGURE 7.2 Circular economy strategies for furniture management.

The portfolio of identified initiatives in the furniture sector, shows that emergent practices related to circular economy have been applied in the stages of design, manufacturing, marketing, use, disposal, and return. In the design and manufacturing phase, *circular design* has been implemented to select high-quality and durable materials with low environmental impact. Additionally, modular designs have been generated to facilitate disassembly and repair by parts, thus increasing the product's lifecycle and recyclability (FURN360, 2015). In the marketing and use phase, customers are provided with information about the materials and composition of the furniture through eco-labels to encourage *awareness* and conscious consumption. Furthermore, business models such as interior and furniture leasing are being implemented, putting the PaaS strategy into practice. Considering this strategy, furniture design can be improved to facilitate subsequent recycling, prolong product lifespan through maintenance and repair services, and promote the reuse and remanufacturing of used furniture pieces (Besch, 2005; Forrest et al., 2017). In the disposal and return phase, strategies like EPR are applied. Manufacturers are incentivized to consider environmental aspects throughout the furniture's life cycle, from its design (implementing modularity and resource efficiency) to its end-of-life stage. They are responsible for waste management and final disposal of the product (FURN360, 2015).

3.2 Characterization of circular economy strategies

Based on the circular economy strategies identified for managing furniture, we assessed their feasibility for implementation in universities and educational institutions. To achieve this, a multi-criteria evaluation was carried out, considering the following criteria.

- ***Administrative criteria*** pertain to the institutional and administrative readiness to fulfill the objectives of the strategy. This includes contractual aspects, stakeholder relationships, and specification of technical requirements.
- ***Operational criteria*** focus on the execution of the strategy, including its implementation, control/monitoring to ensure traceability of performance, and evaluation to identify, quantify, and assess the outcomes.
- ***Cost criteria*** encompass the total expenses associated with implementing the proposed strategy. This includes initial investment, logistics, and operational costs.
- ***Contribution criteria*** consider the potential benefits that the strategy's implementation can offer in the environmental, social, and innovation domains. The social domain considers factors such as creating societal value and promoting fair trade, while the innovation domain considers factors such as technology adoption.
- ***Legal criteria*** correspond to the regulatory framework that governs the proper implementation of the strategy. This may include guidelines on green public procurement or strategic plans mandated by the institutions.

The evaluation was conducted by a group of four experts who have direct involvement in furniture management at universities. The experts utilized a Likert scale ranging from 1 to 5, where 1 represents unsuitability and 5 represents high suitability. Fig. 7.3 shows the evaluation results of the identified strategies based on the described criteria.

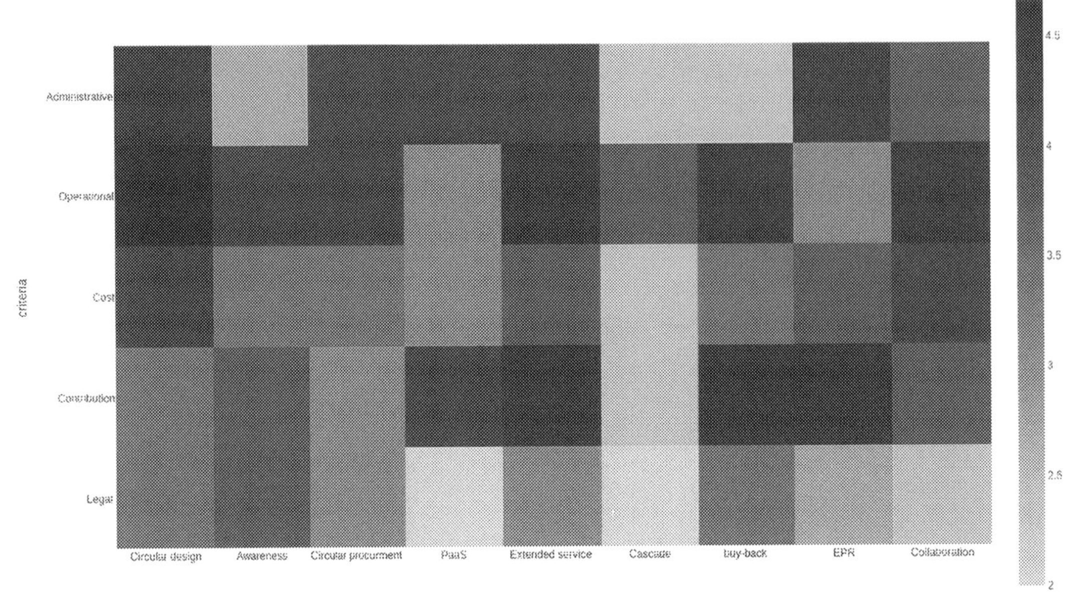

FIGURE 7.3 Strategies evaluation.

The first fact to highlight is that the legal framework consistently emerges as a crucial criterion when considering the implementation of circular strategies for furniture management in universities. This holds particular significance for public universities and institutions, as regulations are typically more stringent in those contexts. In contrast, the operational criteria are not deemed critical for most of the strategies. However, they are of concern for strategies that have already gained traction in national-level discussions (such as extended producer responsibility) or within the institutional context (such as PaaS).

Regarding the cascade use strategy, most of the criteria are considered critical, aligning with the prevailing understanding of this strategy as a complex combination of various circular business models. Conversely, strategies such as extended services, circular design, and circular procurement are generally regarded as suitable across most of the criteria. In terms of contribution, strategies involving not only environmental impact but also innovation, such as EPR, buy-back schemes, and PaaS, are highly evaluated.

3.3 Strategies prioritization

While the prioritization of strategies may vary depending on the specific university or institution, a suggested ranking has been established based on expert opinions. To accomplish this, a paired comparison using the analytic hierarchy process (Saaty, 2008) was employed to calculate weights for each criterion, representing their respective importance. Table 7.1 displays the calculated weights. These weights align with the prevailing notion

TABLE 7.1 Evaluation criteria weights.

Administrative	Operational	Cost	Contribution	Legal
0.19	0.13	0.22	0.19	0.27

that cost and the legal framework typically pose as barriers to the implementation of changes within systems and processes.

The overall importance value of each strategy is calculated as the weighted sum of the average evaluation for each criterion, considering their associated weights. Fig. 7.4 illustrates the prioritization of the strategies. The highest scores, indicating the strategies considered most suitable for implementation, are those that pose fewer challenges to the existing furniture management model. For instance, strategies focusing on improved and circular designs, as well as extended services, may not require significant additional commitments from the university or institution utilizing the chairs. In contrast, strategies that entail the adoption of new circular business models, habit changes, and the utilization of new technologies are considered less suitable for implementation. That is the case of strategies such as circular procurement, PaaS, and the use in cascade.

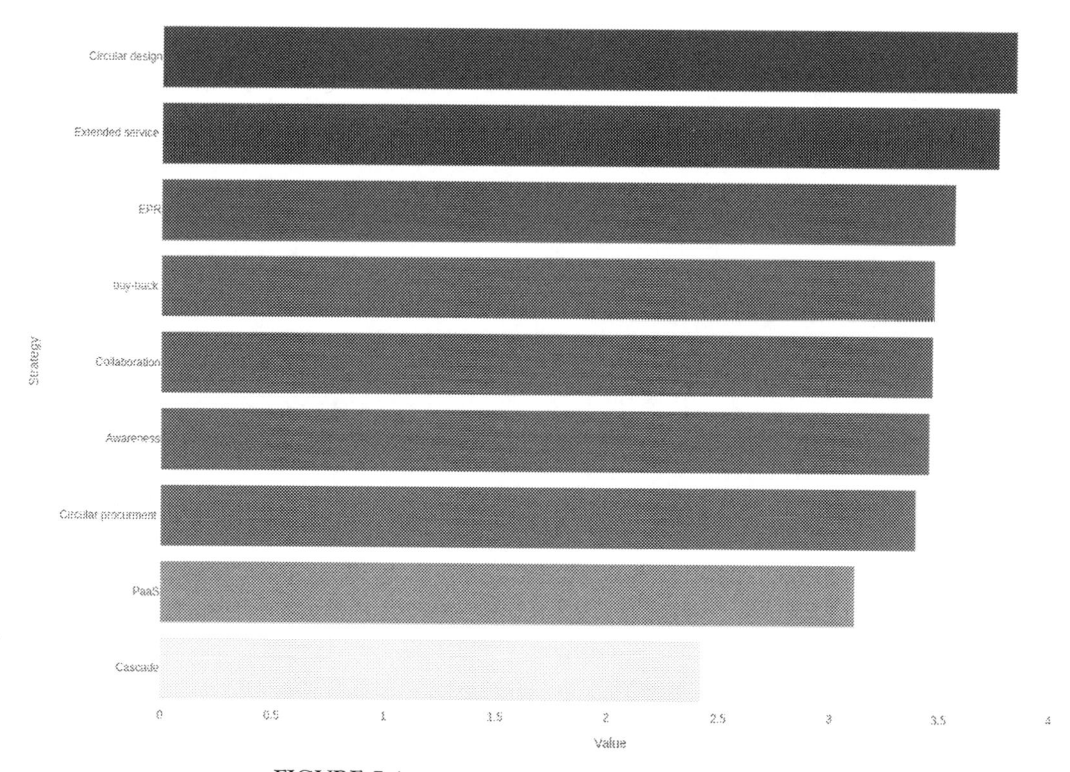

FIGURE 7.4 Circular economy strategies prioritization.

While the previous prioritization of strategies has outlined a path for their initial adoption, we acknowledge the importance of those strategies that, despite receiving lower priority, possess significant potential to instigate structural changes within the institutional furniture management system. This is exemplified by the PaaS strategy, which, in addition to requiring the commitment of various stakeholders, necessitates the alignment of capabilities and the implementation of technological enablers. Therefore, the remainder of this chapter focuses on the PaaS strategy, identifying elements that are crucial for configuring it, proposing a methodology to address those elements, and exemplifying the potential impacts through a case study at the Universidad de Antioquia.

4. Product as a service strategy for furniture at universities

Across different industries and for different products, customers are forsaking ownership of a range of items they used to purchase. They are opting instead to buy access to—and the performance of—these items on either a short or long-term basis. The PaaS business model helps companies capitalize on these new customers' mindsets (Lacy and Rutqvist, 2015).

4.1 Previous work

The literature about PaaS in the furniture sector is scarce, few papers and technical reports have addressed this topic.

A study conducted by Youngjin et al. (2016) proposed new value ideas for PaaS models. The authors suggested an approach that is based on the concepts of general needs (GNs) for the PaaS lifecycle introduced by Kim et al. (2012), as well as business system evolution patterns (BSEPs) described by Mann (2007). In their approach, the researchers identified customer GNs that are already being fulfilled by existing PaaS offerings. They then generated advanced PaaS ideas from an evolutionary perspective by mapping the existing PaaS onto BSEPs. Finally, they selected PaaS ideas that could address the unfulfilled or insufficiently considered GNs. As a result, they generated 16 new ideas centered around the PaaS concept. One of the generated ideas, for example, involves offering customized services based on information obtained through monitoring, such as the amount of dust, humidity, noise, and temperature. This monitoring would be performed using automatic sensors, enabling the provision of tailored services based on the specific needs identified by the collected data.

Similarly, Kristensen and Remmen (2019) propose a framework for generating sustainable value propositions that consider the three levels of the orientation of a PaaS (i.e., product-oriented, service-oriented, and performance-oriented). The framework includes variables such as price, investment, convenience, design, and resource management, aiming to strike a balance between the economy, environment, and society. The proposed framework is applied to a case study in the school furniture sector in Denmark. The results demonstrate that the performance-oriented perspective has the highest potential for generating added value, requiring integrated and collaborative planning between suppliers, schools, teachers, and students. Likewise, Pieroni et al. (2020) reveal the effect of configuring new PaaS models on value-generation outcomes for customers and profits for the company. The research is conducted in two Nordic companies that offer furniture usage through a leasing scheme, which includes maintenance service and subsequent return at the end of the contract. The results identify important characteristics for the success of PaaS models, such as value proposition, customer segmentation, economic growth for the company, and potential resource decoupling, facilitating modularity, restoration, and subsequent reuse or recycling. Additionally, some barriers to PaaS implementations were identified, including negative initial cash flow, financing, and durability of furniture products.

4.2 Configuration of a PaaS system for furniture management

Considering the significant impact that the appropriate design of a PaaS model has on its performance, a methodology for the selection and configuration of its various factors is discussed below. The proposed methodology addresses these factors across three dimensions: actors, level of orientation, and product/service characteristics. To address these dimensions, the methodology provides answers to three guiding questions.

- ¿Which actors will be involved in the PaaS strategy?
- ¿What level of orientation is desired to be achieved?
- ¿What characteristics should the PaaS strategy satisfy for each actor to achieve the desired level of orientation?

Specifically, the design of the strategy considers defining the actors who will be involved in the implementation of the business model. Subsequently, the desired level of PaaS orientation is determined for the PaaS strategy. These two defined dimensions can have multiple combinations and based on the choices made in the actor and orientation dimensions, the elements of the product–service relationship that will be developed in the strategy are established. These elements are materialized through the inclusion of the product and service characteristics.

The factors considered in each of the three dimensions were obtained from the literature review of the circular economy strategies applied to the furniture sector.

4.2.1 Actors involved in the PaaS strategy

Fifteen actors involved and interested in the implementation of the PaaS strategy were identified. These actors were described as stakeholders and entities that interact and contribute to the development of PaaS. They were grouped into four categories: customers, suppliers, government, and collaborative networks. Subcategories were established for the Customer and Supplier actors. Customers were classified, based on their commercial nature, into public entities, private entities, and direct consumers. Suppliers were subcategorized as furniture manufacturers and furniture restorers.

4.2.2 Orientation level

The levels of orientation in the PaaS strategy are referred to by different names in the literature, and three main orientations are identified that represent the level at which tangible products and intangible services are combined to fulfill a market need. These orientations are defined as product orientation, service orientation, and system orientation (Kristensen and Remmen, 2019; Lacy and Rutqvist, 2015; Pieroni et al., 2020; Yang et al., 2018).

Product-oriented, where advice and consultations are provided regarding the use and characteristics of the product; **Use/service/solution-oriented**, where product leasing, product rental, sharing, and pooling arrangements, are generated; and **Results/system-oriented**, in which payment is made per unit of service or functional outcome. The difference between these perspectives lies in the transition from product to service, gradually shifting from transactional negotiations to collaborative relationships or partnerships among stakeholders (Kristensen and Remmen, 2019). Fig. 7.5 presents an outline of the levels of orientation in the PaaS strategy based on Yang (2018).

4.2.3 *Product and service characteristics*

Regarding the characteristics that integrate and constitute the PaaS strategy, a review of the literature identified 54 of them. Those characteristics were referred to as variables, factors, criteria, needs, and characteristics to consider in PaaS implementations. Based on their definitions and similarities 20 characteristics were defined out of the 54 elements identified. Moreover, those characteristics were classified between service characteristics (11) and product characteristics (9) as shown in Table 7.2.

4.3 Product and service characteristics validation

A twofold approach was employed to characterize and validate the product and service characteristics. Initially, a panel of 10 experts directly involved in furniture management at various institutions ranked these characteristics based on their importance. The results, presented in Figs. 7.6 and 7.7, indicate that higher values correspond to a greater level of importance assigned to each characteristic.

While there is not a unanimous consensus among all the experts regarding the importance of these characteristics, some valuable insights can still be gleaned from their responses. Concerning service characteristics, maintenance, and repair receive consistently high rankings

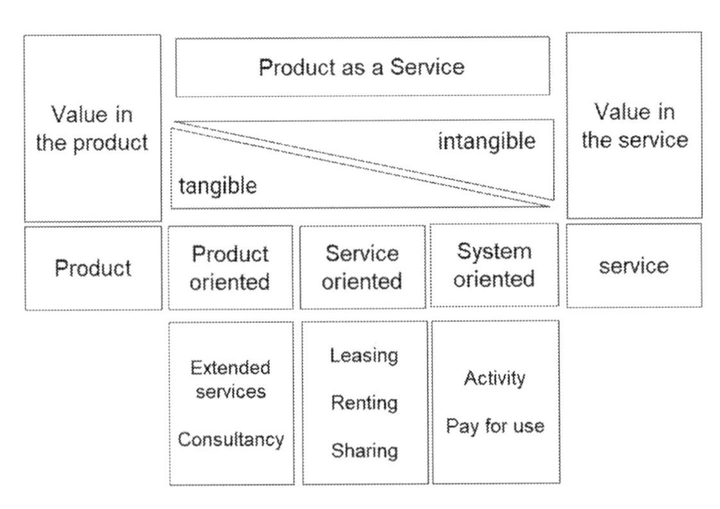

FIGURE 7.5 Levels of orientation in the PaaS strategy.

TABLE 7.2 Service and product characteristics of CE strategies for furniture.

Service characteristics	
CS1 service duration	The period for which a chair supply contract and associated services are provided.
CS2 warranty period	The duration for which the chair supplier guarantees the promised quality to the customer.
CS3 delivery time	The responsiveness in fulfilling a chair order promptly.
CS4 maintenance	Additional preventive or corrective maintenance service that extends the chair's lifespan, ensuring its functionality and availability.
CS5 repair	Minor repair services, such as parts replacement, painting, and screw replacement, that extend the chair's lifespan, maintaining its functionality and availability.
CS6 replacement	Timely replacement of chairs that have significant damage, rendering them nonfunctional.
CS7 recycling opportunity	Utilizing the chair's components as recycling material at the end of its lifespan.
CS8 logistic requirements	Logistic conditions facilitating chair supply management, including transportation, storage, collection, recycling, and/or final disposal.
CS9 usage advisory and training	Basic advisory and training provided by the supplier regarding proper use and maintenance of the chairs.
CS10 collaboration scenarios	Development of joint procurement and chair management strategies involving different stakeholders.
CS11 sustainability training and awareness	Training and awareness initiatives for customers and suppliers regarding the importance of implementing sustainable and environmentally friendly strategies.
Product characteristics	
CP1 cost	Acquisition cost, leasing cost, or implementation cost of services.
CP2 upgradability	The ability of the chair to change its appearance to fit new environments.
CP3 uniformity	Similar physical appearance among chairs.
CP4 ergonomics	Adaptation of the chairs to provide greater comfort for students.
CP5 material composition	Inclusion of environmentally friendly materials and recycled materials in the chairs.
CP6 lifecycle time	The period during which the chair remains in good condition and functional.
CP7 modularity	Design that facilitates disassembly, replacement, repair, and reuse of chairs by individual parts.
CP8 design	Aesthetic characteristic of the chair that makes it appealing to the customer.
CP9 compliance with current regulations	Adherence to the set of laws or regulations governing the furniture sector and its stakeholders.

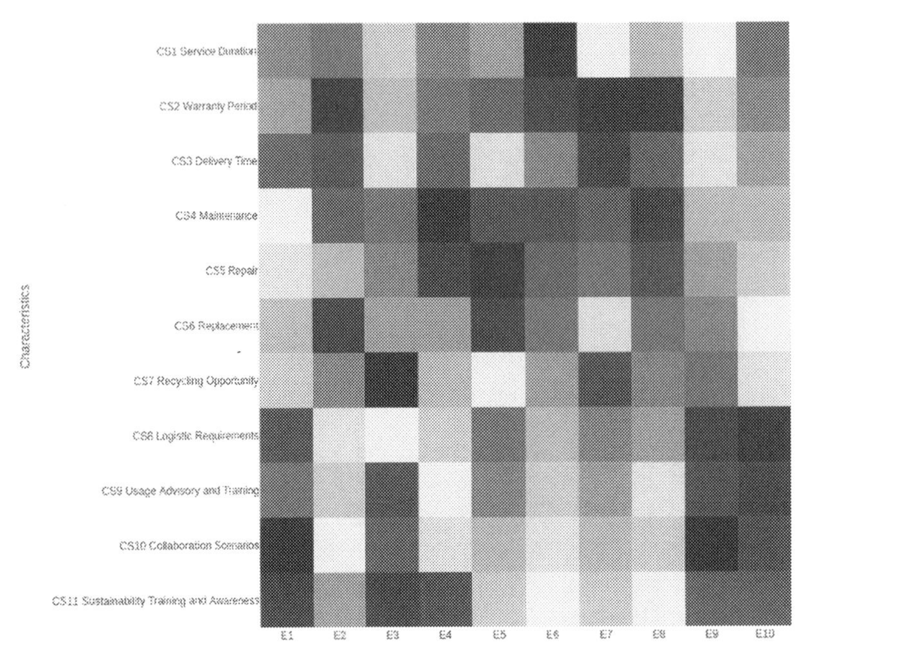

FIGURE 7.6 Service characteristics of a CE strategy for furniture.

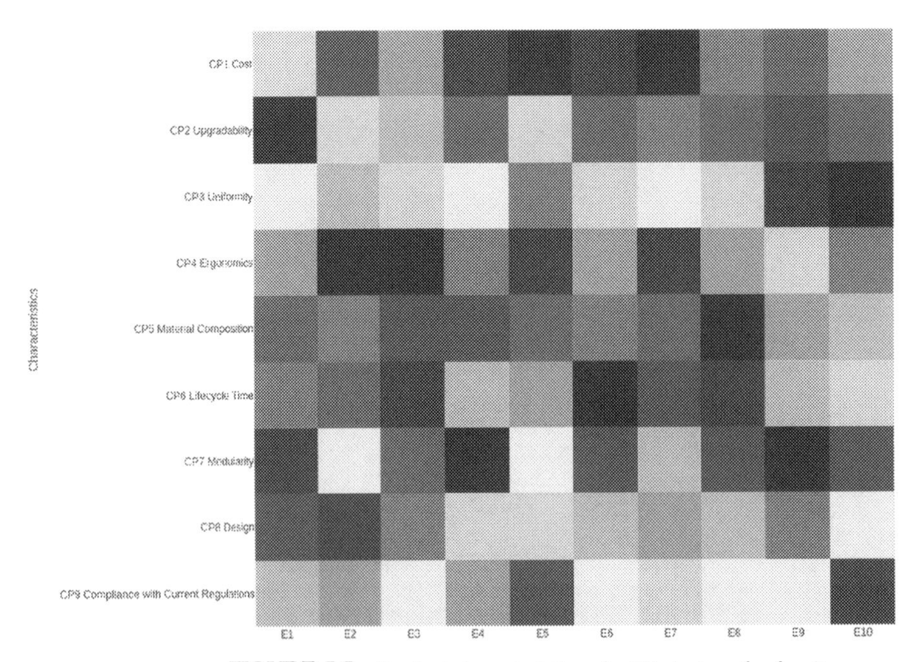

FIGURE 7.7 Product characteristics of a CE strategy for furniture.

TABLE 7.3 Product and services characteristics for different roles and orientation levels.

Client perspective			Supplier perspective		
Product-oriented	Service-oriented	Results-oriented	Product-oriented	Service-oriented	Results-oriented
CP1 cost	CP1 cost	CP1 cost	CP1 cost	CS1 service duration	CP1 cost
CP4 ergonomics	CS1 service duration	CP5 material composition	CP5 material composition	CS2 warranty period	CP5 material composition
CP5 material composition	CS4 maintenance	CP6 lifecycle time	CP6 lifecycle time	CS5 repair	CP9 compliance with current regulations
CP6 lifecycle time	CS5 repair	CS4 maintenance	CP8 design	CS7 recycling opportunity	CS1 service duration
CP8 design	CS6 replacement	CS7 recycling opportunity	CP9 compliance with current regulations	CS11 sustainability training and awareness	CS10 collaboration scenarios

among most experts, while usage advisory, opportunities for collaboration, and training and awareness are not as prominently valued. Turning to product characteristics, cost emerges as the primary consideration in most cases, followed by ergonomics and material composition. Uniformity and compliance with current regulations, on the other hand, do not appear to be deemed particularly important product attributes.

For the second analysis, we ask the experts to assume two different roles: client and supplier. In each of those roles, the experts point out the five most important product or service characteristics when implementing a PaaS system at each of the three possible levels (i.e., product-oriented, use/service/solution-oriented, and results/system-oriented). Table 7.3 summarizes the communalities we observed on the characteristics chosen by the different experts.

The cost is a prevalent product characteristic in nearly all system configurations, whether viewed from the perspective of a client or a provider. This aligns with the current furniture management practices adopted by most organizations. As the model's focus shifts toward service and results-oriented aspects, the significance of maintenance and repair as service characteristics increases. When analyzing PaaS from the supplier's standpoint, collaboration, training, and awareness emerge as key factors in defining the system.

4.4 System design matrix

Based on the definition of the three dimensions and the factors pertaining to each of these dimensions, we have constructed a matrix to identify the relevant product and service characteristics corresponding to various combinations of the orientation level of the PaaS system and the involved actors. An excerpt from the resulting matrix is shown in Fig. 7.8. For

ORIENTATION LEVEL PSS	CLIENTS PUBLIC COMPANY	CLIENTS PRIVATE COMPANY
Product oriented	**CP1** Cost **CP3** Uniformity **CP4** Ergonomics **CP6** Lifecycle Time **CS2** Warranty Period	**CP1** Cost **CP3** Uniformity **CP4** Ergonomics **CP6** Lifecycle Time **CP8** Design **CS2** Warranty Period
Service oriented	**CP1** Cost **CP3** Uniformity **CP4** Ergonomics **CP5** Material Composition **CP7** Modularity **CS1** Service Duration **CS4** Maintenance **CS5** Repair **CS6** Replacement **CS7** Recycling Opportunity	**CP1** Cost **CP3** Uniformity **CP4** Ergonomics **CP5** Material Composition **CP7** Modularity **CP8** Design **CS1** Service Duration **CS4** Maintenance **CS5** Repair **CS6** Replacement **CS7** Recycling Opportunity
Sistem oriented	**CP1** Cost **CP5** Material Composition **CP8** Design **CP9** Compliance with Current Regulations **CS1** Service Duration **CS3** Tiempo de entrega **CS6** Replacement **CS9** Usage Advisory and Training **CS10** Collaboration Scenarios	**CP1** Cost **CP2** Upgradability **CP4** Ergonomics **CP5** Material Composition **CP8** Design **CP9** Compliance with Current Regulations **CS1** Service Duration **CS3** Tiempo de entrega **CS6** Replacement **CS9** Usage Advisory and Training **CS10** Collaboration Scenarios

FIGURE 7.8 System design matrix.

example, the matrix highlights that in a PaaS system oriented toward the product, the furniture manufacturer finds certain characteristics such as cost, supply time, and guarantee time to be particularly significant. The complete matrix and its description are provided as a supplementary material.

5. Case study: Faculty of engineering—University of Antioquia

This section addresses a case study in which the supply chain and operations management of classroom furniture at the University of Antioquia is modeled and simulated. First, we describe and characterize the current supply chain and operations management system using information obtained from semistructured interviews, process observation, and document and information requests. Then, we develop a simulation model of the current system and a proposed system that considers a PaaS strategy. The modeled systems are compared based on performance metrics, such as supply cost and resource utilization.

5.1 Current system description

The diagnosis and description of the current supply chain and operations management system was developed based on the current operations of two of the departments of the university: the Faculty of Engineering on the main campus and the Regionalization Unit (RU-UdeA) located at a satellite campus in the eastern region of Antioquia. We conducted four semistructured interviews with stakeholders associated with the purchasing, supply, logistics, and infrastructure areas of the University. Additionally, we interviewed the supplier of chairs for the Faculty of Engineering over the past 4 years. In that interview.

The Engineering Faculty at the University of Antioquia serves approximately 8500 students across three different buildings. To support academic and teaching activities, approximately 2400 chairs are distributed among the classrooms in these buildings. The faculty has an annual budget of $30,000 for furniture renewal, which is equivalent to an annual supply of 1000 chairs. For example, during 2017 and 2018, a total of 1924 chairs were purchased to replace those that were deteriorated or malfunctioning. The RU-UdeA caters to approximately 2500 students who utilize two buildings with a chair infrastructure similar to that of the Engineering Faculty.

In the current furniture supply chain and operations management system, three processes were identified: procurement, usage and inventory, and final disposal. Those processes, shown in Fig. 7.9, obey a classical linear consumption.

In the initial procurement process, chairs are sourced from a supplier. This involves the infrastructure department submitting a proposal, which then gains approval from the relevant dependency. Subsequently, the technical specifications or criteria for the required chairs are documented. Based on these specifications, a set of criteria is developed, and a weighted evaluation is conducted to select the supplier that best fulfills the requirements. Notably, during the selection of chair suppliers, it is observed that the offers concerning the technical characteristics of the chairs are highly comparable among all bidders. As a result, the final selection is often inclined toward the criterion of lower cost.

In the subsequent process, the furniture is put into use, and the inventory department regularly assesses the condition of the chairs. The University takes charge of minor furniture repairs, procuring replacement backrests, brackets, seats, and fasteners as required. At the conclusion of each academic semester, an inventory check ensures that classrooms meet the minimum requirement for functional chairs. This assessment informs the identification

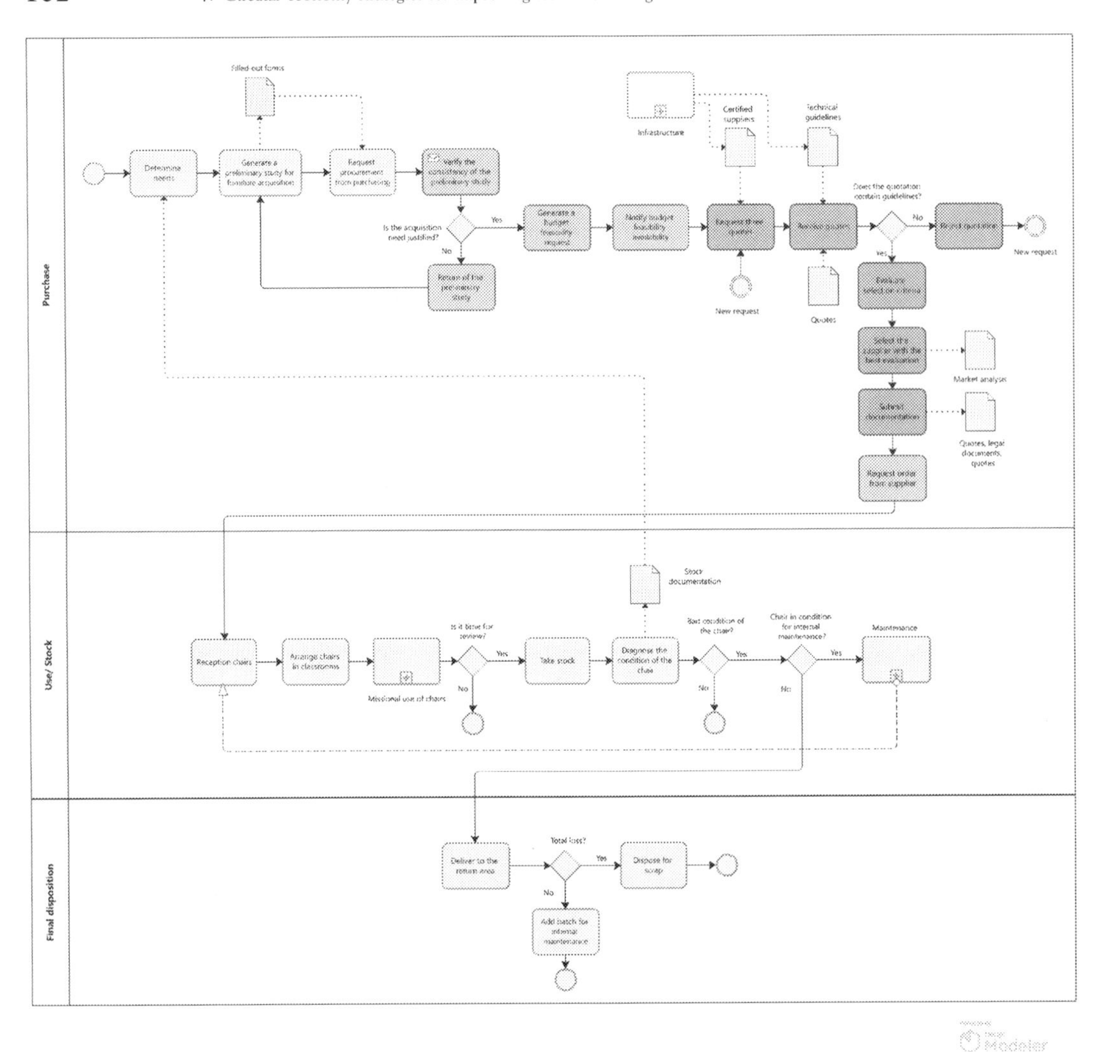

FIGURE 7.9 Processes of the furniture supply and operations management system.

of chairs in need of basic repairs, those that are beyond repair, and the necessity for chair replacements. Due to the relatively lower cost of classroom chairs compared to other infrastructure assets, the University does not maintain a one-to-one traceable inventory. Instead, a general institutional inventory is kept, and quantities in each building or department are monitored.

Finally, in the third process, chairs that are in poor condition and cannot be effectively repaired according to established standards are directed to a disposal area. Here, a decision is made whether the chair is in an acceptable condition to be offered internally within the University or auctioned as scrap to external parties.

5.2 Modeling

We propose an alternative furniture supply chain and operation management system based on a PaaS strategy, which we modeled using system dynamics. The model aims to comprehend the system's behavior and the impact of transitioning from a transactional model of chair procurement to a PaaS model with a service-oriented approach. In the proposed system, chairs are utilized by the customer under a rental model with repair and maintenance services overseen by the supplier. The system dynamics simulation model is typically constructed based on a dynamic hypothesis (Rahmandad and Sterman, 2012; Sterman, 2004), which, in our case, can be summarized as follows:

> The chairs currently in use are primarily located within the classrooms. Regular inventory checks are conducted to assess whether these chairs can continue in use or are deemed defective and unsuitable (defective rate). Among the defective chairs, a certain percentage (repair rate) can be internally repaired within the faculty, involving minor repairs. Once restored, these chairs are reintroduced into the active inventory. Another percentage (irreparable rate) represents chairs that are irreversibly damaged and cannot be repaired internally. Such chairs need to be replaced. The replacement process involves procuring new chairs from the supplier, an operation conducted either annually or when there is a significant shortage of chairs. The number of chairs acquired corresponds to the estimated requirement for furnishing the classrooms and meeting the students' needs. It is expected that the number of chairs in use aligns with the total number of required chairs. Consequently, acquisitions are made to replace those that cannot be repaired.

The causal diagram in Fig. 7.10 illustrates the structure and behavior of the current supply and operations management system as described by the dynamic hypothesis. Arrows connect different variables and are accompanied by a sign (+ or −) indicating the positive or negative polarity of the causal influence between these variables. A positive sign signifies that an

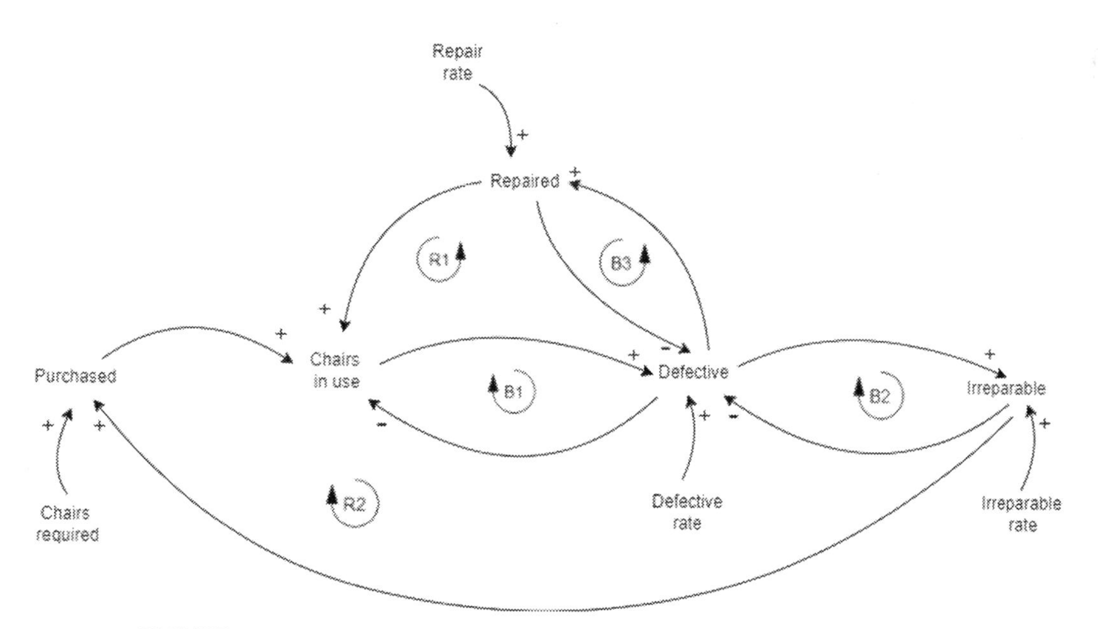

FIGURE 7.10 Causal diagram of current supply and operations management system.

increase in the variable from which the arrow originates leads to an increase in the variable to which the arrow points. Conversely, a negative sign indicates an inverse relationship. Additionally, cycles and closed loops are identified, categorized as either reinforcing (R) for growth or balancing (B) for equilibrium (Rahmandad and Sterman, 2012).

The greater the number of chairs in use, the greater the number that deteriorates and becomes part of the defective group. As these defectives are removed from the classrooms, the greater the number of defectives, the smaller the number of chairs in use. This relationship creates a balancing loop B1, as indicated in the causal diagram. On the other hand, the greater the number of defective chairs, the greater the number that can be repaired. With more repairs, more chairs become available for use, leading to a reinforcing loop R1. Another noteworthy cause—effect relationship represented in the causal diagram is the reinforcing loop R2 generated between the variables irreparable and acquired. The more irreparable chairs there are, the more chairs need to be acquired to meet the requirement of chairs.

The outlined cause—effect relationships present an opportunity for improvement, not only in terms of reducing the number of purchased chairs and their subsequent environmental impact but also in diminishing procurement costs, leading to significant economic benefits. The university stands to capitalize on this opportunity, given its status as a major buyer for most of its suppliers, affording it substantial bargaining power in the acquisition of furniture and other supplies. Therefore, we introduce a classroom furniture supply and operation management system based on a PaaS strategy, following the three-dimensional configuration detailed in Section 4. Firstly, the University is envisioned as the primary client, providing the basis for the system's design. Secondly, to facilitate a smooth transition from the current system, a service-oriented approach is adopted. Under this model, the university would not purchase chairs directly but instead engage in a rental agreement with the supplier. The supplier would ensure the availability of the requisite quantity of chairs for use. The service package includes repair, refurbishment, and replacement services. The repair service entails the replacement of modular parts such as the backrest, seat, or bracket. The refurbishment service covers major corrective repairs, such as welding of the metal structure, while the replacement service guarantees that the rented chairs remain accessible in the classrooms. The causal diagram for the proposed model is illustrated in Fig. 7.11.

The current and the proposed system were implemented as a system dynamics model using the software Powersim Studio 10. Table 7.4 summarizes the main components that were considered for the implementation based on the causal diagrams described before. The parameters represent the input information to the model, accumulator levels represent the state of the system, while flows are the decoupling rates that enable dynamics and imbalance in the system (Sterman, 2004).

5.3 Experiment results

To evaluate the proposed system, we define three different scenarios.

- Scenario 1 (Sc1): A base scenario that considers the current system.
- Scenario 2 (Sc2): A scenario based on the proposed model following a PaaS strategy with similar rates of repairment as in the base scenario.
- Scenario 3 (Sc3): A scenario based on the proposed model following a PaaS strategy with larger rates of repairment than in the base scenario.

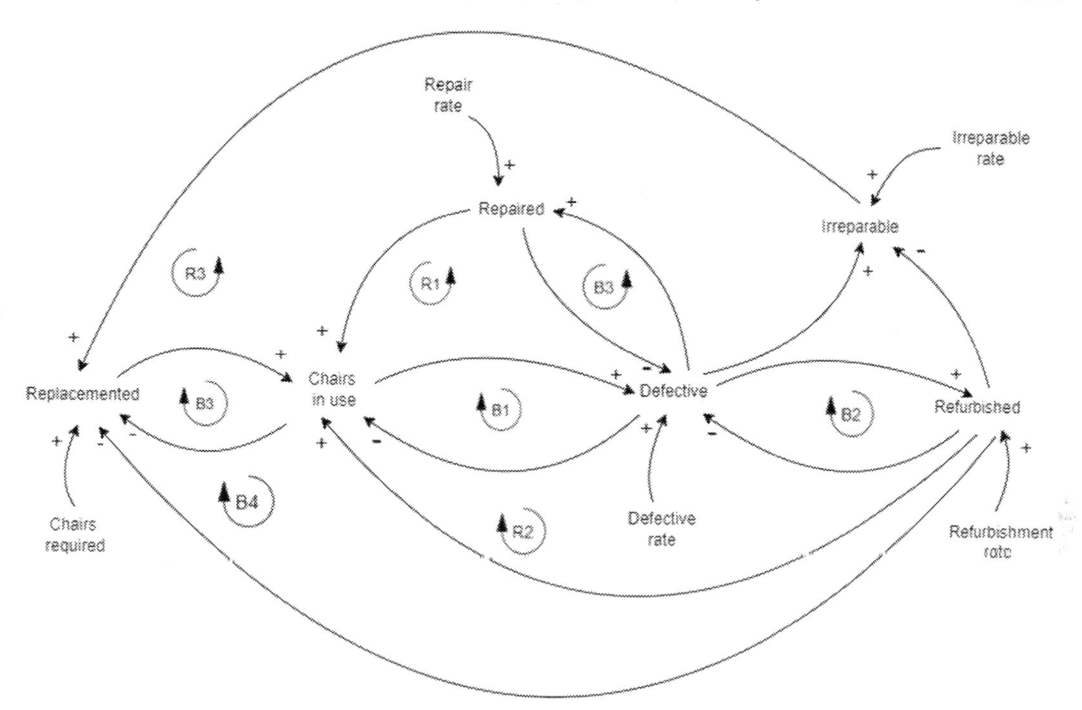

FIGURE 7.11 Causal diagram of proposed supply and operations management system.

TABLE 7.4 Components of the dynamic simulation model.

		Current	Proposed
Parameters			
Defective rate	Proportion of chairs in used that got damaged	X	X
Repairment rate	Proportion of damaged chairs that can be put back to use with small repair	X	X
Refurbish rate	Proportion of damaged chairs that can be put back to use with a large repair or refurbishment		X
Nonrepairable rate	Proportion of damaged chairs that are permanently damaged and cannot be put back to use	X	X
Chairs requirement	Total number of chairs that are required	X	X
Flows			
Damaged chairs	Chairs in use that got damaged	X	X
Repaired chairs	Damaged chairs that are repaired and get back to use	X	X
Refurbished chairs	Damaged chairs that are refurbished and get back to use		X

(*Continued*)

TABLE 7.4 Components of the dynamic simulation model.—cont'd

		Current	Proposed
Nonrepairable chairs	Damaged chairs that cannot be repaired and are disposed	X	X
Replenishment chairs	New chairs that are acquired to replace damaged chairs	X	X
Purchased chairs	New chairs that are acquired to fulfill new demand	X	X
Levels			
Total chairs in use	Total number of chairs in use at the classrooms either owned by the university (current) or rented from a supplier (proposed)	X	X
Total damaged	Total number of damaged chairs either they are repaired or disposed	X	X
Total repaired chairs	Total number of damaged chairs that can be put back to use with small reparations	X	X
Total refurbished chairs	Total number of damaged chairs that can be put back to use with extended reparations or refurbishment		X
Total nonrepairable chairs	Total number of chairs that are permanently damaged and cannot be put back to use	X	X
Total purchased chairs	Total number of chairs purchased to fulfill the demand. They can be bought by the university (current) or by the renting supplier (proposed)	X	X

The final scenario operates under the assumption that higher repair rates would result in a decrease in the requirement for extensive repairs or refurbishments, as well as a reduction in the number of disposed chairs. Experiments are conducted for each scenario, varying the rates of defectiveness, repair, refurbishment, and irreparability. Table 7.5 outlines the range of values considered for each case. For all the defined scenarios, 100 iterations are carried out, with a simulation span of 5 years, using monthly periods.

TABLE 7.5 Scenarios parameter setting.

	Scenario 1		Scenario 2		Scenario 3	
Parameter	**min (%)**	**max (%)**	**min (%)**	**max (%)**	**min (%)**	**max (%)**
Repairment rate	6	12	6	12	6	24
Refurbish rate	–	–	3	5	1.5	2.5
Nonrepairable rate	2	35	14	17	7	8.5
Defective rate	3	4	23	34	15	35

TABLE 7.6 Correlations between performance measurement and parameters.

	Scenario 1		Scenario 2			
Parameter	Cost	Acquisition	Cost	Acquisition	Dual cost	Dual acqui.
Repairment rate	0.35	0.03	0.90	0.16	4.80	4.60
Refurbish rate	−0.00	−0.02	0.05	0.06	4.86	4.63
Nonrepairable rate	0.93	0.99	0.42	0.98	4.83	4.63
Defective rate	0.70	0.44	0.94	0.30	3.82	3.66

TABLE 7.7 Variation of performance measurement among the different scenarios.

	Total cost		Total acquisitions	
Scenario	Average (US)	Variation (%)	Average (chairs)	Variation (%)
Scenario 1	135,712		5890	
Scenario 2	99,317	27	4238	28
Scenario 3	80,742	41	3316	44

An initial analysis focuses on the correlations between the input parameters of the model and the performance measures, namely, the total acquisition cost and the overall number of chairs purchased. The findings, as presented in Table 7.6, demonstrate a robust correlation between both performance measures and the rate of nonrepairable chairs. This suggests that an increase in the rate of nonrepairable chairs leads to higher costs and more acquisitions. Moreover, the defective and repair rates also exhibit a strong correlation with the performance measures, particularly with the cost. The dual variable represents the percentage change in performance for each unit of change in the parameter. Notably, a one-unit shift in the rate of refurbishment triggers a 4.86% and 4.63% change in the total cost and acquisition cost, respectively.

Based on the experiments, the average changes in the performance measures relative to the baseline scenario are computed, as shown in Table 7.7. These variations demonstrate the potential impact of the model adjustments compared to the baseline scenario. The transition scenario (Scenario 1) and the PaaS scenario (Scenario 2) result in substantial reductions in both costs and the total number of chairs purchased.

6. Conclusions

Based on the literature review, nine Circular Economy (CE) strategies were identified as suitable for implementation by universities dealing with a substantial volume of furniture and frequent product replacement. Furthermore, five criteria were established for evaluating the relevance of these strategies. During an expert assessment exercise, the legal, cost, and administrative criteria, commonly observed in the bidding processes of public organizations, emerged as prominent factors.

In evaluating the strategies, a preference is observed for approaches where the institution acts as the client and owns the goods, reflecting current management trends. This preference sidelines more disruptive strategies such as circular sourcing, cascading, and PaaS. Notably, the transition to the PaaS strategy is specifically addressed.

PaaS represents a promising approach for managing classroom furniture in universities, facilitating the integration of economic and environmental sustainability principles. Configuring the PaaS strategy involves identifying key implementation criteria that surpass the conventional paradigm of retail product procurement. This chapter identified and validated the most pertinent product and service characteristics crucial to configuring a PaaS strategy for managing the supply chain and operations of university chairs. Key product characteristics include cost, material composition, ergonomics, modularity, and life cycle time, while service characteristics encompass warranty time, maintenance, repair, and delivery time.

The proposed methodology for configuring PaaS for furniture management in universities is validated through a case study at the University of Antioquia. Employing a system dynamics model, the analysis estimates the impacts on costs and the total number of chairs purchased. The results demonstrate the potential of the PaaS model, indicating significant reductions in both performance measures. However, it is essential to consider the specific implementation peculiarities in each case and their potential cost implications.

Acknowledgments

Pablo Maya-Duque gratefully acknowledges the support of the Colombian General Royalty System within the framework of the BPIN project 2020000100497 *"Alianza Circular, Mejoramiento de la gestión de empaques y envases a través de estrategias de economía circular en la industria de Antioquia."*

Supplementary material

Supplementary data related to this article can be found online at https://doi.org/10.1016/B978-0-443-21651-0.00009-7.

References

Alhola, K., Ryding, S.O., Salmenperä, H., Busch, N.J., 2019. Exploiting the potential of public procurement: opportunities for circular economy. Journal of Industrial Ecology 23 (1), 96–109.
Besch, K., 2005. Product-service systems for office furniture: barriers and opportunities on the European market. Journal of Cleaner Production 13 (10–11), 1083–1094.
Campbell-Johnston, K., Vermeulen, W.J., Reike, D., Brullot, S., 2020. The circular economy and cascading: towards a framework. Resources, Conservation and Recycling X (7), 100038.
De Angelis, R., Howard, M., Miemczyk, J., 2018. Supply chain management and the circular economy: towards the circular supply chain. Production Planning & Control 29 (6), 425–437.
Ellen MacArthur Foundation, 2015. Towards the Circular Economy: Accelerating the Scale-Up across Global Supply Chains.
Forrest, A., et al., 2017. Circular Economy Opportunities in the Furniture Sector. European Environmental Bureau, Brussels, Belgium.

FURN360, 2015. The Furniture Industry and the Circular Economy.
Kim, S., Yoon, B., 2012. Developing a process of concept generation for new product-service systems: a QFD and TRIZ-based approach. Service Business 6 (3), 323–348.
Kirchherr, J., Reike, D., Hekkert, M., 2017. Conceptualizing the circular economy: an analysis of 114 definitions. Resources, Conservation and Recycling 127, 221–232.
Kristensen, H.S., Remmen, A., 2019. A framework for sustainable value propositions in product-service systems. Journal of Cleaner Production 223, 25–35.
Lacy, P., Rutqvist, J., 2015. The product as a service business model: performance over ownership. In: Waste to Wealth: The Circular Economy Advantage. Palgrave Macmillan UK, London, pp. 99–114.
Mann, D., 2007. Hands-on Systematic Innovation for Business and Management. IFR Consultants Limited.
OECD, 2016. Extended Producer Responsibility: Updated Guidance for Efficient Waste Management. OECD Publishing, Paris.
Pieroni, M.P.P., McAloone, T.C., Pigosso, D.C.A., May 2020. Business model innovation for circular economy: integrating literature and practice into a process model. In: Proceedings of the Design Society: DESIGN Conference, vol. 1. Cambridge University Press, pp. 2119–2128.
Rahmandad, H., Sterman, J.D., 2012. Reporting guidelines for simulation-based research in social sciences. System Dynamics Review 28 (4), 396–411.
Saaty, T.L., 2008. Decision making with the analytic hierarchy process. International Journal of Services Sciences 1 (1), 83–98.
Sterman, J.D., 2004. Business dynamics. Systems thinking and modeling for a complex world. Interfaces 34 (1).
Van Oppen, C., Croon, C., Bijl de Vroe, D., 2018. Circular Procurement in 8 Steps. Ecodrukkers/De Toekomst. The Netherlands.
Yang, M., Smart, P., Kumar, M., Jolly, M., Evans, S., 2018. Product-service systems business models for circular supply chains. Production Planning & Control 29 (6), 498–508.
Youngjin, P., Kim, M., Yoon, J., 2016. Generating new product-service system concepts using general needs and business system evolution patterns: a furniture PSS case. Industrial Engineering & Management Systems 15 (2), 181–195.

CHAPTER

8

A multiobjective maximal covering/ p-dispersion model for reverse vending machine location

Sebastián Bedoya, Jaime Luis Mejía, Pablo Maya-Duque and Juan G. Villegas

Departamento de Ingeniería Industrial, Facultad de Ingeniería, Grupo Analytics and Research for Decision Making, Universidad de Antioquia, Medellín, Colombia

1. Introduction

Packaging and containers are designed to contain and store products to protect or help preserve them while facilitating their transport and display. Although in many cases they are necessary, packaging and containers have become a significant source of contamination and waste generation, and their effective management implies a series of political, economic, technological, logistical, and design challenges that should be addressed. For instance, in the European Union, each person generates about 180 kg of waste from packaging and containers per year, which means they are responsible for about 36% of municipal solid waste. Additionally, a large part of the virgin materials used in the European Union, 40% of plastics and 50% of paper, are destined for packaging and containers. The situation in Colombia is similar, but the amount of waste from packaging and containers taken to landfills or poorly disposed of in the ecosystem is probably higher due to lower recycling and recovery rates. In particular, regarding the use of plastics, the primary material used for containers and packaging in Colombia, about 700,500 tons are put on the market each year. Of this quantity, only 32% is collected, but only 3% of the material reaches a closed circularity: that is, the container and packaging return to being a container or packaging again (Rojas, 2023). In recent years, most efforts have focused on reducing the weight and material content of packaging, particularly plastics, and on developing new materials. However, large amounts of packaging, such as composite materials, still need to be managed by current collection and treatment systems. Additionally, the need for changes

Evolution and Trends of Sustainable Approaches
https://doi.org/10.1016/B978-0-443-21651-0.00015-2

in the management of packaging and containers interacts with industry trends such as electronic commerce and digitization, which makes their control more complex. Governments have responded to this challenge by implementing policies such as extended producer responsibility or regulating the employment of single-use plastics. In this context, initiatives with disruptive ideas have become relevant regarding the design of packaging or the redesign of supply chains, considering circular business models that minimize waste generation and encourage the intensive use of materials.

Particularly, extended producer responsibility (EPR) is an environmental policy promoted by the Organization for Economic Cooperation and Development (OECD) in which *"a producer's responsibility for a product extends to the postconsumer stage of its life cycle"* (OECD, 2016), that is, producers are responsible for planning, organizing, and financing postconsumer waste management, from its collection to its treatment, to extend its useful life. This environmental policy framework aims to (1) generate incentives for producers to incorporate environmental considerations into the design of their products, processes, and services; (2) displace responsibility (physical and economic, totally or partially) from the territorial entities to the producer.

As Fig. 8.1 illustrates, different actors are involved in an EPR system for packaging and containers. The government is the one that issues the normative basis of the system within the framework of which the producers design and update their management plans, awareness strategies, and the alliances that promote the management of the end-of-the-life cycle of their products. Producers can address this task individually or collectively through producer responsibility organizations or collective plans that, with the financing of the companies, help the collective of producers meet the established EPR goals. This is particularly meaningful for small and medium-sized companies that do not have the logistical and commercial capacity to deal with large-scale processes such as waste collection.

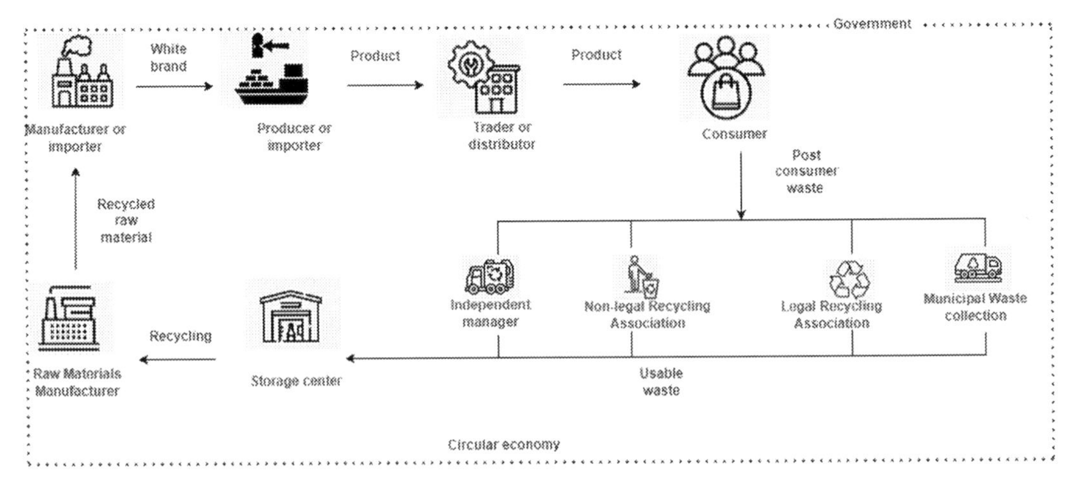

FIGURE 8.1 Actors in an extended producer system.

Once there is a postconsumer package or container, it must be collected. In Colombia, this is mainly done by individuals or associations related to recycling groups. In this stage, the material is collected, classified, and sorted to be stored at the collection centers. From there, the material will be integrated into the chain again through transformation processes. In the transformation stage, certifications of material recovered and used are issued according to the environmental waste management plans.

As another option, collection strategies using reverse vending machines (RVMs) appear within this context. An RVM is a mechatronic device where consumers exchange beverage containers (mainly bottles and cans) for monetary benefits like coupons, movie tickets, and deposit refunds (Sambhi and Dahiya, 2020). RVM use encourages consumers to dispose of plastic bottles in appropriate places that enter the reverse logistic flow. RVMs are simple to use and require limited resources to operate. Moreover, they offer a cost-effective recycling solution for improving recycling rates since they have the potential to reach a wider audience when located at shopping centers, supermarkets, airports, metro/bus stations, and other public places. Thanks to global environmental awareness, the market of RVMs is projected to grow at a 12.2% yearly rate during the 2021–28 period (Dataintelo, 2023; Research and Markets, 2022). Most current research on RVMs focuses on their cost-effective design and automation (Alfarisi et al., 2022; Sambhi and Dahiya, 2020; Zia et al., 2022).

On the other hand, a facility location problem consists of selecting locations for facilities or equipment to serve a set of customers in the best possible way (Saldanha-da-Gama, 2022). These problems appear ubiquitously in the design of transportation systems and supply chains. Facility location applications for green and reverse logistics usually study the trade-off between environmental (e.g., total emissions of the supply chain) and economic objectives (e.g., the total cost of the supply chain) (Alumur and Bektaş, 2019; Velázquez-Martínez and Fransoo, 2017). Therefore, multiobjective facility location models are particularly interesting in green/reverse facility location applications (Farahani et al., 2010). Mono and multiobjective facility location problems in reverse logistics for plastic recycling have focused on the location of collection and classification centers (Sheriff et al., 2017), processing facilities (Bing et al., 2015), and the design of closed-loop supply chains (Papen and Amin, 2019; Soleimani et al., 2016).

Despite its popularity, the literature on reverse logistics for plastic waste collection and recycling has neglected the study of RVM location (Valenzuela et al., 2021). Likewise, the location of RVM has neither been studied in the strategic design of waste reverse networks (Van Engeland et al., 2020) nor in the literature on waste bin location (Rossit and Nesmachnow, 2022). To fill this gap, in this book chapter, we present a multiobjective facility location model that can be used to assist companies in the selection of the sites to locate their (new) RVMs. Moreover, to illustrate the model's applicability, we present a case study conducted in a Colombian startup company dedicated to the fabrication and commercialization of RVMs in an innovative circular business model.

The rest of this chapter is organized as follows. Section 2 presents the multiobjective facility location model notation and formulation. The details and results of the case study are given in Section 3. Finally, Section 4 summarizes the main findings and presents possible extensions of our work.

2. Multiobjective facility location model

As pointed out by O'Connor et al. (2010) and Rossit and Nesmachnow (2022), the location of recycling containers is a critical factor in affecting their level of usage, and therefore, user accessibility to collection points is of paramount importance in the implementation of recycling programs. Poor location decisions diminish the material entering the formal waste collection system, reducing the amount of appropriately deposed or valorized waste. Similarly, Rossit and Nesmachnow (2022) highlight that the distance between bins and their relation with the collected volume has not been sufficiently studied in the waste collection literature.

RVMs serve at the same time as waste collection points for beverage containers, as marketing points for companies offering monetary benefits, and as environmental education channels (Stoyanov, 2015). Therefore, in the location of RVMs, conflicting objective functions arise: the maximization of collected waste/consumers visiting them and the total cost of operating the system while keeping a minimum distance constraint between located RVMs. Therefore, in this work, we propose a simple multiobjective facility location model aimed at maximizing the usage of RVMs in two dimensions (quantity of recycled material and the number of reached consumers) and minimizing the total operation cost of the system.

In a closely related work, Erfani et al. (2018) consider maximizing the inhabitants' usage of waste bins in a case study in an Iranian urban district. However, they do not consider the existing trade-off in our RVM location application nor use a multiobjective model to represent the location decision. In a similar study, Toutouh et al. (2020) proposed a multiobjective bin location problem with three objectives: collected volume, total investment, and user accessibility. These authors resorted to evolutionary algorithms to solve the resulting optimization problem. By contrast, the size of the instances solved in our case study enabled the use of optimization software as a solution alternative.

The proposed model combines the elements of maximal covering, uncapacitated facility location, and p-dispersion problems (Current et al., 2002). A description of the elements of the multiobjective location model follows. The set S represents the candidate sites where new RVMs can be installed. For each location $s \in S$, there are three important quantities: an estimation of the monthly amount of material that can be collected v_s, a potential number of monthly consumers interacting with an RVM located at the site t_s, and the monthly operation cost of a machine located at this site c_s. The values of these three quantities vary from site to site and can be estimated based on historical information and the characteristics of a given location using statistical models (Walk et al., 2020). For instance, the amount of material tends to be larger in RVMs near low-income quarters. Conversely, RVMs located at sites with restricted access, like universities or company headquarters, have smaller quantities than sites with open access, like shopping centers or public buildings. Similarly, the operating cost of an RVM varies from site to site. Sometimes, a company or university willing to host an RVM is willing to pay for their operating costs. The cost of the electricity needed to operate the machine is lower or higher depending on the location. Additionally, different formats can be used depending on the location of the RVM; a completely automated RVM or a human-assisted kiosk are the main alternatives.

In the model, the binary decision variables X_s represent the location of an RVM in a candidate site $s \in S$ ($X_s = 1$), or not ($X_s = 0$). Additionally, to prevent RVMs from being very close to each other, we use a p-dispersion-like constraint that includes the distance between pairs of sites $s, s' \in S$ ($d_{s,s'}$), a minimum distance threshold (d_{min}) between RVMs to be located, and a large positive quantity (M). Using this notation, it is possible to formulate the following multiobjective optimization model:

$$Max\ z_1 = \sum_{s \in S} t_s X_s \tag{8.1}$$

$$Max\ z_2 = \sum_{s \in S} v_s X_s \tag{8.2}$$

$$Min\ z_3 = \sum_{s \in S} c_s X_s \tag{8.3}$$

Subject to:

$$d_{min} + \left(M - d_{s,s'}\right)X_s + \left(M - d_{s,s'}\right)X_{s'} \leq 2M - d_{s,s'},\ \forall s, s' \in S : s \neq s' \tag{8.4}$$

$$X_s \in \{0, 1\}\ \forall\ s \in S \tag{8.5}$$

The first objective function (z_1) represents the total number of consumers having interaction with the RVMs, the second one (z_2) the total weight of the plastic bottles collected in the entire system, and the third one (z_3) the operation cost of the installed RVMs. The main constraint of the optimization model in Eq. (8.4), borrowed from the p-dispersion problem (Current et al., 2002), indicates that a minimum distance between the RVMs to be located must be kept. Finally, Eq. (8.5) defines the decision variables as binary.

Note that constraint (4) is required in the model to prevent RVMs from being very close to each other in the optimal solution, a phenomenon that often occurs in facility location models where market share maximization is sought (e.g., competitive facility location models (Méndez-Vogel et al., 2023)). The model (1)–(5) was implemented under an extended ϵ-constraint approach to handle the multiple objectives (Mesquita-Cunha et al., 2023) as follows:

$$Min\ z_3 = \sum_{s \in S} c_s X_s - \rho \sum_{k=1}^{2} \frac{z_k}{r_k} \tag{8.6}$$

Subject to:

$$\sum_{s \in S} t_s X_s \geq \epsilon_1 \tag{8.7}$$

$$\sum_{s \in S} v_s X_s \geq \epsilon_2 \tag{8.8}$$

$$d_{min} + \left(M - d_{s,s'}\right)X_s + \left(M - d_{s,s'}\right)X_{s'} \leq 2M - d_{s,s'}, \forall s, s' \in S : s \neq s' \tag{8.9}$$

$$X_s \in \{0, 1\} \ \forall \ s \in S \tag{8.10}$$

where ϵ_1 and ϵ_2 are lower bounds set for the first two objective functions that are iteratively increased using a uniform grid to obtain efficient solutions of problem (1)–(5). In this mono-objective model, parameter ρ takes very small values (usually between 10^{-3} and 10^{-6}) to maximize the two other objective functions lexicographically once the minimum cost is achieved. Additionally, to normalize the second term in the objective function, parameters r_1 and r_2 represent the range value for the total visits (z_1) and total collected material (z_2), respectively. We resort to the Mosel modeling language, to solve problem (6)–(11) using a grid of 100 points, with 10 intervals for each ϵ constraint. The solution of this model takes a few seconds in a standard laptop computer; a running time consistent with the integer-friendly behavior of facility location problems (Daskin, 2008). An Excel version of the model with additional constraints on the number of RVMs to be located and budget limits was transferred to the company as a spreadsheet-based decision support system for its planning process. We used the open-solver add-in of Excel to deploy this version of the model (Mason, 2012).

3. Case study

Several countries, including Colombia, have promoted the circular economy model due to its economic, environmental, and social benefits. The last Colombian governments have issued several regulations and policies to promote the circular economy, as illustrated by the timeline in Fig. 8.2. A particular focus on EPR-related laws is presented in this figure.

In 2009, product return plans were introduced into the medicines and lead batteries supply chains. In 2010, the National Sustainable Consumption Policy was implemented, and selective collection systems for waste batteries, computers, light bulbs, and used tires were established. In 2013, the postconsumer pesticide product return system was implemented. In 2016, a circular economy approach was proposed in the National Policy for the Integrated Management of Solid Waste. Finally, in 2018, the National Circular Economy Strategy was presented, and incentives were established to promote migration toward a circular approach. This approach will continue in the plan drawn up by the government for 2022–26. In 2018 and 2020, significant advances were made in the environmental management of packaging and containers by regulating the extended responsibility model for this type of waste (Resolution 1407 of 2018, further modified by Resolution 1342 of 2020). This environmental policy scheme establishes that producers are obliged to maintain an updated plan for the environmental management of their waste for its best use and reincorporation into the company's production cycle. Likewise, the

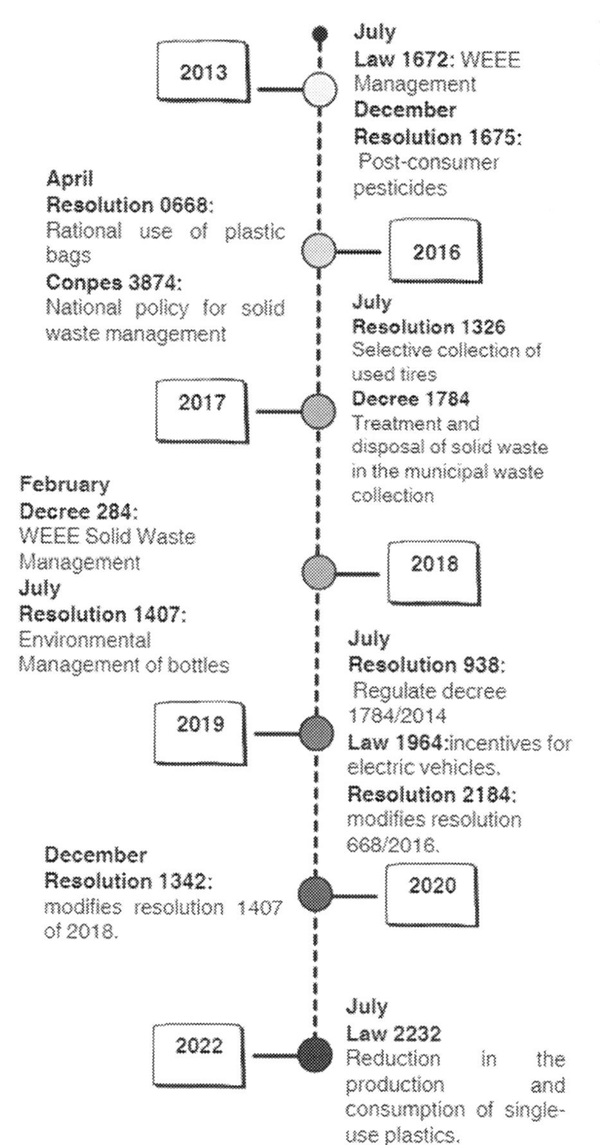

FIGURE 8.2 Evolution of the circular economy laws and policies in Colombia.

Ministry of Environment and Sustainable Development published the National Plan for its sustainable management. It seeks to implement a responsible consumption of plastic and natural resources, establishing goals of minimum use of 25% of plastic products until reaching 100% recyclable and reusable plastics. That plan is complemented by Law 2232 of 2022, which gradually establishes measures to reduce the production of single-use plastics.

3.1 The company and the data

The operation of the company under study corresponds to a circular economy business model supported in an application where users are the fundamental actors in the recycling of plastic containers. In their business model, users register in a web platform of the company. When they deposit a bottle at an RVM of the company, they receive in exchange a series of points that they can redeem for movie tickets, mobile voice and data services, spa, public services, clothing, and tours, among others. In addition, they offer an advertising service on the screens of the points for brand positioning through ecological campaigns. Through the screens, the mobile application, and its website, the company mobilizes people for intelligent recycling through education, innovation, and love for the planet, preventing bottles from reaching landfills, thus aiming at citizen participation as a fundamental element of the process.

There were a series of interviews to understand the company's business model, the planning methodology, types of format used (fully automated RVM or kiosk), criteria for selecting locations, growth projections, collection process, and information analysis on the operation and the users. After knowing the actual inner processes of the company, there was a request for the data to implement model (1)–(5). For instance, its current and candidate collection points, number of visits, volume collected per point, and estimation of installation and operation costs. Additionally, there was a request for qualitative data about the recycling collection centers, such as type of access (indoor or outdoor), format type (kiosk or machine), georeferencing (latitude and longitude), and the socioeconomic stratum of the sites. Mainly, the company was interested in analyzing the location of their RVM in 35 points in Medellín, the second-largest Colombian city. The candidate locations include shopping centers, universities and technical education institutions, hospitals, public parks, convention centers, and specific downtown buildings. Using the information provided by the company, we analyzed different scenarios and the trade-off between the three objectives.

3.2 Results and discussion

Initially, we analyze the effect of the minimum distance constraint on the model's results. Three scenarios were run with values of d_{min} in the set $\{0 \text{ km}, 1 \text{ km}, 2 \text{ km}\}$. Different results were obtained, Table 8.1 presents summary statistics of these scenarios: the number of efficient solutions obtained, and the maximum values for the coverage objective functions (z_1, z_2), cost (z_3) and number of RVM located in the extreme point of the efficient frontiers obtained with the three values of d_{min}.

TABLE 8.1 Scenarios with different values of d_{min}.

Scenario	$d_{min} = 0$	$d_{min} = 1$ km	$d_{min} = 2$ km
Number of solutions in the efficient frontier	19	12	11
Maximum number of RVM located	35	21	19
Maximum number of visits (z_1)	33,628	26,928	23,590
Maximum collected quantity [kg] (z_2),	17,180	13,758	12,180
Maximum cost [COP$] (z_3),	56,236,000	33,836,000	31,236,000

Table 8.1 shows that if the minimum distance between RVMs is considered, it is no longer feasible to locate them in all the candidate sites envisaged by the company, decreasing from the 35 original sites to a value near 20 RVMs (a 42% reduction approximately) when the distance limit is between 1 and 2 km. This reduction in the number of RVMs located also represents a proportional reduction in the cost of about 40%. Although this reduction in the number of RVM in the system also represents a decrease in the total number of customers interacting with the machines and the amount of material collected, the values obtained with this distance constraint seem more realistic. Very close RVMs could cannibalize their market share, and therefore, the estimation provided by the values of t_s and v_s could be unreasonably optimistic. Specifically, Fig. 8.3 presents efficient solutions with 19 RVMs obtained

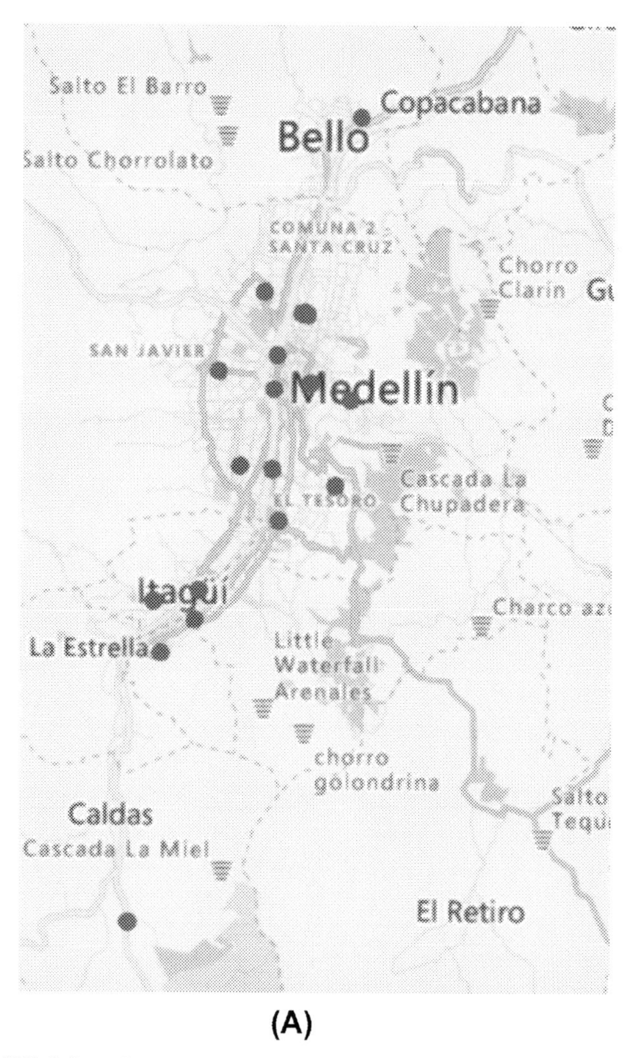

FIGURE 8.3 Efficient solutions with 19 RVMs and different d_{min} thresholds.

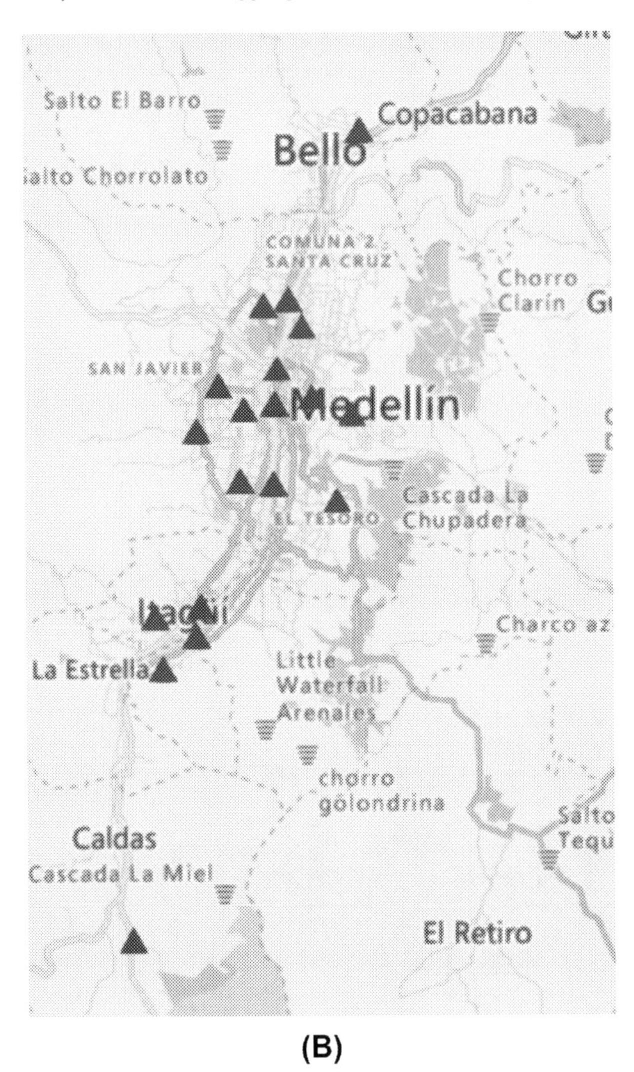

(B)

FIGURE 8.3 cont'd

with the optimization model for the three scenarios. The comparison of the three solutions shows that in the scenario without the minimum distance threshold (Fig. 8.3A), three RVMs are crowded in the northern part of the city, cannibalizing each other market. By contrast, if a minimum distance between RVMs is imposed, a better spread of them is obtained. Particularly, when the minimum distance threshold reaches 2 km, some RVMs are located in northern and southern areas of the city that were not covered in the previous two scenarios (comparing the solutions of Fig. 8.3B and C).

Using a minimum distance of 1 or 2 km seems reasonable for further analysis as the values of collected material and number of visits could be more realistic. Given that the objective

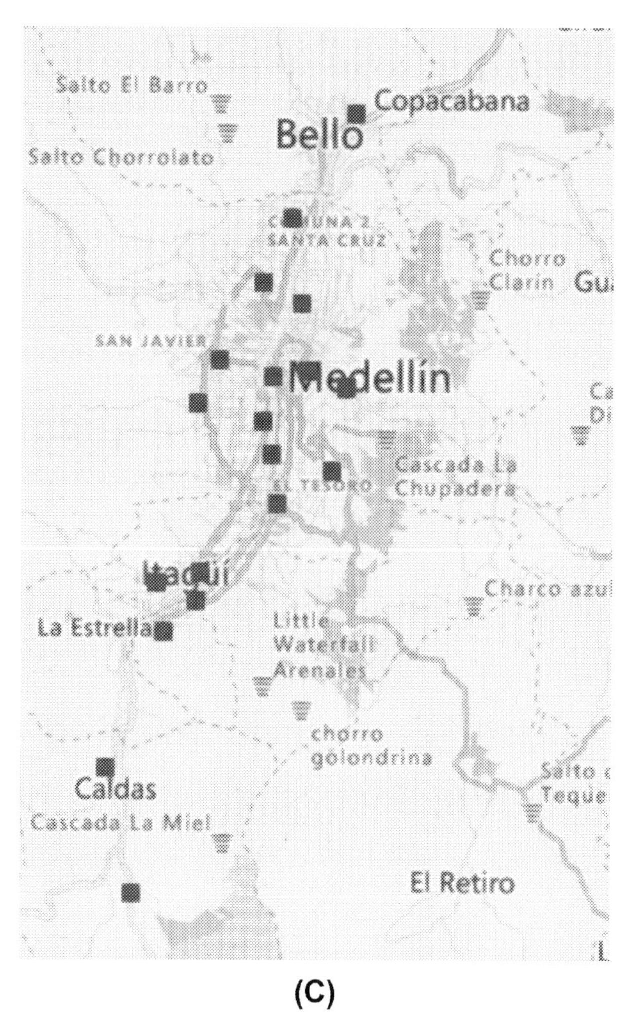

(C)

FIGURE 8.3 cont'd

functions are similar with these values of d_{min} (12% absolute difference approximately), we decided to continue the analysis using a value of 2 km to obtain locations that are better spread in the city.

Table 8.2 presents the details of the solutions of the resulting efficient frontier. As it can be seen, a solution with more than 19 RVMs is infeasible when $d_{min} = 2$ km. Moreover, we only found 11 solutions in this efficient frontier. This phenomenon is mainly due to the number of intervals used. We put a limit of 10% of the total visits and material as the initial ϵ value for both constraints and increase them in steps of 10%. Consequently, solutions with very few RVMs did not appear. Moreover, a large value of the ϵ constraint for one of the objectives also rises the other one automatically due to the high correlation between the total material

TABLE 8.2 Solutions for the efficient frontier with $d_{min} = 2$ km.

Solution	Total visits (z_1)	Total material (z_2) [kg]	Total cost (z_3) [COP$]	Number of RVMs
1	3637	1727	2,954,000	3
2	6806	3482	5,908,000	5
3	10,264	5172	9,216,000	7
4	13,722	6862	12,524,000	9
5	13,842	6974	12,878,000	9
6	17,060	8440	15,832,000	11
7	16,891	8617	17,058,000	11
8	20,219	9981	19,494,000	12
9	20,587	10,348	20,440,000	13
10	23,180	12,171	28,308,000	17
11	23,590	12,180	31,236,000	19

and total visits. Therefore, several duplicated solutions were discarded when building the final efficient frontier.

As said before, the efficient frontier shows that the number of total visits and total material collected in the system are highly correlated. Nonetheless, there are solutions with the same number of RVMs and similar costs, offering a different location pattern of the RVMs and, therefore different amounts of material collected and number of visits (e.g., solutions 4 and 5 in the frontier). On the other hand, some solutions highly differ in cost despite having the same number of RVMs (e.g., solutions 6 and 7 in the frontier).

Now, we turn the attention to the trade-off between cost and the total visits/total collected material; Figs. 8.4 and 8.5 present the bidimensional views of the efficient frontier to depict such trade-offs. These figures show that at the beginning, the growth of the number of visits and collected materials followed a rather linear behavior until solution 9 in the frontier. Once this solution is reached, decreasing returns are obtained, and to increase the visits and the collected material several RVMs must be added, leading to a significant jump in the total cost. For instance, changing from solution 9 to 10 requires an increase of 38% in the cost to obtain only a 13% increase in the total visits and 18% in the collected material. More striking is the 10% increase in cost from solution 10 to 11 to obtain a negligible increase in total visits (1.8%) and collected material (0.1%). Finally, in this bidimensional view, some points seem dominated as they show a slight decrease in the objective function depicted in the figure. However, in the other objective function, they offer an increase compared to the previous solution and therefore are nondominated solutions. This is, for instance, the case of solution 7.

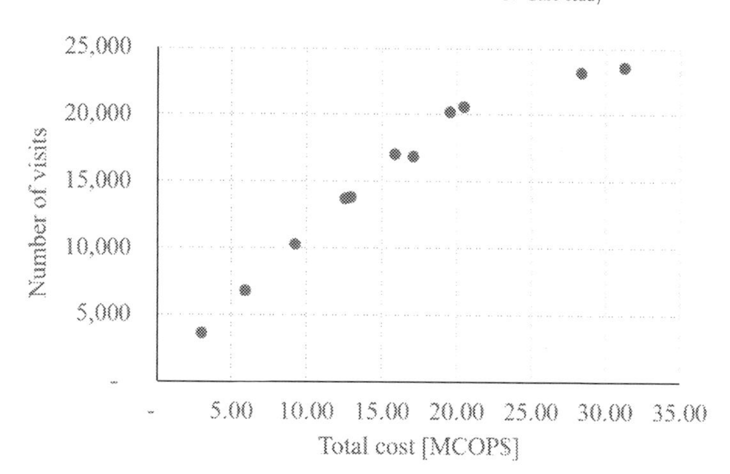

FIGURE 8.4 Trade-off between the total cost and the total visits.

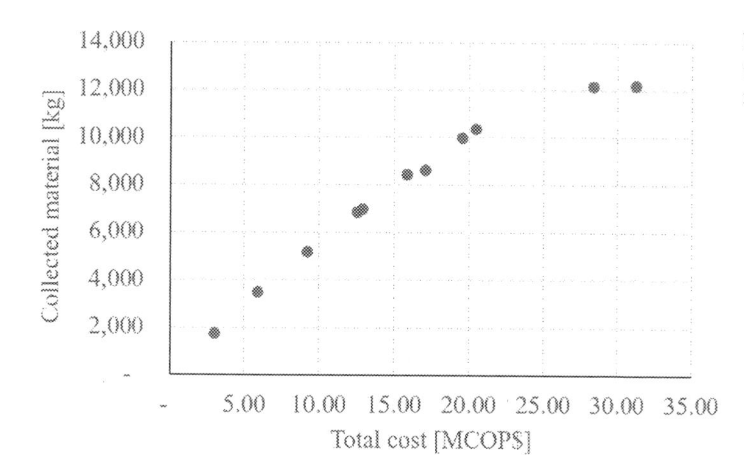

FIGURE 8.5 Trade-off between the total cost and the total collected material.

3.3 Managerial insights

Finally, we analyze the attractiveness of the different candidate sites for the location of the RVMs. To do so, we count the number of appearances of them in the solutions of the frontier. A naïve form to rank the sites could be to order them by the values of t_s and v_S. Additionally, these values could be used in a multicriteria sorting method to support the selection of the sites (Alvarez et al., 2021). However, using these criteria would ignore the interaction of the RVMs that results from the minimum distance threshold constraint. Fig. 8.6 presents in descending order the number of times that each candidate site appears in the solutions of the efficient frontier. Half of the candidate sites were never selected for the location of RVMs (18/35). On the other hand, two of them appeared in all solutions of the frontier.

For the sake of confidentiality, the names of the sites were anonymized in this figure. However, with the names provided in the table (indicating the type of location), it is clear that

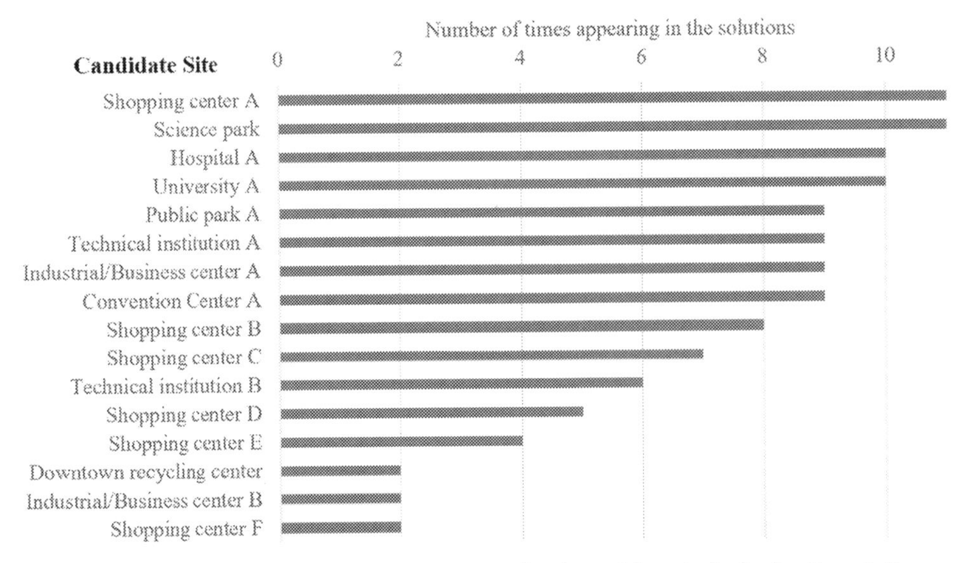

FIGURE 8.6 Count the number of appearances of each candidate site in the frontier solutions.

shopping centers are the most preferred locations for the RVMs as one of the top locations is of this type, and five others (from a total of 11 candidate location of this type) appeared in several solutions. Other free access spaces like public parks and science parks are also very attractive for the location of RVMs. Business, industrial, and convention centers are also attractive locations for RVMs, as the three candidate sites of this type were selected at least once in the solutions of the frontier. Remarkably, with a few exceptions, universities and technical institutions are not selected frequently as RVM locations by the model. Despite being 10 out of the 35 candidate sites, only three appear in the solutions of the efficient frontier. An interesting exception is University A, located in a zone of the city without other nearby candidate sites, which was selected in almost all solutions of the frontier. This last behavior can be explained by the restricted access to these locations (mainly to students, professors, and employees) that generates a smaller public for the RVMs and, therefore, less visits and collected material. Two sites in the list of candidates deserve a final discussion. The only hospital in the list of candidate sites is selected in many frontier solutions due to its location in an area of the city without other nearby candidate sites. Conversely, a recycling collection center in the city's downtown is selected in a few solutions.

4. Conclusions and future research

Reverse vending machines are an attractive option for material collection in plastic packaging recycling. Their suitability as collection points and educational/marketing devices offers an interesting alternative to designing reverse logistics networks for plastic packaging. This unique synergy between material collection and user education and engagement is an attractive attribute for using RVM in implementing EPR policies. Nonetheless, the location of these devices is seldom studied in the literature.

Therefore, in this chapter, we introduce a simple multiobjective facility location problem to support the decision-making process for companies interested in the use of these machines in their extended producer responsibility strategies. The proposed model aims to maximize users' visits and collected materials while minimizing costs and considering a minimum distance threshold between reverse vending machines as the main constraint. From the many multiobjective optimization modeling options, we decided to use the ϵ-constraints method to explore the trade-off between the three objectives of the problem.

Using this model in a case study for a startup company operating in Medellín (Colombia) illustrates its appropriateness as a decision support tool. The analysis of the case study reveals that ignoring the minimum distance between machines could result in crowded locations with several nearby machines. By contrast, using different distance thresholds allows the control of the spread of the RVM in the city. As is common in multiobjective facility location models, the efficient frontier offers solutions with a different trade-off in the objective functions. For example, having too many machines leads to diminishing returns as the cost of locating extra machines outweighs the minimal increase in visits and collection.

The analysis of the frequency of appearance of the candidate sites in the solutions of the efficient frontier unveils the attractiveness of shopping centers for the location of RVMs when compared against educational institutions and other types of locations. Likewise, some isolated points are also attractive for the location of these machines as they cover areas of the city that any other candidate location could not serve.

Finally, the proposed model offers several possibilities for future research. The multiobjective approach could be changed to other methods like the scalarization of the objectives or goal programming (Guggeri et al., 2023). Likewise, the model itself can be extended to explicitly include the customers and their preferences in the selection of the RVMs. For instance, discrete choice models can be embedded in calculating the volumes and visits to each candidate site following the approach presented in Haase and Müller (2014). Likewise, solving large-scale instances of this problem that require heuristics and metaheuristic approaches is also a possible research direction.

Acknowledgments

Pablo A. Maya gratefully acknowledges the support of the Colombian General Royalty System within the framework of the BPIN project 2020000100497 "*Alianza Circular, Mejoramiento de la gestión de empaques y envases a través de estrategias de economía circular en la industria de Antioquia*"

References

Alfarisi, L., Mulyana, E., Faroqi, A., 2022. Reverse vending machine for exchange empty containers to money. In: 2022 8th International Conference on Wireless and Telematics (ICWT), pp. 1–4. https://doi.org/10.1109/ICWT55831.2022.9935455.

Alumur, S.A., Bektaş, T., 2019. Green location problems. In: Laporte, G., Nickel, S., Saldanha Da Gama, F. (Eds.), Location Science. Springer International Publishing, pp. 591–610. https://doi.org/10.1007/978-3-030-32177-2_20.

Alvarez, P.A., Ishizaka, A., Martínez, L., 2021. Multiple-criteria decision-making sorting methods: a survey. Expert Systems with Applications 183, 115368. https://doi.org/10.1016/j.eswa.2021.115368.

Bing, X., Bloemhof-Ruwaard, J., Chaabane, A., Van Der Vorst, J., 2015. Global reverse supply chain redesign for household plastic waste under the emission trading scheme. Journal of Cleaner Production 103, 28–39. https://doi.org/10.1016/j.jclepro.2015.02.019.

Current, J., Daskin, M., Schilling, D., 2002. Discrete network location models. In: Drezner, Z., Hamacher, H.W. (Eds.), Facility Location. Springer Berlin Heidelberg, pp. 81–118. https://doi.org/10.1007/978-3-642-56082-8_3.

Daskin, M.S., 2008. What you should know about location modeling. Naval Research Logistics 55 (4), 283–294. https://doi.org/10.1002/nav.20284.

Dataintelo, 2023. Global Reverse Vending Machine Market. Dataintelo. https://dataintelo.com/report/global-reverse-vending-machine-market/.

Erfani, S.M.H., Danesh, S., Karrabi, S.M., Shad, R., Nemati, S., 2018. Using applied operations research and geographical information systems to evaluate effective factors in storage service of municipal solid waste management systems. Waste Management 79, 346–355. https://doi.org/10.1016/j.wasman.2018.08.003.

Farahani, R.Z., SteadieSeifi, M., Asgari, N., 2010. Multiple criteria facility location problems: a survey. Applied Mathematical Modelling 34 (7), 1689–1709. https://doi.org/10.1016/j.apm.2009.10.005.

Guggeri, E.M., Ham, C., Silveyra, P., Rossit, D.A., Piñeyro, P., 2023. Goal programming and multi-criteria methods in remanufacturing and reverse logistics: systematic literature review and survey. Computers & Industrial Engineering 185, 109587. https://doi.org/10.1016/j.cie.2023.109587.

Haase, K., Müller, S., 2014. A comparison of linear reformulations for multinomial logit choice probabilities in facility location models. European Journal of Operational Research 232 (3), 689–691. https://doi.org/10.1016/j.ejor.2013.08.009.

Mason, A.J., 2012. OpenSolver - An Open Source Add-in to Solve Linear and Integer Progammes in Excel. Operations Research Proceedings 2011. Springer, Berlin Heidelberg, pp. 401–406. https://doi.org/10.1007/978-3-642-29210-1_64. http://opensolver.org.

Méndez-Vogel, G., Marianov, V., Lüer-Villagra, A., 2023. The follower competitive facility location problem under the nested logit choice rule. European Journal of Operational Research 310 (2), 834–846. https://doi.org/10.1016/j.ejor.2023.03.008.

Mesquita-Cunha, M., Figueira, J.R., Barbosa-Póvoa, A.P., 2023. New $\epsilon -$ constraint methods for multi-objective integer linear programming: a Pareto front representation approach. European Journal of Operational Research 306 (1), 286–307. https://doi.org/10.1016/j.ejor.2022.07.044.

O'Connor, R.T., Lerman, D.C., Fritz, J.N., Hodde, H.B., 2010. Effects of number and location of bins on plastic recycling at a university. Journal of Applied Behavior Analysis 43 (4), 711–715. https://doi.org/10.1901/jaba.2010.43-711.

OECD, 2016. Extended Producer Responsibility: Policy Highlights- Guidance for Efficient Waste Management. OECD publishing. https://www.oecd.org/environment/waste/Extended-producer-responsibility-Policy-Highlights-2016-web.pdf.

Papen, P., Amin, S.H., 2019. Network configuration of a bottled water closed-loop supply chain with green supplier selection. Journal of Remanufacturing 9 (2), 109–127. https://doi.org/10.1007/s13243-018-0061-y.

Research and Markets, 2022. The Worldwide Reverse Vending Machine Industry Is Expected to Reach $736.9 Million by 2030. https://www.globenewswire.com/en/news-release/2022/05/16/2443719/28124/en/The-Worldwide-Reverse-Vending-Machine-Industry-is-Expected-to-Reach-736-9-Million-by-2030.html.

Rojas, T., March 9, 2023. Pacto por los Plásticos en Colombia: Acelerando la economía circular del plástico. Tecnología Del Plástico. https://www.plastico.com/es/noticias/pacto-por-los-plasticos-en-colombia-acelerando-la-economia-circular-del-plastico.

Rossit, D.G., Nesmachnow, S., 2022. Waste bins location problem: a review of recent advances in the storage stage of the Municipal Solid Waste reverse logistic chain. Journal of Cleaner Production 342, 130793. https://doi.org/10.1016/j.jclepro.2022.130793.

Saldanha-da-Gama, F., 2022. Facility location in logistics and transportation: an enduring relationship. Transportation Research Part E: Logistics and Transportation Review 166, 102903. https://doi.org/10.1016/j.tre.2022.102903.

Sambhi, S., Dahiya, P., 2020. Reverse vending machine for managing plastic waste. International Journal of System Assurance Engineering and Management 11 (3), 635–640. https://doi.org/10.1007/s13198-020-00967-y.

Sheriff, K.M.M., Subramanian, N., Rahman, S., Jayaram, J., 2017. Integrated optimization model and methodology for plastics recycling: Indian empirical evidence. Journal of Cleaner Production 153, 707–717. https://doi.org/10.1016/j.jclepro.2016.07.137.

Soleimani, H., Seyyed-Esfahani, M., Shirazi, M.A., 2016. A new multi-criteria scenario-based solution approach for stochastic forward/reverse supply chain network design. Annals of Operations Research 242 (2), 399–421. https://doi.org/10.1007/s10479-013-1435-z.

Stoyanov, D., 2015. Sustainable marketing: a global benchmark perspective on the vending industry. Review of Integrative Business and Economics Research 4 (2), 1–19.

Toutouh, J., Rossit, D., Nesmachnow, S., 2020. Soft computing methods for multiobjective location of garbage accumulation points in smart cities. Annals of Mathematics and Artificial Intelligence 88 (1–3), 105–131. https://doi.org/10.1007/s10472-019-09647-5.

Valenzuela, J., Alfaro, M., Fuertes, G., Vargas, M., Sáez-Navarrete, C., 2021. Reverse logistics models for the collection of plastic waste: a literature review. Waste Management & Research: The Journal for a Sustainable Circular Economy 39 (9), 1116–1134. https://doi.org/10.1177/0734242X211003948.

Van Engeland, J., Beliën, J., De Boeck, L., De Jaeger, S., 2020. Literature review: strategic network optimization models in waste reverse supply chains. Omega 91, 102012. https://doi.org/10.1016/j.omega.2018.12.001.

Velázquez-Martínez, J.C., Fransoo, J.C., 2017. Green facility location. In: Bouchery, Y., Corbett, C.J., Fransoo, J.C., Tan, T. (Eds.), Sustainable Supply Chains, vol. 4. Springer International Publishing, pp. 219–234. https://doi.org/10.1007/978-3-319-29791-0_9.

Walk, J., Hirt, R., Kühl, N., Hersløv, E.R., 2020. Half-empty or half-full? A hybrid approach to predict recycling behavior of consumers to increase reverse vending machine uptime. In: Nóvoa, H., Drăgoicea, M., Kühl, N. (Eds.), Exploring Service Science, vol 377. Springer International Publishing, pp. 107–120. https://doi.org/10.1007/978-3-030-38724-2_8.

Zia, H., Jawaid, M.U., Fatima, H.S., Hassan, I.U., Hussain, A., Shahzad, S., Khurram, M., 2022. Plastic waste management through the development of a low cost and light weight deep learning based reverse vending machine. Recycling 7 (5), 70. https://doi.org/10.3390/recycling7050070.

CHAPTER

9

Closed-loop supply chain and extended producer responsibility (EPR): A literature review

Milagros Marizcurrena[1], *Victoria Morás*[1], *Guillermo Ulery*[1], *Daniel Rossit*[2] *and Pedro Piñeyro*[1]

[1]Universidad de la República, Montevideo, Uruguay; [2]Department of Engineering, INMABB, Universidad Nacional del Sur (UNS)-CONICET, Bahía Blanca, Argentina

1. Introduction

The mitigation of industrial activities on the environment is a central topic in the different research agendas of countries and the UN. Achieving sustainable management of productive resources and reducing the volume of waste represent two of the greatest challenges for industrial systems and for the scientific community that proposes to contribute. One of the main strategies to achieve this double impact is the recovery of products once they have ended their useful life (Govindan et al., 2015). In this way, they are prevented from being discarded, reducing the amount of waste, and the consumption of raw materials is also reduced, by using a large part of the already industrialized materials in the recovered products (Matsumoto et al., 2016).

The problem of recovery of used products has become an increasingly popular topic (Sadrnia et al., 2021), both its research and applications having grown. Guide and Van Wassenhove (2009) identified that, in 1990, although the practice was used, research was scarce. Global interest began to grow by identifying the economic potential, both in increasing profits and in reducing costs, while for Europe, the focus on this issue began due to regulations imposed by the European Union on the end of the life cycle of the products (Pazhani et al., 2021). Subsequently, it was identified that the economic benefit is not the only one, but that there are also benefits on the environment: reduction of waste, gas emissions and

Evolution and Trends of Sustainable Approaches
https://doi.org/10.1016/B978-0-443-21651-0.00007-3

optimization of resources, and the added value that the product adopts, positively affecting the image that is provided to the final customer (Guide and Van Wassenhove, 2009).

Recovering the products also implies carrying out a proper management of logistics operations. Reverse logistics involves all activities related to the collection, recovery, and/or proper disposal of used products. Considering the consumer society we live in today, where the consumption of resources and the generation of waste are increasing, manufacturers are increasingly seeking to integrate reverse logistics operations into their supply chain, thus seeking to minimize the environmental impact caused by their companies (Li et al., 2018). The worldwide increase in environmental regulations has intensified the importance of reverse logistics, adding the fact that it also leads to greater profitability by reducing transportation, inventory, and storage costs (Ilgin et al., 2010). It is for the points aforementioned that it is possible to find many research works in the literature, focused on the recovery of used products and remanufacturing (Tornese et al., 2019).

The objective of this chapter is to present a state of the art on the planning of production and flow of products in closed supply chains. Throughout this work, the current situation of the topic is exposed, informing the reader about the ideas, standards, and methods used today along with the benefits achieved by this type of practices. Similarly, gaps and aspects that have not been considered in the articles surveyed in this literature review are identified, ensuring that the topic is presented to the reader in the most complete way possible. For the preparation of this chapter, the international regulatory and legal framework was taken into account, as well as that of Uruguay, as an example of a developing South American country and to be able to reveal how the issue is found in these cases. In addition, an analysis of scientific documents was also carried out, seeking to contrast the realities presented by the different authors regarding the recovery of used products and the design of closed supply chains with remanufacturing options.

The rest of the chapter is organized as follows. In Section 2, an environmental regulatory framework is provided along with information on the current EPR situation in Uruguay and in the world regarding the recovery of used products. In Section 3, the motivation and why we chose to study this topic are explained. Section 4 first describes the systematic literature review procedure carried out for this research. Second, the analysis of the most relevant articles found is included, classifying them into two large categories: those that deal with production planning in closed supply chains and those that address the flow of the supply chain with product recovery. Finally, the conclusions are addressed in Section 5.

2. Extended producer responsibility context

2.1 Extended producer responsibility

Extended producer responsibility (EPR) seeks to involve producers/manufacturers in the recovery process of the products they generate. This implies that producers/manufacturers assume the management of the waste they generate, including the associated costs. Thus, producers must deal with the treatment or disposal of the product once its life cycle has ended (OECD, 2021). To achieve this objective, there are different techniques that producers can adopt, as recovering end-of-life or end-of-use products.

On the other hand, because waste volumes are increasing in all the countries and societies, governments decide to take these responsibilities policies to mitigate the negative impact of waste. In the following, main parts of these regulations will be explored and detailed.

2.2 Current environmental regulations

In the current international framework, numerous countries have begun to take concrete actions with regard to sustainability issue, which is why there is increasing number of governments that are choosing to implement environmental regulations that regulate the use of natural resources and promote the implementation of recovery activities for used products. In particular, the recovery of used products is one of the methods most sought to be adopted in the region.

As already mentioned in the introduction, product recovery is a problem and an issue that has become very important for some decades now, but it was important to highlight the regulations linked to this issue.

Uruguay, as a country belonging to MERCOSUR, is directly affected by Law 17,712, which was approved on November 27, 2003, as the "MERCOSUR Framework Agreement on the Environment," and has as its objective the protection of the environment and the sustainable use of natural resources. The countries involved in said law must seek the creation of national regulations regarding environmental matters and consider the different realities of the country. They must also collaborate to ensure safe working conditions and promote environmental education.

This agreement confirms the principles of the Rio Declaration on Environment and Development, which was the manifesto developed after the United Nations Conference on Environment and Development in Rio in 1992. This recognizes the integral and interdependent nature of the planet, so that the protection of ecosystems, and for this, sustainable production, is declared as the interest of all countries and all human beings.

On the other hand, there are ISO 14000 standards, which are standards for Environmental Management Systems, which provide limit values and forms of evaluation on the quality of air, water and soil, and smoke emissions from vehicles. These not only seek the preservation of the environment but also specify environmental goals and specific parameters, so that it becomes achievable and measurable, so that environmental management can be carried out by companies in a simpler way.

2.3 Motivation

The world has been dominated by a mass production and consumption model, in which goods are generally produced from raw materials, used, and at the end of their useful life, they end up being discarded as waste. In the past decades, societies have constantly increased the consumption of goods significantly, which has also led to an increase in the generation of waste, thus producing problems in the environment and in the health of the population.

Every year, 380 million tons of plastic are produced in the world for various uses, of which 12.7 million reach the oceans and end up harming the lives of more than 700 species (Greenpeace). In turn, of the more than 1032 tons of electronic waste that are discarded each year, less than 17% can be treated (Greenpeace).

However, lately, the importance of responsible product management has begun to be observed, generating a positive impact both from an environmental and social point of view and from an economic point of view. Manufacturers are coming under increasing pressure from governments and the creation and updating of standards and laws that seek to minimize environmental impact. Similarly, the influence of interest groups, activists, and society itself is also increasing, thus leading producers to seek to design their supply chains in a way that is as environmentally friendly as possible. However, this is not the only reason why product recovery, subsequent treatment, and the design of a closed supply chain are so important today, adopting these practices also has environmental and economic advantages (Blackburn et al., 2004). For manufacturers, from an economic point of view, remanufacturing is an industrial practice that allows them to recover the lost value of the product, offering various benefits such as cost savings by reducing the amount of material they purchase and store, as well as the energy consumed in the production stages (Piñeyro and Viera, 2009). From the same point of view, society and consumers also benefit through the creation of jobs, since, in general, remanufacturing requires a greater proportion of man-hours than the manufacture of a new product (Thierry et al., 1995; Guggeri et al., 2023). In turn, companies end up offering a product with the same features as a new one, but at a lower price (Piñeyro and Viera, 2009). Many remanufacturing companies even argue that remanufactured products can have higher quality, since quality control systems are usually higher than those of new products (Matsumoto et al., 2016; Piñeyro and Rossit, 2020). Similarly, by consuming remanufactured products with a service cycle shorter than the usual life cycle of the products, customers will always have more frequent access to products of the latest technology and design (Guide and Van Wassenhove, 2009). From an environmental point of view, product recovery and remanufacturing specifically play an important role in minimizing the amount of discarded products and reducing air and water pollution, by reducing greenhouse gas emissions. greenhouse gases, other toxic pollutants, and the use of raw materials and energy (Guggeri et al., 2023).

At the same time, remanufacturing provides manufacturers with other benefits that are not so explicit. First, it internalizes revenue by preventing resellers of damaged products or waste merchants from taking advantage of the company's own products. Second, it allows companies to offer a new range of products and services, outside the standard consumption ranges, thus accessing new consumers (Matsumoto et al., 2016). Third, the design of products for remanufacturing processes improves the productivity and general quality of the products offered by the company, by optimizing the reverse engineering used. Finally, adopting this type of recovery practices for used products and everything that entails allows companies to anticipate future environmental legislation and taxes, gaining competitiveness over those companies that are not proactive on this issue (Guide and Van Wassenhove, 2009).

Despite this, today, remanufactured products are still a relatively small portion compared with newly manufactured products (Xu and Feng, 2014).

3. Systematic literature review procedure and article filtering

This section presents the search for articles for this review. For performing the reviewing, the Systematic Literature Review method (Briner and Denyer, 2012) was adopted. Briner and

Denyer (2012) propose a series of steps to carry out this review, which, in a general way, consist of defining the theme and types of study, then determining those studies that are of interest according to the criteria taken, and finally, extract the relevant information about each study of the defined topic, and write a report with the contributions of each study.

On this occasion, to carry out the systematic review, it is necessary to define specifically the topic to be studied. To this end, iterative searches were carried out with different keywords in the three databases considered: Scopus, Springer, and Science Direct. By reaching an adequate scope of articles with the defined requirements, the keywords were established: "programming" AND "remanufacturing" AND "closed loop Supply chain" AND NOT "fuzzy" AND NOT "stochastic" AND "EPR." That is, we are looking for those items that consider a closed supply chain, with remanufacturing and reference to the extended responsibility of the producer with a deterministic approach. From this search, the number of articles taught are provided in Table 9.1.

Next, a verification was carried out to ensure that all articles have mathematical programming models, that they are deterministic, and that the model defines the flow of the products with the location of the centers established, leaving out those that define the location of the different centers.

After this verification, a total of 37 articles were confirmed that met all the criteria (Table 9.2).

TABLE 9.1 Number of articles in each journal considering the keywords: programming, remanufacturing, closed-loop supply chain, nonfuzzy nonstochastic, and EPR.

Journals	Number of articles found
Scopus	67
SpringerLink	5
ScienceDirect	37
TOTAL	109

TABLE 9.2 Number of articles in each journal considering the detailed keywords, which have a mathematical model and are deterministic.

Journals	Number of articles that meet the criteria
Scopus	34
Springer	1
ScienceDirect	2
TOTAL	37

Subsequently, an in-depth reading of the articles was carried out to determine: objective or objectives of the model, if it is multilevel, multiperiod, multiproduct, the resolution method, if it plans production or if it designs the supply chain, if the model considers environmental impact, inspection, quality levels, EPR, or outsource services.

Regarding the different particularities of the models, in Fig. 9.1, how many meet each characteristic (multiproduct, multiperiod and multilevel) of the 37 total items is presented. To continue, regarding the topic, two major issues were found in the articles: those that were focused on production planning in closed supply chains and those focused on the flow in CLSCs (Fig. 9.2).

Then, once these two themes were detected, different categorization criteria for the articles were established. On the one hand, for the issue of production planning, they were classified according to: production planning in CLSCs, environmental impact, EPR, inspection, and outsourcing of services. In Fig. 9.3, the number of articles that correspond to each category can be seen.

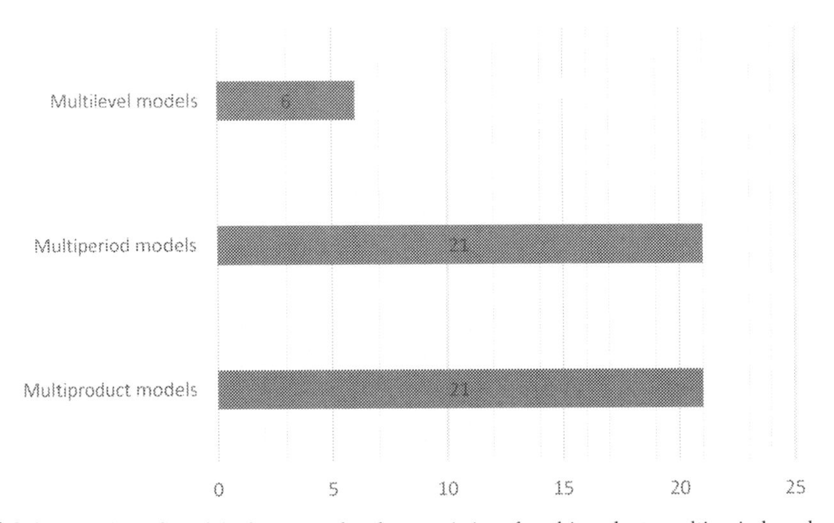

FIGURE 9.1 Number of models that meet the characteristics of multiproduct, multiperiod, and multilevel.

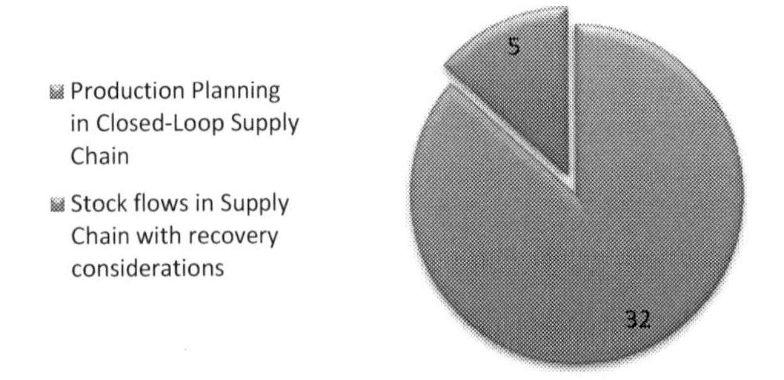

FIGURE 9.2 Production planning and supply chain design proportion of articles.

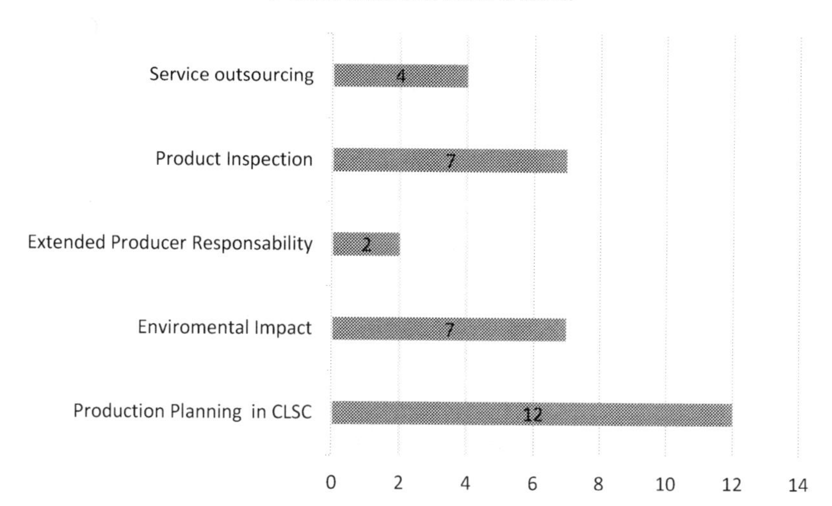

FIGURE 9.3 Categorization of the subject of production planning in closed supply chains.

4. Review of academic articles of interest

For the review of the articles, they are classified into two subsections: one subsection dedicated to those articles that plan production in closed supply chains, and the other subsection to those articles that focus on the flow of this chain without planning production. This division distinguishes those articles in which along with the supply chain, the internal operations of the agents involved are also analyzed, from those articles in which the supply chain is analyzed without analyzing the internal operations of any particular agent.

4.1 Production planning in closed supply chains

Production planning consists of determining the aspects related to the acquisition of resources and raw materials together with the production activities required to satisfy consumer demand in the most efficient or economical possible way. Together with scheduling (at the operational level), production planning (capacity at the tactical level) forms production management in the supply chain (Rowshannahad et al., 2018). Then, in the following works, this activity is considered in closed supply chains, which take into account the reverse flow of materials and decisions on transfers to collection centers, inspection, cleaning, transformation, reuse, among others, are incorporated. Therefore, considering production planning within a closed supply chain consists of considering the "upstream" returns within resource, material, and production planning decisions to satisfy customer demand.

4.1.1 Production planning in closed-loop supply chains

This subsection is dedicated to that articles that plan production in closed supply chains. Initially, those articles that address direct problems (i.e., with any special feature) are presented. Then, the rest of the articles that tackled peculiar decisions or cover topics of another nature such as environmental impact, inspection or outsourcing of services, or legislation on extended producer responsibility will be presented.

4.1.1.1 Single product systems

A tour of the works is carried out, beginning with those that involve a single product, then with those that pose the problem for more than one product, and finally, those that are multilevel.

Chekoubi et al. (2018) study the integrated production-inventory-routing problem (IPIRP), by developing a linear programming model that aims to determine the optimal batch of units to be remanufactured, along with the optimal batch of units to be produced and the quantity of these to be stored. In this way, it seeks to minimize costs and satisfy the needs of customers. The problem is a dynamic lot sizing with remanufacturing, coupled with a vehicle routing problem with pickup and delivery, and can be seen as an extension of the classic "Enabled dynamic lot sizing with reconditioning."

It is considered that there is a central repository that includes both types of articles, unlike the one presented by Zhang et al. (2011), who consider two independent deposits. In the latter, there is also a dynamic batch sizing problem, known as capacitated production planning (CPP). These problems lead to a program for the collection and processing of used products, which come from customers, with remanufacturing, repair, recycling, or disposal of waste as a processing option.

Both works (Chekoubi et al., 2018; Zhang et al., 2011) assume that the quality of remanufactured products is the same as those produced from new raw materials, so that the demand can be satisfied with either of these two.

The fact that such demand can be satisfied by both new and remanufactured items is linked to the definition of remanufacturing, since the remanufactured product must have the same quality as a new product, but the former must be less expensive (Tornese et al., 2019). Contrary to this definition, Steeneck and Sarin (2017) and Zeballos (2021) propose that the remanufactured product and the new product are different, there being a distinction in demand, price, and quantities produced and transported.

Moreover, (Steeneck and Sarin, 2017) consider that they are substitutes for each other, which means that in case one increases its price, the demand for the other will increase. They also consider that the demand for remanufactured products may not be satisfied because they are composed of parts that may or may not be available depending on the amount of products that have been disassembled.

Both papers (Steeneck and Sarin, 2017; Zeballos, 2021) raise issues whose objective is to maximize profits in a closed supply chain. The first paper (Steeneck and Sarin, 2017) focuses on the problem of deciding which is the best policy to take according to the parameters taken: only recovery, remanufacturing and recovery, only remanufacturing or not collecting products at all, solving also the quantity of products to collect and the quantity of products at each destination. The second (Zeballos, 2021) proposes a closed supply chain in which the customer, after using the product, delivers it to the centers where it is broken down into raw materials. From there, these have three possible destinations: to be discarded, sold to third parties or sent to the recycling center to be used for remanufacturing, and returned to the market to be sold.

Based on the analysis conducted, (Steeneck and Sarin, 2017) mention the incidence that the value of the recovery ratio parameter has on the results of the model since, in case it is very small, the reverse flow is not profitable and in case it is large, the more reduced the costs are.

In view of this fact, (Zeballos, 2021) proposes that the proportion of products that are recovered from the modular or traditional designs should be a variable with a minimum and a maximum value as parameters. In turn, for materials, there is also a recovery variable, whose value is fixed for some materials, but for others, it is determined on the basis of a minimum and a maximum value. On the other hand, there is a last parameter that takes the value that two related materials should have.

As in these works, Hariga et al. (2017) raise their model to an economic objective, which in this case is to minimize costs, to optimize inventory levels and replacement decisions, as well as the sequence of newly manufactured and remanufactured batches in supplier stage. This work develops a model that jointly optimizes the production sequence of multiple manufacturing and remanufacturing batches, the number and size of both batches, as well as the scheduling decisions for delivery to the subsequent buyer. It should be noted that for this work, consignment stock (CS) was considered as a stock policy, that is, systems in which the products are stored by the customer but belong to the seller until they are used by the buyer. This type of stock policy is increasingly common.

4.1.1.2 Multiple products

The works related to production planning in closed supply chains of multiple products simultaneously are presented in the following.

Sifaleras and Konstantaras (2015) propose a multiproduct model that seeks to determine how much to store, produce, and remanufacture to satisfy the demand for each of the products and in turn minimize total costs. To do this, they studied the multiproduct dynamic lot sizing problem with remanufacturing activities (MDLSRP). To solve the model, the authors propose the use of a metaheuristic algorithm called general variable neighborhood search (GVNS).

Similarly, Amin and Zhang (2014) considered a multiproduct problem, with the particularity that they included two models in their study. The first one seeks to determine the quantities to be produced, where the production centers will be located (within a range of possible locations), and the flow of products in the chain, minimizing the costs, while his second mathematical model is an extension of the first, thus leading to a multiobjective problem. This seeks, in addition to minimizing costs, minimizing operating times and the number of defective products. The authors suggested that there could be different types of recovery technology in collection centers, each with its own processing cost. From this, there is the particularity that an installation cost of the different technologies in the collection centers is included.

Within the articles that consider multiple products, it is worth highlighting two papers that take into account aspects that are not included in other papers studied. On the one hand, the article by Hashemi et al. (2014) seeks to maximize benefits in an integrated multiproduct manufacturing and remanufacturing system of the aerospace industry, in which products are returned for revision cycles or after a certain amount of flight hours. The article considers from the reverse flow, after the inspection and cannibalization stages, how many components are discarded, remanufactured, transformed, or repaired. In this, it is considered that only the manufacturing stage is associated with a certain percentage of defects, and that all the other processes do not (remanufacturing, repair, transformation). The fact that the model gives priority to using the components of the returned products is highlighted, and if they are not enough to cover the demand, the new products and their components are taken into account to satisfy it.

On the other hand, in their work, Zhang et al. (2012) propose the use of a Lagrangian relaxation to create subproblems, which are then solved using polynomial algorithms created for the case. The authors mention that an advantage of this method is the fact that, for large-scale problems, the exact solving methods fail or complete in extremely long solving times, while this one manages to give a margin less than the optimal value in a way effective. However, it is clear that for simple problems, the method used by researchers is complex and may be inefficient.

4.1.1.3 Multilevel systems

In this section, multilevel production systems were considered. This approach aims to consider within the same model the different levels of the hierarchical structure. That is, they are used when the variables to be studied are grouped into sets, which in turn belong to another group at a higher level. This relationship between the data, which is called "nested," must be considered when carrying out the model since the variables depend on the class to which they belong, but also on the group to which their class belongs.

Rowshannahad et al. (2018) propose a closed-loop manufacturing system with recovery of by-products, coproducts, and/or components, whose by-products in the reverse flow go to a separate remanufacturing line. In other words, two coupled processes are studied: one for manufacturing and the other for remanufacturing, since the final product is made up of a nonremanufacturable raw material and another remanufacturable one. Since it is considered that the final product uses two intermediate products, which must be acquired externally or remanufactured, the problem can be determined as a multilevel problem, because the demand for the latter depends on the demand for the final products. It must be taken into account that the remanufacturing process can only be repeated a limited number of times, due to the chemical characteristics of the material, which is an aspect that until now had not been seen in the rest of the articles.

As in this work, Ghafarimoghadam et al. (2016) propose a multilevel that allows determining the level of inventory of both manufactured and remanufactured products. As a difference, the decision, within a range of possible locations, of which manufacturing/remanufacturing centers are convenient to have available is included in their model. The "one-way substitution" is considered, which establishes that the demand for remanufactured products can be satisfied with manufactured products at remanufactured prices, but not vice versa. This ends up positively affecting both manufacturing plants that have the challenge of selling remanufactured products and customers who prefer to buy cheaper products of acceptable quality. After presenting the model, the article proposes the use of a robust optimization method to face the inherent uncertainty and be able to isolate the solution from the fluctuations in the input data, both demand and returns as well as transportation costs.

On the other hand, Subulan and Tasan (2013) propose the Taguchi design of experiments (DOE) method to validate the resolution of their model and sensitivity analysis. The model consists of optimizing all the stages of the CLSC. The model contemplates reverse flows between final consumers, retailers, collection centers, and remanufacturing plants, concluding in remanufactured products sold in wholesale centers. The model has three particularities. First, nonfulfillment of demand by both wholesalers and retailers is allowed, but a cost is included for delays at the end of the planning horizon, and retailers do not receive early deliveries of products. Second, the times that the products take to be transported from one node to another in the network are taken into account, and they are all assumed to be the same. Third, the collection of the products takes place in the retail centers, and the number of returns that occur in a certain period corresponds to a fraction of the demand that the wholesalers satisfy in the previous period.

4.1.2 Environmental impact

Continuing in the line of works that focus on the planning of production in closed supply chains, those that, in some way, have explicitly considered the impact on the environment will be studied further. It has the particularity that most of these works investigate the reduction of carbon dioxide, contemplating different activities. On the one hand, these works are presented and, on the other hand, those that took other criteria. The focus of these works in this aspect is linked to the fact that in recent years, the international community has become considerably aware of the importance of environmental issues, which have come to occupy a priority place on the world agenda. Concern about environmental deterioration has led multiple actors to put pressure on companies to take environmental impacts into consideration in their production processes, from nongovernmental organizations and the media to consumers, employees, unions, and shareholders.

4.1.2.1 CO_2 emission reduction

Sarkar et al. (2017) included carbon emissions in their analysis. In this case, the minimization of CO_2 is specifically focused on the transportation activity. The authors propose in this case a multisector supply chain with a logistics third party providing transport and collection services. It is suggested that packaging is one of the aspects that produces more waste and thus pollution in terms of transportation, so to improve sustainability. If only the economic impact is considered, it is considered that remanufacturing is usually the best alternative. This case is studied with a mixed integer nonlinear programming (MINLP) model, in which the total cost is minimized with the simultaneous optimization of the remanufacturing rate, the capacity of the containers, the necessary number of containers, retailers' shipping sequence, and cycle time. Under these conditions, it is concluded that omitting transportation costs, in addition to producing negative environmental impacts, leads to nonoptimal solutions or local optimums, which can cause economic losses. Similarly, it is specified that the total cost is not highly sensitive to those of transportation and carbon emissions, and that the information on these costs directly affects the remanufacturing rate. For the situation analyzed, it was determined that an exclusively remanufacturing production policy is not adequate since it is very expensive, regardless of the relationship between the costs of remanufacturing and manufacturing with new raw materials. This happens even more when the transport distances are long.

Unlike Sarkar et al. (2017), who consider the carbon dioxide emissions emitted by transport, Kwak (2018) focuses on that emitted by manufacturing through a restriction in which it imposes an upper bound of CO_2 emissions produced by said activity. The author identifies that the exponential technological advance that we are experiencing has a direct impact on the works presented to date on the subject. The author understands that the original design of the products with the passage of time no longer meets the needs of customers, so constant adaptation is necessary. Based on this concern, a mixed integer programming model (MIP) is proposed that considers an improvement in product design, and optimal solutions are reached on product design, sale price, and product planning production under the objectives of maximizing profits and market share. To determine when the design change is necessary, variables called "Generational Difference" are used. These variables are a relative measure that indicates, in technological terms, how old a specific part is compared with the most innovative part. The three variables are differentiated according to the product being considered. The author proposes five different scenarios for which he solves the model, and under the conditions taken and applied to laptops, the author concludes that the decision to have new and refurbished products with parts resale outweighs the decision to have new only both in profits as in market share.

Langarudi et al. (2019) propose an MILP model that seeks to minimize costs and emissions from remanufacturing and transportation. The authors get involved with the theme of car batteries and propose a network with reverse logistics that extracts and recycles lead, plastic, and sulfuric acid. The fact of recycling these components prevents them from polluting the environment, and on the other hand, sulfuric acid is sold to third parties for the production of fertilizers, chemicals, and sodium sulfate (used in paper and glass industries), so even a profit is made on this side.

Tornese et al. (2019) delve into the study of pallets, which are essential for the transport of all types of merchandise and used in most supply chains (Inbound Logistics Latam, 2016). In 2018, it was estimated that the consumption of pallets internationally exceeded 3980 million units per year, and that in South and Central America, consumption exceeded 149 million (Carmona, 2018), so it is understood that pallet waste will tend to increase and, if not managed efficiently, can have a significant impact on the environment. Tornese et al. (2019) address the economic and environmental benefits of having a preventive repair schedule for two types of pallets, proposing a multiobjective, multiproduct integer linear optimization model. The trade-off is analyzed between: the preventive repairs to the pallets carried out by the company, which have as a consequence the decrease in the number of times that the pallet is sent to a remanufacturing center, which reduces transport and operations, therefore the costs and emissions, or the opportunity losses generated by these preventive policies, related to a shorter life of the products, since they are repaired before it is necessary. In this case, emissions minimization is geared toward transportation and remanufacturing activities and the opportunity for lost emissions. The latter is defined as the proportion of repair emissions that are lost in anticipation of remanufacturing activities. As conclusions, the results are compared with the best possible case, that is, the optimal repair schedule. Under these conditions, the authors highlight that the fact of implementing a repair schedule that reduces emissions ends up increasing the economic cost by 4%. On the other hand, when a repair schedule is adopted that reduces costs, it ends up increasing CO_2 emissions by up to 6%. This means that, regardless of the type of recommended repair schedule that is used, significant losses will not be seen in either of the two objectives in the case study.

Finally, in this section, Chekoubi et al. (2020) approached a problem of the production-inventory-routing problem, considering the economic policy "cap-and-trade." This seeks to minimize carbon dioxide emissions and thus reduce the effects of this compound on the environment, by limiting companies that produce this substance. In this way, the country defines the carbon release limits that companies must respect as a result of their productive activity and also establishes the fine for not complying with said limits. On the other hand, companies can, if they wish, buy more emission allowances in a market where other companies sell their allowances if they release less CO_2 than the limit. Therefore, this chapter deals again with production, inventory, and distribution decisions, but also including the costs involved in the CO_2 released by transportation, production, and remanufacturing, and also any other type of emissions released during storage. In addition, the fact that there is a maximum of emissions that can be released for the sum of all activities is included as a restriction.

4.1.2.2 Other environmental impact alternatives

Following are those articles that consider the impact on the environment but with other alternatives that do not include CO_2 emissions, presenting a focus on waste. According to a World Bank report carried out in 2018, 2010 million tons of municipal solid waste are generated annually in the world, and at least 33% of them are not managed without risk to the environment, and the growth of the population and economic development will cause the amount of waste worldwide to increase by 70% in the next 30 years (World Bank, 2018).

Spaltini et al. (2021) delved into the circular economy, together with the economic impact on Industry 4.0. The evolution of circular economy, in addition to being based on the traditional 3R's (Reduce, Reuse and Recycle), it is currently considered to be made up of 6R's, in which the three strategies of Recover, Redesign, and Remanufacture are also added. For this reason, the circular economy is expected to provide new jobs, and in this way, sustainability is also promoted. However, the great paradigm of CE is that it requires great changes in industrial practices and in the consumption patterns of society. The authors propose a model in which they seek to maximize economic gains under the costs of new raw material, disassembly, primary processing and printing costs, logistics and assembly costs, and also environmental benefits. This last term is based on waste. By considering the environmental gains, the researchers highlight the fact that Industry 4.0 is not only compatible with but also favors the circular economy.

Finally, Zhang and Meng (2021) focused, in addition to performing the reverse flow to close the supply chain, on shared participation. More specifically, they sought to create a model that allows finding the optimal shareholding between the retailer and the manufacturer while maintaining a certain independence (less than half of the shares). The authors include the environmental impact seeking to maximize profits, which are affected by "environmental costs." Under the conditions of the model, they conclude that the greater shared participation can mean benefits for the client by finding lower prices.

4.1.3 *Extended producer responsibility (EPR)*

This section highlights the use of EPR legislation. These policies seek to involve and hold the manufacturer responsible for the waste from their production, either physically or financially. That is, they must deal with the treatment or disposal of the product after being used by the final consumer (Subulan and Tasan, 2013), to make their manufacture more sustainable.

As the amount of waste is increasing around the world, more and more governments are adopting this policy, but each opting for the form that provides the best results from an environmental perspective, also considering the economic impact that this measure can have on business. Some producers decide to focus on redesigning their products to prevent waste, while others choose to recycle or remanufacture their used products (Subulan and Tasan, 2013). This is why there is an increasing recovery of used products.

For example, Trochu et al. (2015) incorporate in their model a legislation on EPR that was applied in Canada in 2010 and whose objective is that companies contribute to the costs that municipalities have due to the recovery of waste. To create this regulation, it was necessary to estimate rates for 1 kg of raw material collected by government services, which also depend on the types of raw materials (that is, each material has a cost of being collected, measured in kilograms). Companies must annually declare the amount of each raw material that was used in their productions, and thus, a cost is obtained that until now had not been considered in any article. In this case, it is also considered that the production of the products can be carried out using two lists of materials, one traditional and the other with more ecological, lighter, and more expensive raw materials (so that the resulting final product will also be lighter). This is why the trucks that transport the products have different capacities in terms of volume and weight that they can transport. Continuing with the differentiation between the products, the ecological product will be manufactured using a "greener" process, with specialized employees for this. However, the demand can be satisfied both with the traditional product and with its ecological version. This situation was analyzed with data in the context of fast-moving consumer goods, since the production of these generates a large number of containers, packaging and printing and millions are consumed daily. For this reason, many companies in the industry spend thousands of dollars in compensation, and in this way, the impact of the legislation is observed (Trochu et al., 2015).

On the other hand, Krikke (2011) also chooses to include EPR legislation in his model, with the difference that this time it is applied to waste electrical and electronic equipment (WEEE). In these, the contribution of the supply chain to the carbon footprint of the life cycle of these products is very high, because the materials used to produce them are very harmful to the environment, unlike the work (Trochu et al., 2015), in which the importance of product recovery lies in the massiveness produced. Due to the substitution effect (of a part of the supply chain going forward), according to the author, the recovery of used products usually improves the ecological footprint of the processes, since recovered products and materials are used as components to create new products. This development in which there is substitution is called "closed-loop recovery," and to maximize this substitution, it is sought that the recovered items return to the original supply chain. However, it is mentioned that this event increases the flow of goods, since a reverse flow is added to the chain, which can have a negative impact on the environment. The other way to recover products, when they do not return to the original supply chain and there is no substitution or CLSC, is by selling them in a cascading market: lower segments parallel to primary markets (Krikke, 2011).

In this work, the author focuses on the way in which the products are recovered and on the destination that these have after their useful life. In the case of (Krikke, 2011), decision modeling techniques were used to analyze a phenomenon, and it was concluded that the viability of recovery in the supply chain depends to a large extent on the design of the network; that is, the distance in which the points of the chain are found. In addition to what I already assumed, the carbon footprint of CLSCs is strongly dependent on the substitution effect. On the other hand, in Trochu et al. (2015), the authors do not mention what happens to the waste when it is collected by the municipality.

Krikke shows, for this case, what is the structure of the supply chain to maximize the substitution effect. First of all, the arrangement is important, that is, the choice about recovery options for the reverse chain. For example, you can opt for the reuse of the product, the component, or the material, apply the cascade option, or without any type of recovery (but waste). This depends on the quality and volume of the returned products. On the other hand, the design of the network must be analyzed: the location of the warehouses to which the recovered products will be sent, and the location of the dispatch centers, as well as the transport between these sites (they may or may not be centralized).

4.1.3.1 Inspection

One of the stages in the recovery of used products is inspection, from which the status of the returned product and/or its components is known and decisions are made about its destination. This section will focus on works that have explicitly dealt with this procedure. Although the inspection can be carried out in different stages, the works reviewed mainly propose early inspection after the return of the used product, bringing direct cost savings.

Yang et al. (2017) highlight how awareness of manufacturers' responsibility is increasing and propose a model in which they designate inspection as one of the stages that used products that are returned to the CLSC will go through. This case study seeks to plan the production of a single product in a single period by making decisions under the objectives of maximizing profits and minimizing CO_2 emissions. In this work, it is the final consumers themselves who transport the products to the distribution/collection centers, where they will be later inspected, and it will be defined if the parts of the product can be used for remanufacturing, or if they have to be discarded. This is a difference from what Srinivasa and Khan (2018) propose where inspection is used one step later. The proposed CLSC is for customers to deliver EOL products to retailers, and retailers send them to the center where the parts are disassembled, inspected, and sorted according to whether they can be reused, salvaged, or scrapped. If this model is carried out, transport costs may have a greater weight than in Yang et al. (2017) because all the products used are transferred, including one more node in the network, without taking into account the option that much of the product can be discarded.

As in Srinivasa and Khan (2018) and Sadrnia et al. (2021), propose an inspection in the disassembly center, and as in Yang et al. (2017), they seek to minimize costs and minimize CO_2 emissions, but with the particularity that they are also interested in maximizing the social impact through greater opportunities for job. This last objective is peculiar as it is not considered in any of the works reviewed, and it is also interesting how it manages to target three different areas of decision: economic, environmental, and social. This arises as a consequence of the search for a "triple bottom line sustainability approach" (Sadrnia et al., 2021). The authors understand that the decision to have a sustainable supply chain ends up being a strategic option for companies, based on government regulations, pressure from environmental groups, and the benefits of recycling (Sadrnia et al., 2021), and based on this is that they seek sustainability in the CLSC.

Finally, Pazhani et al. (2021) contribute to this series of works that mention the inspection with two models that differ in the design of the chain. A more traditional one is proposed, with a structure called "REMIND," and another model with a "RENDER" design. The latter has the particularity that the retailers send the used products to the hybrid center, and they are inspected there. The term "hybrid center" is used because in the outward flow, these centers are used as a warehouse, and in the reverse flow, it is used as an inspection center. Regarding the resolution, the authors point out that solving a large-scale problem with integer variables is very difficult for them, so they resort to relaxation to find solutions. The authors are based on the theorem of unimodular matrices that establishes that a matrix A is unimodular if and only if the determinants of each submatrix of A are -1, 0 or 1, and from said matrix, it is possible to find conditions that allow to find the optimal solution of the problem.

4.1.3.2 Returned product quality levels

Nenes and Nikolaidis (2012) address a profit maximization problem, which seeks to optimize the obtaining of returns from a cell phone model, assuming that the returned products come from everywhere and in different quantities and qualities. There is the particularity that depositions do not exist since all returned products can be remanufactured regardless of their quality level, and even the best quality return has to go through remanufacturing processes. On the other hand, it is also considered that the company can choose to sell in lower markets at a low price those products that are returned, but it was not decided to carry out any remanufacturing activity.

Contrary to what was discussed by Nenes and Nikolaidis (2012) and Torkaman et al. (2017) in this case present a model that seeks to minimize configuration, inventory, manufacturing, and remanufacturing costs. The number of quality levels is arbitrary, and these affect the remanufacturing time and, therefore, said costs. The authors mention that production planning has been investigated since the beginning of the 20th century and establish that one of the greatest complexities of this problem is the configuration time of the machinery (change of tools and cleaning). Therefore, his motivation is to develop a model that is multistage with complex configurations and dependent on each other to be able to apply in industries that are not as simple as the automotive industry. A production line is considered in which there is a machine at each stage with limited capacity. In turn, the model receives used products, and it is considered that both they and those manufactured are remanufactured and that they have defects. One of the methods proposed by the authors to solve the model is the "Rolling horizon" method, which is used in those cases where there is uncertainty about the values that will adapt the parameters in future periods (Torkaman et al., 2017). On the other hand, the second resolution method uses the technique of searching for neighboring solutions until finding the optimal one ("Neighborhood search"), just like the article (Sifaleras and Konstantaras, 2015).

As in Torkaman et al. (2017) and Keshavarz-Ghorbani and Pasandideh (2021) have the objective of minimizing costs, but in this case, they chose to develop a nonlinear multilevel model that determines production and remanufacturing, defining a minimum level of acceptance of the quality of the products that are returned to the chain. It should be noted that the authors considered that within the production and remanufacturing processes, there is a percentage of defective products, this being a random rate that follows a known distribution function. Unlike the other articles discussed in this section, in Keshavarz-Ghorbani and Pasandideh (2011), due to the extensive computational time required to solve the problem, four methods are proposed to carry out the resolution of the model. The first of these is the Fibonacci algorithm, an iterative algorithm that seeks to find the extrema of the objective function and is useful for relatively small problems. With this method, depending on the desired error tolerance, the number of iterations to be carried out is defined, then the initial search interval is divided into three equal parts, and until the stopping criterion is reached, the iteration is repeated, rapidly approaching and cornering the optimal value. The other three tested algorithms are genetic algorithms (GA), invasive weeds optimization (IWO) and moth-flame optimization (MFO) algorithm. As conclusions, for the proposed model and with the conditions taken, in Keshavarz-Ghorbani and Pasandideh (2011), it is detailed that the Fibonacci algorithm, for small problems, achieves better values of the objective function, compared with the four metaheuristic algorithms. However, when it comes to medium-to-large-size problems, the computational time required is so great that they were not able to solve it.

4.1.3.3 Outsourcing of services

A possible decision to take to close the supply chains is the outsourcing of services. Companies may not have the necessary resources to carry out the reverse flow and choose to hire a third party to provide them. The works that include this practice in their research are presented further.

Masoudipour et al. (2017) present a problem in which a company is considering the advantages of hiring a third party to collect and process used products. This company offers three possible methods of recovery: repair, remanufacturing, and recycling for another supply chain. The authors consider the same center both to distribute new or remanufactured products and to collect used products. To define the destination of the collected products, in this problem, the inspectors of the collection and distribution centers are asked to indicate a number between 0 and 18 that reflects the current quality of the product. Then, based on the established quality and the minimum quality parameters for repair (q_{rep}: Minimum quality necessary for repair) and remanufacturing (q_{rem}: Minimum quality necessary for remanufacturing), the destination of the used product is defined. If it is greater than q_{rep}, the product is repaired and returned to the market, if its value is between q_{rep} and q_{rem}, it is remanufactured and then returned to the market, and if it is less than q_{rem}, it is recycled and sent to a secondary supply chain. On the other hand, the quality levels also determine how much the incentive will be returned to the customer (being 50% if the product is repaired, 30% if it is remanufactured, or 5% if it is recycled) and the remanufacturing costs.

Similarly, Hashemi et al. (2016) propose a model that seeks to maximize total benefit, in which the remanufacturing of aerospace products is outsourced. The authors consider that, even though this option incurs higher costs, the outsourced company has a more reliable "lead time" because it is specialized in this type of products/components. The authors suggest that there is a lot of uncertainty in returns, so to consider this variation in the demand for reverse flow components, they consider different scenarios. Each of them has a percentage of random disassembly defects with a certain probability of occurrence.

On the other hand, unlike the two works just mentioned where the option of entirely outsourcing one or several activities was considered, Kammoun et al. (2013) propose a reverse flow in which the used product is disassembled, and if there is not enough capacity to process the components, a third party is hired to do it. To achieve this, a goal programming (GP) model is proposed. The model, among other things, defines the quantity of components that are sent to each outsourced remanufacturing plant and the purchases that must be made from suppliers with the objective of minimizing the total costs of purchasing raw materials and remanufacturing.

Finally, unlike the works in this section, Kannan et al. (2017) raise the problem from the perspective of the third party, who provides logistics services to recover products at the end of their useful life in the most efficient way possible. This company seeks to recover the maximum possible value from the final products, through recycling, remanufacturing or both. To do this, a balanced disassembly line is established for each of the different types of products. The authors focus on a product recovery network that begins with the collection of returned products. For this, the decision is made to open warehouses in areas close to homes, and the items are consolidated there, to be later inspected. According to their quality, these are classified into relatively new products that will go to a secondary market, into items that need to be disassembled and their components used independently to fix other products (in the service center or as spare parts), or into items that are completely discarded. The objective of the authors is to determine the optimal allocation of products in the disassembly lines, in addition to determining the optimal flow of products and components to the facilities, under the condition of maximizing revenue.

4.2 Supply chain flow with product recovery

In this section, the works that focus on the design of the supply chain, mainly, considering material flows between the different stages of this chain are presented further.

The impact of e-commerce on supply chains has been growing in recent years, changing the way transactions are generated and material flows managed. Chen et al. (2019) rightly state that this impact also occurs in closed-loop supply chains, so to take full advantage of these technologies, chains must be designed that consider e-commerce as a flow channel itself, and not just as a channel, complement to the traditional one. Therefore, they will develop a CLSC model that considers two sales channels: online and offline. The offline channel has the particularity that the products do not have to be sent to the distribution center, but are sent by a third-party company directly from the place where they are remanufactured to the customers. This chapter tries to study how the two channels of sale and recycling affect the CLSC and also understand how the incentive prices given to customers when returning their products influence the channel chosen by each customer. It is important to clarify that the sale prices of the products do not depend on the channel through which they are sold. The authors comment that the resolution of the model and a sensitivity test remain pending to reach numerical results on the dilemma posed.

Later, Yoda et al. (2020) also focus on the issue of supply chain design and planning with an integer linear programming (ILP) model that seeks to maximize profits and the recovery ratio, both from reuse and recycling. Unlike (Chen et al., 2019) in which a given system is analyzed, in this case, they focus on proposing a procedure to implement remanufacturing. Step 1 is to determine the bill of materials and estimate the costs of assembly, disassembly, disposal, and treatment and also the estimation of the recovery rate. Then, in step 2, the ILP is formulated and solved, and finally, in step 3, the results obtained are analyzed, understanding the effects of remanufacturing, the useful years of the product, and the remanufacturing prices. The ILP model presented takes into account the life span of the product, the cost of reuse and recycling of each component, disassembly costs, purchase of new components and sales of remanufactured products, used parts, and recycled materials. From the resolution of the model applied to PCs, the authors reach some interesting conclusions about the incidence of some parameters in the decisions made by the model. For example, the authors show that for these cases when the recovery rate is high, remanufacturing is an effective option. It is also highlighted that of the remanufactured products, the PC is one of those that can obtain the highest profits. However, when you can't make enough profit from remanufactured products, it pays both economically and environmentally to make a profit from reused components.

Within these simpler cases of optimization of supply chain flows, it is worth highlighting the study carried out by Alqahtani and Gupta (2015). This is the method that they incorporate to solve the MILP that seeks to determine the flow of products in the network, since it is a method of linear physical programming (LPP, for its acronym in English). This method has the advantage that it allows the decision-maker to express his preferences on the criteria values without taking the traditional forms of weighting, but rather it is expressed in different degrees of desirability. The authors consider cost and revenue in this chapter.

On the other hand, there are some works that analyze supply chain design in a more comprehensive manner than previous works, mainly, they consider decisions about the quantity and locations of the facilities, as well as the flow of materials in the network. Among the works that address this problem, Elahi and Franchetti (2014) stand out, who in turn add considerations for reducing network waste. On the other hand, Wang et al. (2012) present a model that deepens the one published by (Elahi and Franchetti, 2014), since it not only considers waste but also takes into account CO_2 emissions and aims to examine the "trade-off" between economic costs and ecological effects in a CLSC. Therefore, the proposed model is a multiobjective MILP that seeks to minimize costs, emissions, and waste generation. The collection of returns considered is calculated by area and has a positive relationship with respect to private consumption. When considering waste, all waste that is generated throughout the entire CLSC is taken into account. The authors define the amount of waste generated as the amount of products returned from markets to recycling centers minus the amount of products sent from recycling centers to remanufacturing plants and disposal centers. On the other hand, the minimization of carbon emissions considers both transportation and remanufacturing operations. Finally, in the conclusions, it is highlighted that under these conditions, by increasing the rate of return, recovery rate, or the replacement cost rate, an optimum can be found between the economic and ecological objectives. In turn, decision-makers may have a variety of options with the same CO_2 emissions but with different costs and different waste generation. Each of these options may have a different network design, but they are all optimal solutions, which provide both an economic and an ecological evaluation when making decisions.

5. Conclusions

This chapter presents a review of the literature on the problem of production and flow planning in closed supply chains, with special focus on the remanufacturing of used products. The environmental problem of materials recovery was addressed, highlighting the importance for the current and future societies. Similarly, the environmental regulatory context that approaches that problem was briefly presented, both in Uruguay and internationally. Within the relevant literature, it was possible to identify that there is a growing trend in recent years in the amount of research on this topic. This is due to the increase in regulations, and awareness and interest in reducing environmental impact, thus seeking to counteract the increase in consumption and waste, a consequence of the traditional production system.

To carry out the review of the articles of interest, a systematic procedure was carried out, in which certain different scientific databases were considered, certain keywords and a specific period were defined. From the studies reviewed, some main conclusions can be pointed out. On one hand, in many cases, results were compared between a scenario with product recovery and without product recovery, and it was possible to demonstrate both the economic benefits that this activity entails, such as a greater market share and a better customer perception toward a more efficient company, sustainable.

On the other hand, it is also highlighted that those authors who focused only on the flow of the supply chain mention that it would be convenient to extend the model to also consider production planning. Also, it was detected that those articles that consider the environmental impact apply it from CO_2 emissions.

The computational burden of the models used in the articles was also an issue worth mentioned, since those that turned out to be too complex to solve were relaxed or solved with metaheuristics. Within these resolution techniques, the ones that were found most often were the genetic algorithm (GA) and the so-called general variable neighborhood search (GVNS).

Finally, although progress was noted in the research and in the models that were reviewed throughout the period taken (2020–21), great potential is detected on this topic since the authors repeatedly suggest possible extensions of their models, highlighting a great terrain that has not yet been explored.

References

Alqahtani, A.Y., Gupta, S.M., 2015. Multicriteria optimization for the delivery of products across multiple periods in a reverse supply chains environment. In: Applications of Management Science. Emerald Group Publishing Limited, pp. 3–18.

Amin, S.H., Zhang, G., 2014. Closed-loop supply chain network configuration by a multi-objective mathematical model. International Journal of Business Performance and Supply Chain Modelling 6 (1), 1–15.

Blackburn, J.D., Guide Jr, V.D.R., Souza, G.C., Van Wassenhove, L.N., 2004. Reverse supply chains for commercial returns. California Management Review 46 (2), 6–22.

Briner, R.B., Denyer, D., 2012. Systematic review and evidence synthesis as a practice and scholarship tool. Oxford Handbook of Evidence-Based Management: Companies, Classrooms and Research 112–129. https://doi.org/10.1093/oxfordhb/9780199763986.013.0007.

Carmona Marín, R., Hernández González, C., 2018. Logística inversa de los desechos de pallets de madera usadas en el Puerto de Cartagena de Indias: "caso Barrio Martínez Martelo, Sector La Cuchilla". Monograph, Universidad de San Buenaventura.

Chekoubi, Z., Trabelsi, W., Sauer, N., April 2018. The integrated production-inventory-routing problem in the context of reverse logistics: the case of collecting and remanufacturing of EOL products. In: 2018 4th International Conference on Optimization and Applications (ICOA). IEEE, pp. 1–6.

Chekoubi, Z., Trabelsi, W., Sauer, N., Majdouline, I., April 2020. The integrated production-inventory-routing problem with remanufacturing under carbon cap-and-trade. In: 2020 IEEE 6th International Conference on Optimization and Applications (ICOA). IEEE, pp. 1–7.

Chen, C., Zhang, G., Xia, Y., 2019. Remanufacturing network design for dual-channel closed-loop supply chain. Procedia CIRP 83, 479–484.

Elahi, B., Franchetti, M., June 2014. A new optimization model for closed-loop supply chain networks. In: 2014 IEEE International Technology Management Conference. IEEE, pp. 1–9.

Ghafarimoghadam, A., Karimi, A., Mousazadeh, M., Pishvaee, M.S., 2016. A robust optimisation model for remanufacturing network design problem with one-way substitution. International Journal of Services and Operations Management 24 (4), 484–503.

Govindan, K., Soleimani, H., Kannan, D., 2015. Reverse logistics and closed-loop supply chain: a comprehensive review to explore the future. European Journal of Operational Research 240 (3), 603–626.

Guggeri, E.M., Ham, C., Silveyra, P., Rossit, D.A., Piñeyro, P., 2023. Goal programming and multi-criteria methods in remanufacturing and reverse logistics: systematic literature review and survey. Computers & Industrial Engineering 185, 109587.

Guide Jr, V.D.R., Van Wassenhove, L.N., 2009. The evolution of closed-loop supply chain research. Operations Research 57 (1), 10–18.

Hariga, M., Aśad, R., Khan, Z., 2017. Manufacturing-remanufacturing policies for a centralized two stage supply chain under consignment stock partnership. International Journal of Production Economics 183, 362–374.

Hashemi, V., Chen, M., Fang, L., 2014. Process planning for closed-loop aerospace manufacturing supply chain and environmental impact reduction. Computers & Industrial Engineering 75, 87–95.

Hashemi, V., Chen, M., Fang, L., 2016. Modeling and analysis of aerospace remanufacturing systems with scenario analysis. The International Journal of Advanced Manufacturing Technology 87, 2135–2151.

Ilgin, M.A., Gupta, S.M., 2010. Environmentally conscious manufacturing and product recovery (ECMPRO): a review of the state of the art. Journal of Environmental Management 91 (3), 563–591.

Inbound Logistics Latam, 18 julio 2016. "Pallets, el poder está detrás de la cadena de abastecimiento", Portal Inbound Logistics Latam Magazine. Available at: https://www.il-latam.com/blog/articulos-centrales/pallets-el-poder-destras-de-la-cadena-de-abastecimiento/. (Accessed 21 September 2021).

Kammoun, A., Benaissa, M., Chabchoub, H., May 2013. Planning model in reverse logistic. In: 2013 International Conference on Advanced Logistics and Transport. IEEE, pp. 446–451.

Kannan, D., Garg, K., Jha, P.C., Diabat, A., 2017. Integrating disassembly line balancing in the planning of a reverse logistics network from the perspective of a third party provider. Annals of Operations Research 253, 353–376.

Keshavarz-Ghorbani, F., Pasandideh, S.H.R., 2021. Optimizing a two-level closed-loop supply chain under the vendor managed inventory contract and learning: Fibonacci, GA, IWO, MFO algorithms. Neural Computing & Applications 33, 9425–9450.

Krikke, H., 2011. Impact of closed-loop network configurations on carbon footprints: a case study in copiers. Resources, Conservation and Recycling 55 (12), 1196–1205.

Kwak, M., 2018. Optimal line design of new and remanufactured products: a model for maximum profit and market share with environmental consideration. Sustainability 10 (11), 4283.

Langarudi, N.R., Sadrnia, A., Sani, A.P., 2019. Recovering lead, plastic, and sulphuric acid from automobile used batteries by mathematical reverse logistics network modelling. Progress in Industrial Ecology, an International Journal 13 (1), 63–83.

Li, Y., Kannan, D., Jha, P.C., Garg, K., Darbari, J., Agarwal, N., 2018. Design of a multi echelon product recovery embeded reverse logistics network for multi products and multi periods. Annals of Operations Research 1–22.

Masoudipour, E., Amirian, H., Sahraeian, R., 2017. A novel closed-loop supply chain based on the quality of returned products. Journal of Cleaner Production 151, 344–355.

Matsumoto, M., Yang, S., Martinsen, K., Kainuma, Y., 2016. Trends and research challenges in remanufacturing. International Journal of Precision Engineering and Manufacturing-Green Technology 3, 129–142.
Nenes, G., Nikolaidis, Y., 2012. A multi-period model for managing used product returns. International Journal of Production Research 50 (5), 1360–1376.
Organization for Economic Co-operation and Development, 2021. Extended Producer Responsibility. Disponible en. https://www.oecd.org/env/tools-evaluation/extendedproducerresponsibility.htm.
Pazhani, S., Mendoza, A., Nambirajan, R., Narendran, T.T., Ganesh, K., Olivares-Benitez, E., 2021. Multi-period multi-product closed loop supply chain network design: a relaxation approach. Computers & Industrial Engineering 155, 107191.
Piñeyro, P., Rossit, D.A., December 2020. Simultaneous lot-sizing and scheduling with recovery options: problem formulation and analysis of the single-product case. In: International Conference of Production Research–Americas. Springer International Publishing, Cham, pp. 102–112.
Piñeyro, P., Viera, O., 2009. Inventory policies for the economic lot-sizing problem with remanufacturing and final disposal options. Journal of Industrial and Management Optimization 5 (2), 217–238.
Rowshannahad, M., Absi, N., Dauzère-Pérès, S., Cassini, B., 2018. Multi-item bi-level supply chain planning with multiple remanufacturing of reusable by-products. International Journal of Production Economics 198, 25–37.
Sadrnia, A., Sani, A.P., Langarudi, N.R., 2021. Sustainable closed-loop supply chain network optimization for construction machinery recovering. Journal of Industrial and Management Optimization 17 (5), 2389–2414.
Sarkar, B., Ullah, M., Kim, N., 2017. Environmental and economic assessment of closed-loop supply chain with remanufacturing and returnable transport items. Computers & Industrial Engineering 111, 148–163.
Sifaleras, A., Konstantaras, I., 2015. General variable neighborhood search for the multi-product dynamic lot sizing problem in closed-loop supply chain. Electronic Notes in Discrete Mathematics 47, 69–76.
Spaltini, M., Poletti, A., Acerbi, F., Taisch, M., 2021. A quantitative framework for Industry 4.0 enabled Circular Economy. Procedia CIRP 98, 115–120.
Srinivasan, S., Khan, S.H., 2018. Multi-stage manufacturing/re-manufacturing facility location and allocation model under uncertain demand and return. The International Journal of Advanced Manufacturing Technology 94 (5–8), 2847–2860.
Steeneck, D.W., Sarin, S.C., 2017. Determining end-of-life policy for recoverable products. International Journal of Production Research 55 (19), 5782–5800.
Subulan, K., Tasan, A.S., 2013. Taguchi method for analyzing the tactical planning model in a closed-loop supply chain considering remanufacturing option. The International Journal of Advanced Manufacturing Technology 66, 251–269.
Thierry, M., Salomon, M., Nunen, J.V., Wassenhove, L.V., 1995. Strategic issues in product recovery management. California Management Review 37 (2), 114–136. https://doi.org/10.2307/411657926.
Torkaman, S., Ghomi, S.F., Karimi, B., 2017. Multi-stage multi-product multi-period production planning with sequence-dependent setups in closed-loop supply chain. Computers & Industrial Engineering 113, 602–613.
Tornese, F., Pazour, J.A., Thorn, B.K., Carrano, A.L., 2019. Environmental and economic impacts of preemptive remanufacturing policies for block and stringer pallets. Journal of Cleaner Production 235, 1327–1337.
Trochu, J., Chaabane, A., Ouhimmou, M., 2015. Tactical planning model for procurement and manufacturing decisions under environmental regulations. IFAC-PapersOnLine 48 (3), 362–367.
Wang, Y.C., Lu, T., Gao, C.H., Zhang, C.H., Chen, C., 2012. Research on remanufacturing closed-loop logistics network design under low-carbon restriction. Applied Mechanics and Materials 159, 224–234.
World Bank, Septiembre 20, 2018. Los desechos: un análisis actualizado del futuro de la gestión de los desechos sólidos. Disponible en: https://www.bancomundial.org/es/news/immersive-story/2018/09/20/what-a-waste-an-updated-look-into-the-future-of-solid-waste-management.
Xu, Y., Feng, W., 2014. Develop a cost model to evaluate the economic benefit of remanufacturing based on specific technique. Journal of Remanufacturing 4, 1–12.
Yang, Y., Huang, Z., Qiang, Q.P., Zhou, G., 2017. A mathematical programming model with equilibrium constraints for competitive closed-loop supply chain network design. Asia Pacific Journal of Operational Research 34 (05), 1750026.
Yoda, K., Irie, H., Kinoshita, Y., Yamada, T., Yamada, S., Inoue, M., 2020. Remanufacturing option selection with disassembly for recovery rate and profit. International Journal of Automation Technology 14 (6), 930–942.

Zeballos, L.J., 2021. Approach for improving the life-cycle profit exploding remanufacturing conditions. Computers & Industrial Engineering 152, 107025.

Zhang, S., Meng, Q., 2021. Electronics closed-loop supply chain value co-creation considering cross-shareholding. Journal of Cleaner Production 278, 123878.

Zhang, J., Liu, X., Tu, Y.L., 2011. A capacitated production planning problem for closed-loop supply chain with remanufacturing. The International Journal of Advanced Manufacturing Technology 54, 757–766.

Zhang, Z.H., Jiang, H., Pan, X., 2012. A Lagrangian relaxation based approach for the capacitated lot sizing problem in closed-loop supply chain. International Journal of Production Economics 140 (1), 249–255.

CHAPTER

10

Solving a routing problem with time windows for an Argentinian case study in solid waste management

Diego Gabriel Rossit[2], Adrián Andrés Toncovich[1] and Franco Bautista Heine[1]

[1]Department of Engineering, Universidad Nacional del Sur, Bahía Blanca, Argentina; [2]Department of Engineering, INMABB, Universidad Nacional del Sur (UNS)-CONICET, Bahía Blanca, Argentina

1. Introduction

The development of urbanization is associated with an increase in the rate of municipal solid waste (MSW) generation, a trend that is not expected to reverse any time soon (Hoornweg et al., 2015).

An adequate MSW system planning must address structural, social, and environmental problems and assume a precautionary principle due to its impact on the health of its inhabitants, identifying threats that may hamper the good quality of life of citizens, as well as potential strengths that can make flourish the forgotten regions of the country. The environmental perspective must necessarily analyze the systemic behavior of the city, based on its flows of both people and inputs, products, and waste. These environmental challenges in cities arise from the need to enhance existing resources and redirect urban expansion, minimizing its ecological cost (Naredo, 2007; Sodiq et al., 2019). It is known that in the vast majority of Argentine cities, the MSW is deficient. In addition, they do not value wasting resources, energy and promoting a culture that is not very committed to the environment in which they live. For this reason, there is an increasing interest in cities in the efficient management of the MSW system to mitigate the environmental, social, and economic impact of this growing volume of waste (Cavallin et al., 2020; Rossit et al., 2020).

Evolution and Trends of Sustainable Approaches
https://doi.org/10.1016/B978-0-443-21651-0.00001-2

1.1 Waste management in Argentina

In 2004, the National Law No. 25916 was passed by the Congress of Argentina to promote the recovery of household waste, understood as such, the use of the resources contained in the waste, without endangering human health and without using methods that may harm the environment, through the implementation of appropriate methods and processes, reducing waste destined for final disposal and minimizing the negative impacts that these wastes may produce on the environment. Since the promulgation of this law, different provincial and municipal initiatives have been launched in Argentina to meet the proposed objectives (Rossit et al., 2017).

In Argentina, the management of MSW is under municipal jurisdiction. It is important to highlight that the Argentine municipal system is characterized by being heterogeneous, in the sense that there are multiple types of municipalities with different attributions and responsibilities, as established by each of the provincial constitutions and organic laws. The determination of the municipal regime constitutes a power reserved to the provinces and not delegated to the federal government, for which reason, 24 different regimes coexist (one for each province and one for the Autonomous City of Buenos Aires) that define the typology and the degree of autonomy in various ways. After the incorporation of Article 123 in the National Constitution in 1994, municipal autonomy was enshrined, which resulted in an increase in responsibilities and the municipalities became the first government recipients of the multiple demands of civil society that were previously directed to other levels of government, so the impact of their interventions acquired a central place in the scope of the quality of life of the population. However, the formal delegation of functions was not always accompanied by the transfer of material, economic, and technical resources necessary to carry them out, and thus, the real capability of the municipalities to implement policies is often limited (Smith and Revell, 2016).

Taking into account the aforementioned, each city began by making the necessary improvements according to the state/province in which their respective waste management system was located (Cavallin et al., 2020). In several cities, these improvements involved the eradication of open-air waste dumps and their replacement with landfills, the implementation of source separated MSW collection systems, and the installation of MSW treatment plants. However, in other large cities of the country, the lack of a comprehensive treatment of MSW remains with, e.g., the use of open-air waste dump as a final disposal method (Esparza, 2021). In addition, the vast majority of Argentine local governments depend on waste management programs and subsidies at the provincial and national levels. In very few cases, municipal resources exceed 30% of the total expenses related to the MSW system. This evidence shows a scenario that presents not only a scarcity of resources but also insufficient capabilities to generate them and face their own problems autonomously.

1.2 Aim and organization of this chapter

In connection with what has been said before, the development of support tools for decision-making based on operations research can help make better decisions in shorter times to favor efficient resource management in Argentina. Thus, the motivation of this work is to propose a tool that contributes to national knowledge about MSW efficient handling and that helps in decision-making for those municipalities that want to evolve toward a process that minimizes waste management costs and maximizes waste recovery. Particularly, in this chapter, a model is proposed to solve the MSW collection problem based on a vehicle routing problem considering capacity and maximum route time constraints. This model is then applied in simulated instances of the city of Bahía Blanca. Preliminary results have shown that the proposed model is capable of solving real scale instances of the city.

This work is structured as follows: Section 2 presents a review of the literature related to the problem addressed. In Section 3, the mathematical model corresponding to the problem is developed. The case study is introduced in Section 4. The experimental work is developed in Section 5. Finally, Section 6 presents the conclusions of the chapter and future lines of research.

2. Review of the related literature

The design of MSW collection routes in a bin-based system is a typical application case of the well-known VRP (vehicle routing problem). Generally, the resolution of a VRP consists of determining the best route, in terms of a given specific criteria, that runs through a certain set of points, taking into account different restrictions (Toth and Vigo, 2002). Among the criteria usually considered, we can mention the minimization of times, costs, and number of vehicles needed. Among other constraints, the limitation of the capacity of the vehicles, the time limits on the routes, the time windows of the places to visit can be listed. VRPs are known problems of great complexity in computational theory, which require a number of integer variables and constraints that grow exponentially with the number of points to visit (Lenstra and Kan, 1981).

The MSW collection problem has been addressed by a significant number of authors using various techniques, both through heuristic and exact methods. Beliën et al. (2012) and Han and Ponce Cueto (2015) provide extensive reviews of these works. The collection phase of the waste management process is one of the major contributors to the costs in the reverse logistic chain of the MSW system especially in developing countries (Di Maria and Micale, 2015). In Latin American countries, the cost of the collection phase is even more important in the MSW system due to the extended usage of the door-to-door system and rear-loading trucks equipped with diesel engines (Benitez et al., 2021), which is associated with high costs mainly due to fuel consumption, truck maintenance services, and crew wages (Sousa et al., 2018).

In the case of Argentina, some works can be found on applications of VRP models to solve the collection route design problem. For example, Bonomo et al. (2012) implement mathematical programming models to plan the collection routes for waste containers in the southern area of the City of Buenos Aires. In Braier et al. (2017), through integer programming models, the collection of recyclable waste is planned in the city of Morón. Fermani et al. (2020) and Rossit et al. (2021) solve VRP with capacity restrictions for cases of the city of Bahía Blanca with exact and heuristic tools. In Mahé et al. (2021), a combined collection route design and collection point sizing problem for small instances in the same city is solved with a Benders decomposition method and the inclusion of tailor-made valid cuts. This work was expanded in Mahé et al. (2023) with the inclusion of new valid cuts. Then, Delle Donne et al. (2021) proposed a model to efficiently assign street sweepers to city blocks and the routes to be followed by collection trucks for leaf bag pickup in the city of Trenque Lauquen. Finally, in Molfese et al. (2022), the analysis of the convenience of constructing a transfer plant was studied for the city of Bahía Blanca.

In this work, it is proposed to continue previous work considering VRP cases in the city of Bahía Blanca that incorporate the time limit restriction due to the working hours of the drivers of the collection vehicles.

3. Mathematical model

The time-limited VRP studied here can be formalized through a mathematical programming model that includes the following definition of sets, parameters, and variables (Table 10.1).

Considering these parameters and variables, the mathematical model of the problem can be expressed using the following Eqs. (10.1)(10.9).

$$min \sum_{i,j\in M', \forall j\neq i, k\in K} x_{ij}^k \, t_{ij} \tag{10.1}$$

TABLE 10.1 Sets, parameters, and variables of the model.

Sets	
K	Set of trips that can be made by collection vehicles.
I	Set of clean points to collect.
Parameters	
C	Collection vehicle capacity.
T	Time limit of the drivers' working day.
c_i	Volume of waste to be collected at the clean point $i\in I$
t_{ij}	Travel time from the clean point $i\in I$ to the clean point $j\in I$
tc_i	Service time in the clean point $i\in I$ (MSW collection).
td_0	Unloading time of MSW accumulated in the vehicle at the depot (transfer plant).
Variables	
x_{ij}^k	Binary variable that has the value 1 if the collecting vehicle during the trip $k\in K$ travels from the clean point $i\in I$ to the clean point $j\in I$, 0 otherwise.
u_i^k	Auxiliary continuous variable for removing subtours.
v_i^k	Auxiliary continuous variable to limit the time per execution.

Subject to:

$$\sum_{i\in I, i\neq 0, j\neq i, k\in K} x_{ij}^{k} = 1, \forall j\in I \tag{10.2}$$

$$\sum_{i\in I, i\neq 0, j\neq i, k\in K} x_{ji}^{k} = 1, \forall j\in I \tag{10.3}$$

$$\sum_{j\in I, j\neq 0} x_{0j}^{k} \leq 1, \forall k\in K \tag{10.4}$$

$$\sum_{i\in I, j\neq i, k\in K} x_{ij}^{k} - \sum_{i\in I, j\neq i, k\in K} x_{ji}^{k} = 0, \forall j\in I \tag{10.5}$$

$$u_{i}^{k} - u_{j}^{k} \leq C\left(1 - x_{ij}^{k}\right) - c_{j}, \forall\ i, j\in I, j\neq i, k\in K \tag{10.6}$$

$$v_{i}^{k} - v_{j}^{k} \leq T\left(1 - x_{ij}^{k}\right) - t_{ij} - tc_{j}, \forall\ i, j\in I, j\neq i, j\neq 0, k\in K \tag{10.7}$$

$$v_{ik} - v_{jk} \leq T + T\left(1 - x_{ij}^{k}\right) - t_{i0} - td_{0}, \forall\ i, j\in I, j\neq i, k\in K \tag{10.8}$$

$$x_{ij}^{k}\in \{0; 1\}, u_{i}^{k}, v_{i}^{k} \geq 0, \forall\ i, j\in I, k\in K \tag{10.9}$$

The proposed objective is to minimize the total travel time during the collection of dry and wet waste, as shown in Eq. (10.1). On the other hand, the restrictions imposed on the model are presented from Eqs. (10.2)–(10.9). Eqs. (10.2) and (10.3) guarantee that each node (clean point) is visited only once by the collection vehicle so that it enters and exits that node only once, except in the case of node 0, which corresponds to the depot (transfer plant), from which it can exit and enter as often as necessary. Eq. (10.4) ensures that a maximum of one exit is made from the depot for each possible route. Eq. (10.5) refers to the conservation of flow in the path while Eq. (10.6) prevents the formation of subtours. Eqs. (10.7) and (10.8) ensure that the time per journey does not exceed the maximum time established for the working day. Eq. (10.9) defines the binary and nonnegative variables of the model.

4. Case study

The city of Bahía Blanca is located in the South of Buenos Aires province, Argentina. According to the last census conducted in 2022, the city has 335,190 inhabitants, being the 30th largest city in the country. The city is considered a relatively developed economic center in its area of influence (World Bank Group, 2016). Moreover, the city has one of the main export ports of Argentina (Port Consultants Rotterdam: Port Vision Bah í et al., 2017) through which a large amount of domestic trade and agricultural and livestock products are channeled (Deckers et al., 2018). Another relevant feature of the city is the presence of prestigious higher education centers and an important industrial sector.

These features caused Bahía Blanca to be an attraction center for regional migration and a regional service center in the last decades. This has had an impact in terms of the generation of MSW. For instance, waste generation rate per inhabitant has increased from 623 g/day in 2006 to 1194 g/day in 2018 (Vazquez et al., 2020).

In most of the city of Bahía Blanca, door-to-door collection is currently carried out of undifferentiated waste that is deposited in a landfill site (Rossit et al., 2019). Therefore, for the city considered in this work, the waste collection service still only has a hygienist approach, that is, the activity does not contemplate the recovery of recyclable materials, and nonetheless, it represents 12% of the municipal budget (Cavallin et al., 2020). Currently, a source sorted door-to-door collection project is being performed as a pilot test to extend the lifespan of the landfill. In addition to the implementation of source separation policy to increase the recycling rate, other policies can be implemented to improve the MSW management of the city. Among these policies, the use of a bin-based system has some benefits over the usage of a door-to-door collection system. A system of differentiated containers for recyclable and nonrecyclable waste would not only help to recover the recyclable fraction of the waste, reducing the environmental impact of its final disposal, but it would also reduce the number of points necessary to visit in the collection routes and, therefore, the distances traveled by the vehicles with respect to a door-to-door system (where the collection vehicle visits each home to collect the waste) such as the one the city currently has. As has been highlighted in other recent Latin American studies, this causes not only a decrease in the logistics costs of the system with respect to the door-to-door collection system (Blazquez and Paredes, 2020; Gilardino et al., 2017) but also a reduction in the environmental impact through the reduction of the emission of greenhouse gasses (Gilardino et al., 2017). For this reason, efforts are being made in this regard by different groups in the city (Cavallin et al., 2020).

The present work is projected to advance one step further, focusing on the logistics of collection. In particular, a model is proposed to approach the design of the routes of the collection vehicles considering the restrictions of capacity and maximum time of the route taking into account the working day.

5. Computational experimentation

The following subsections present the instances used and the results of the computational experimentation.

5.1 Instances description

The instances used represent four real scenarios of the logistics of MSW collection from the city of Bahía Blanca built on the basis of data obtained from a field survey (Cavallin et al., 2020). Two different plans for the disposal of the waste accumulated in the containers are

TABLE 10.2 Description of instances.

Name	Collection frequency (times/week)	Kind of waste	Amount of waste (m^3)	Number of clean points
Instance 1	6	Wet	57.911	69
Instance 2	4	Dry	155.198	69
Instance 3	6	Wet	57.911	79
Instance 4	3	Dry	232.797	79

studied, considering the collection of both fractions of waste: dry and wet. These plans contemplate the option of separating the containers into wet and dry waste while considering the option of using different collection frequencies. The plans differ fundamentally in terms of the frequency with which it is planned to collect dry waste, thus forming the instances that are described in Table 10.2.

The model parameters were estimated as follows. Distances and travel times between clean points were calculated using the Open Source Routing Machine (2021) based on the OpenStreetMap platform OpenStreetMap Contributors (2017). For this purpose, a procedure developed by Vázquez Brust (2018) in the R programming language was used. The service time in each clean point was estimated based on the work developed by Carlos et al. (2019), where the average unloading time for each container and the number and types of containers accumulated in each place are described. The unloading time at the end of the journey in the sanitary landfill site was estimated at 8 min, in line with the estimates available in similar studies in the literature (Owusu et al., 2019). To estimate the operational capacity of the compactor truck, the volumetric capacity indicated by the manufacturer (21 m^3) was multiplied by the compression ratio, which differs depending on the type of waste. A compression ratio of 1.5:1 and 3:1 was estimated for wet and dry waste, respectively, in line with the values found experimentally by other authors for these types of waste (Lavee and Nardiya, 2013; Zbib et al., 2019). Finally, the maximum time of the route was set at 6 h according to the working day following the current legislation. The rest of the data of the instances can be consulted in https://github.com/diegorossit/Intances-Heine-et-al.-2023-.git.

5.2 Implementation and results

The model was implemented using the Pyomo optimization package (Hart et al., 2017) in a Python environment. To solve the optimization model, CPLEX version 22.1.0.0 (International Business Machines Corporation, 2022) was used. The experiments were carried out using a personal computer with an Intel Core i5-3570 @ 3.40 GHz processor with 12 GB of RAM and a 64-bit Windows 10 operating system. A time limit of 3600 s was set for each run.

The results initially obtained are shown in Tables 10.3 and 10.4. To evaluate the influence of the maximum number of routes allowed, different executions of the model were carried out. Table 10.3 corresponds to the establishment of a maximum number of possible routes

TABLE 10.3 Results of the resolution of the instances, $|K| = 30$.

Instance	Travel time (min)	Unloading time (min)	Total route time (min)	Gap (%)	Nº of routes
Instance 1	65.03	16	130.67	34.38	2
Instance 2	78.57	24	213.41	45.69	3
Instance 3	321.69	152	527.41	85.46	19
Instance 4	383.41	184	732.65	87.80	23

TABLE 10.4 Results of the resolution of the instances, $|K| = 5$.

Instance	Travel time (min)	Unloading time (min)	Total route time (min)	Gap (%)	Nº of routes
Instance 1	54.61	16	120.25	21.85	2
Instance 2	76.07	24	210.91	43.85	3
Instance 3	71.51	16	141.23	34.36	2
Instance 4 *(k = 6)*	145.84	40	351.08	67.86	5

of 30, while Table 10.4 corresponds to the establishment of a maximum number of routes equal to 5. It should be noted that the theoretical minimum number of routes based on capacity (total of accumulated residue divided by the capacity of the vehicle) fluctuates between two and four trips for all the scenarios.

$$|K|_{min} = ceil\left(\sum_{i\in I, i\neq 0} c_j / C\right) \tag{10.10}$$

From Table 10.3, it can be seen that the model is capable of solving the required instances. However, it is found that the reported gaps are relatively high in Instances 3 and 4. Similarly, the number of routes in both instances is relatively large when contrasted with respect to the theoretical minimum number of routes. An experiment was carried out using a maximum number of possible routes of 5 (which exceeds the theoretical minimum), the results of which are presented in Table 10.4. In instance 4, $|K| = 6$ had to be used because in the maximum time of 3600 s with $|K| = 5$, no feasible solution was found. With respect to the results of Table 10.3 in terms of the value of the objective function (travel time) for all the instances considered, improvements were obtained. The average improvement in the functional value was 40%, and the greatest improvement was in Instance 3 (78%).

Continuing with the intention of cutting the number of trips, the model was then solved with a modified objective function taking into account the unloading time at the end of each route/trip. That is, Eq. (10.1) was replaced by Eq. (10.11) and, therefore, the resulting model consisted of Eq. (10.11) combined with Eqs. (10.2)(10.9).

$$min \sum_{i,j\in M', \forall j\neq i, k\in K} x_{ij}^{k}\, t_{ij} + \sum_{i\in M', \forall i\neq 0, k\in K} x_{i0}^{k}\, td_{0} \tag{10.11}$$

The results of the computational experiments with this last model are presented in Table 10.5. The results were different, obtaining an improvement of 10% in Instance 1 and a deterioration of 8% in Instance 3 with respect to Table 10.3. On the other hand, a small improvement arises in Instance 2 (<1%) and no change in Instance 4. The numbers of routes/trips remained the same as those from Table 10.3.

5.3 Use of valid inequalities

As it was mentioned earlier, the presented model is a hard computational problem. In particular, it contains many symmetric solutions. Two solutions are said to be symmetric if they have the same objective function value but different variable assignments (Mahé et al., 2023). For example, it is equivalent for the solution algorithm if a given a route is assigned indistinctly to any of the vehicles. One way to reduce the complexity of the problem consists in adding valid inequalities (VIs) to the model. A VI is a constraint that reduces the size of the feasible region of the problem without eliminating all optimal solutions (Dror et al., 1985). In this case, we use the VIs that were proposed in Mahé et al. (2023): (1) the farthest clean point is assigned to the first route; (2) being K an ordered set, a route k cannot be used if route $k-1$ was not used; and (3) the auxiliary variables associated with the load at the beginning of each route (u_0^k) are equal to zero. In addition, the valid inequality proposed by Dror et al. (1994) is included, which requires the activation of the first k routes, where $k = |K|_{min}$ theoretical minimum of trips from Eq. (10.10).

The results of Table 10.6 indicate that the introduction of valid inequalities allowed to improve the results with respect to Table 10.3 in Instances 1 and 2 (18% and 24% improvement in travel time). Similarly, in these solutions the optimality of the solutions could be tested. On the other hand, in Instances 3 and 4, the solutions were worse in terms of the objective function than those in Table 10.2.

Figs. 10.1 and 10.2 show the graphical representation of the two routes obtained for Instance 1, for which the optimal result is reached.

TABLE 10.5 Results of the resolution of the instances, $|K| = 30$. Modified objective function.

Instance	Travel time (min)	Unloading time (min)	Total route time (min)	Gap (%)	N° of routes
Instance 1	58.44	16	124.08	42.67	2
Instance 2	78.38	24	213.22	58.32	3
Instance 3	346.99	152	552.71	90.63	19
Instance 4	383.41	184	732.65	91.76	23

TABLE 10.6 Results of the resolution of the instances, $|K| = 30$ with valid cuts (inequalities).

Instance	Travel time (min)	Unloading time(min)	Total route time (min)	Gap	N° of routes
Instance 1	53.12	16	118.76	0% (521.31 [s])	2
Instance 2	59.72	24	194.56	0% (116.95 [s])	3
Instance 3	423.9	224	701.62	86.57%	28
Instance 4	423.9	224	813.14	83.45%	28

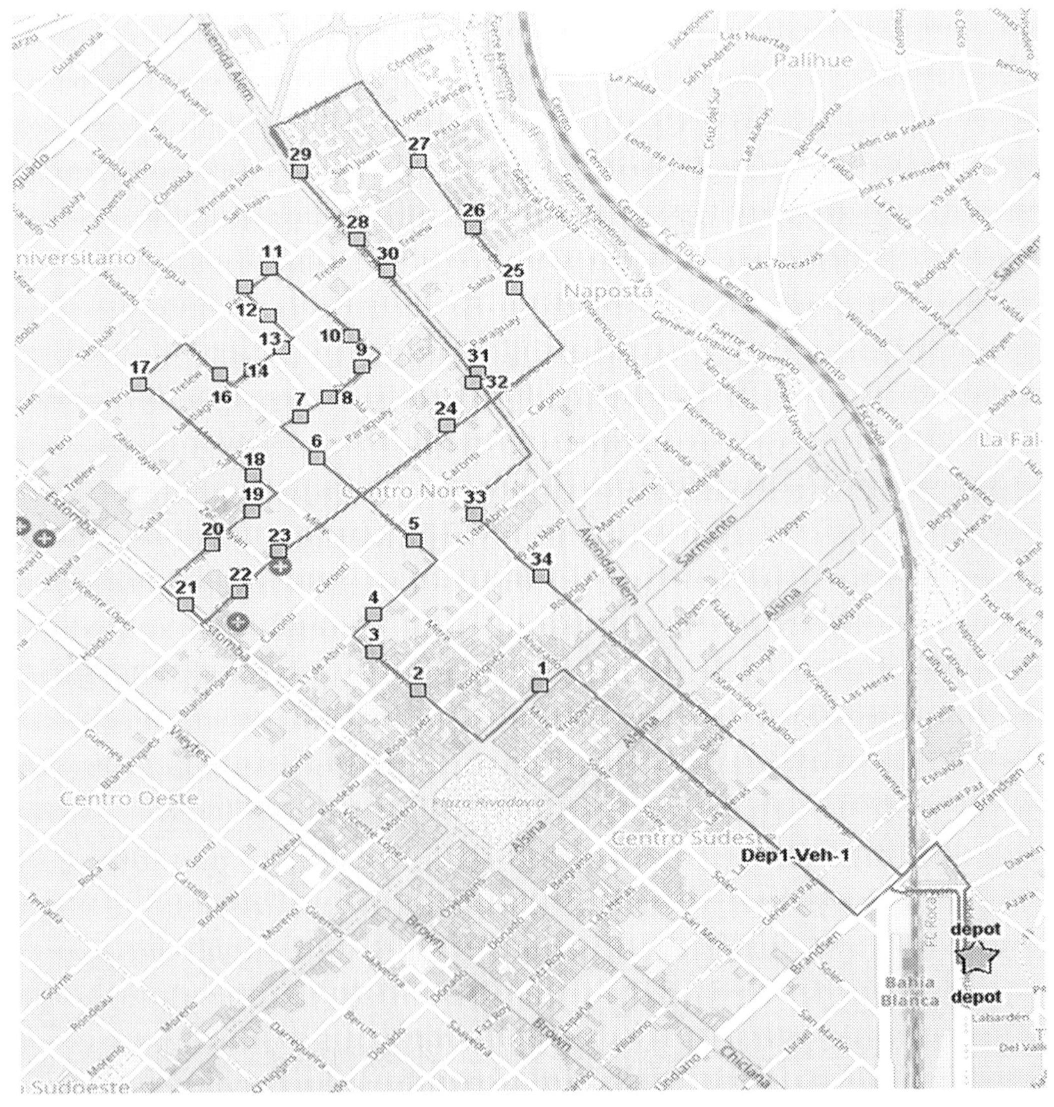

FIGURE 10.1 Collection route N°1 of the optimal solution for Instance 1.

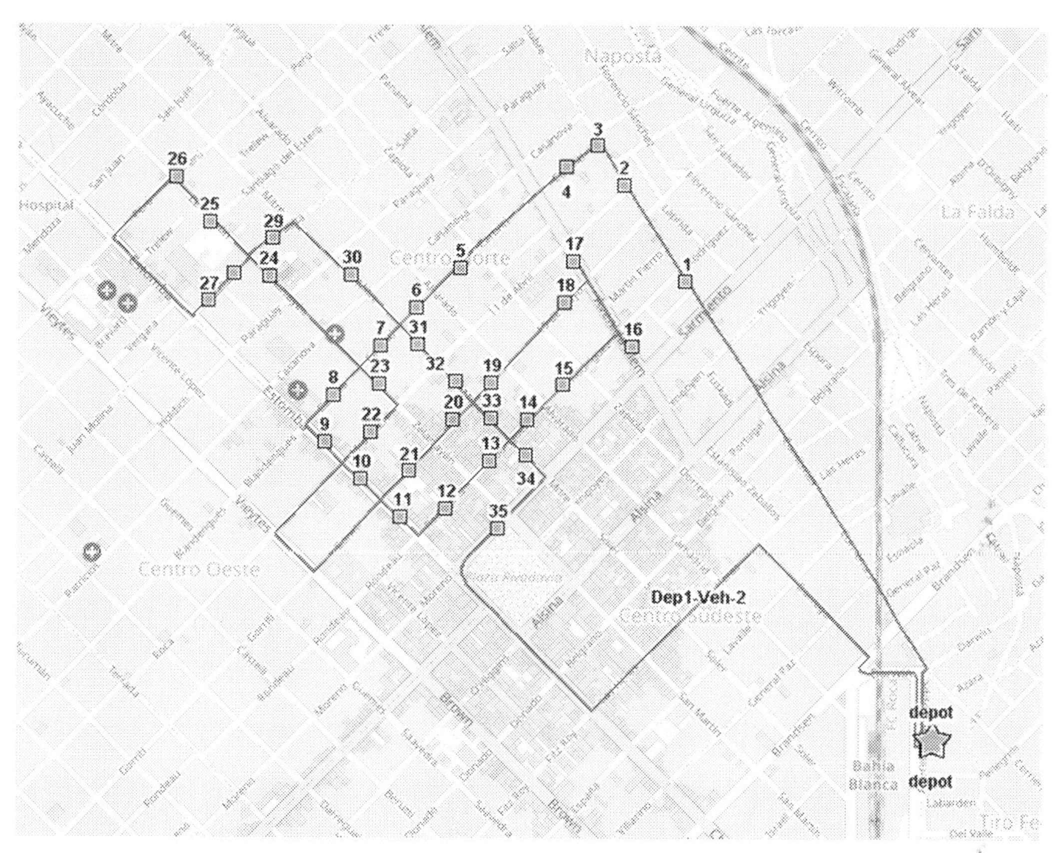

FIGURE 10.2 Collection route N°2 of the optimal solution for Instance 1.

6. Conclusions

The mismanagement of the municipal solid waste (MSW) system can provoke important environmental, economic, and social issues. Therefore, finding new tools to make MSW logistics more efficient is a pressing concern in today's societies, especially in developing countries, which generally lack large budgetary and technical resources. This chapter is focused on the resolution of waste collection problems for the Bahía Blanca area considering capacity and total time of the route restrictions. For this purpose, an integer-mixed programming model is proposed whose implementation is carried out in a Python environment and is solved with the CPLEX software.

The experimentation was carried out on four real instances of the city of Bahía Blanca. Preliminary results indicate that the model was able to present feasible solutions to the problem. Furthermore, two different modalities of the model were evaluated to limit the number of routes. The strategies implemented to decrease the number of routes were: decreasing the potential number of routes and assigning a penalty to the generation of new routes in the objective function. Reducing the potential number of routes was the strategy that provided the best results, allowing not only to reduce the total number of routes but also to obtain a better value of the objective function. Similarly, since it was evident that most of the solutions had large gaps, additional experiments were carried out incorporating valid inequalities to eliminate symmetric solutions in the feasible region and the minimum number of trips. These executions gave inconclusive results, allowing to obtain the optimal result in some instances, but worsening the value of the functional in others.

As a line of work for the future, the application of visual attraction metrics is proposed as additional desired objectives, since it has been noticed that compactness and noncrossing paths are properties that are favored by practitioners of the MSW routing plan. Another future research line is concerned with the resolution of instances through metaheuristic strategies, considering that routing problems can be computationally challenging to solve by means of exact methodologies.

Acknowledgments

The first author of this chapter is grateful for the funding received from the Secretary of Science and Technology of the Universidad Nacional del Sur through a Scholarship for Advanced Students. In addition, the authors wish to acknowledge the funding received through research projects 24/ZJ39 and 24/J084 from the Universidad Nacional del Sur and the research project PICT-2021-I-INVI-00217 from the AGENCIA I+D+i of Argentina. The authors also thank Victoria Herrán Symonds, an industrial engineering graduate from the Universidad Nacional del Sur, who has collaborated with this line of research from the beginning.

References

Beliën, J., De Boeck, L., Van Ackere, J., 2012. Municipal solid waste collection and management problems: a literature review. Transportation Science 48 (1), 78–102.

Benitez, R., Gomez, R., Rivas, P., Botello, J., Huerta, O.F., García, A., Rueda, J., 2021. Optimization of municipal solid waste collection routes in a Latin-American context. Journal of the Air & Waste Management Association 71 (11), 1415–1427.

Blazquez, C., Paredes, G., 2020. Network design of a household waste collection system: a case study of the commune of Renca in Santiago, Chile. Waste Management 116, 179–189.

Bonomo, F., Durán, G., Larumbe, F., Marenco, J., 2012. A method for optimizing waste collection using mathematical programming: a Buenos Aires case study. Waste Management & Research 30 (3), 311–324.

Braier, G., Durán, G., Marenco, J., Wesner, F., 2017. An integer programming approach to a real-world recyclable waste collection problem in Argentina. Waste Management & Research 35 (5), 525–533.

Carlos, M., Gallardo, A., Edo-Alcón, N., Abaso, J.R., 2019. Influence of the municipal solid waste collection system on the time spent at a collection point: a case study. Sustainability 11 (22), 6481.

Cavallin, A., Rossit, D., Herrán, V., Rossit, D., Frutos, M., 2020. Application of a methodology to design a municipal waste pre-collection network in real scenarios. Waste Management & Research 38 (Suppl. l), 117–129.

Deckers, D., van Dijk, M., Grund, M., Hendrickx, G., de Koning, R., Smit, N., 2018. PoDeA, Port Development Argentina. Technical Report. Delft University of Technology, Netherlands.

Delle Donne, D., Di Tomaso, V., Duran, G., 2021. Optimizing leaf sweeping and collection in the Argentine city of Trenque Lauquen. Waste Management & Research 39 (2), 209–220.

Di Maria, F., Micale, C., 2015. Life cycle analysis of management options for organic waste collected in an urban area. Environmental Science and Pollution Research 22, 248–263.

Dror, M., Ball, M., Golden, B., 1985. A computational comparison of algorithms for the inventory routing problem. Annals of Operations Research 4 (1), 1–23.

Dror, M., Laporte, G., Trudeau, P., 1994. Vehicle routing with split deliveries. Discrete Applied Mathematics 50 (3), 239–254.

Esparza, J., 2021. Clasificación y afectación por residuos sólidos urbanos en la ciudad de La Plata, Buenos Aires, Argentina. Revista Internacional de Contaminación Ambiental 37, 357–371.

Fermani, M., Rossit, D., Toncovich, A., 2020. A simulated annealing algorithm for solving a routing problem in the context of municipal solid waste collection. Communications in Computer and Information Science 1408, 63–76.

Gilardino, A., Rojas, J., Mattos, H., Larrea, G., Vázquez, I., 2017. Combining operational research and life cycle assessment to optimize municipal solid waste collection in a district in Lima (Peru). Journal of Cleaner Production 156, 589–603.

Han, H., Ponce Cueto, E., 2015. Waste collection vehicle routing problem: literature review. Promet - Traffic & Transportation 7 (4), 345–358.

Hart, W., Laird, C., Watson, J., Woodruff, D., Hackebeil, G., Nicholson, B., Siirola, J., 2017. Pyomo-optimization Modeling in Python. Springer, Heidelberg.

Hoornweg, D., Bhada-Tata, P., Kennedy, C., 2015. Peak waste: when is it likely to occur? Journal of Industrial Ecology 19 (1), 117–128.

International Business Machines Corporation: User's Manual for IBM ILOG CPLEX Optimization Studio 22.1.0, 2022. International Business Machines Corporation. Available in: https://www.ibm.com/docs/en/icos/22.1.0.

Lavee, D., Nardiya, S., 2013. A cost evaluation method for transferring municipalities to solid waste source-separated system. Waste Management 33 (5), 1064–1072.

Lenstra, J., Kan, A., 1981. Complexity of vehicle routing and scheduling problems. Networks 11 (2), 221–227.

Mahéo, A., Rossit, D., Kilby, P., 2021. A Benders decomposition approach for an integrated bin allocation and vehicle routing problem in municipal waste management. Communications in Computer and Information Science 1408, 3–18.

Mahéo, A., Rossit, D.G., Kilby, P., 2023. Solving the integrated bin allocation and collection routing problem for municipal solid waste: a Benders decomposition approach. Annals of Operations Research 322 (1), 441–465.

Molfese, S., Rossit, D., Frutos, M., Cavallin, A., 2022. Optimization of waste collection through the sequencing of micro-routes and transfer station convenience analysis: an Argentinian case study. Waste Management & Research (in press).

Naredo, J., 2007. La reconciliación virtual entre economía y ecología en el nuevo desarrollismo ecológico. Papeles de Relaciones Ecosociales y Cambio Global 100, 33–51.

Open Source Routing Machine: Routing Service, 2021. Available in: http://project-osrm.org/.

OpenStreetMap Contributors: Maps Download from, 2017. Available in: https://planet.osm.org https://www.openstreetmap.org/.

Owusu, F., Oduro, S., Essandoh, H., Wayo, F., Shamudeen, M., 2019. Characteristics and management of landfill solid waste in Kumasi, Ghana. Scientific African 3, e00052.

Port Consultants Rotterdam: Port Vision Bahía Blanca 2040. Technical Report, 2017. Available in: https://puertobahiablanca.com/vision_portuaria_2040/files/downloads/Vision_BB_2040-EN.pdf.

Rossit, D., Tohmé, F., Frutos, M., Broz, D., 2017. An application of the augmented ε-constraint method to design a municipal sorted waste collection system. Decision Science Letters 6 (4), 323–336.

Rossit, D., Nesmachnow, S., Toutouh, J., 2019. Municipal solid waste management in smart cities: facility location of community bins. Communications in Computer and Information Science 978, 102–115.

Rossit, D., Toutouh, J., Nesmachnow, S., 2020. Exact and heuristic approaches for multi-objective garbage accumulation points location in real scenarios. Waste Management 105, 467–481.

Rossit, D., Toncovich, A., Fermani, M., 2021. Routing in waste collection: a simulated annealing algorithm for an Argentinean case study. Mathematical Biosciences and Engineering: MBE 18 (6), 9579–9605.

Smith, H., Revell, K., 2016. Micro-incentives and municipal behavior: political decentralization and fiscal federalism in Argentina and Mexico. World Development 77, 231–248.

Sodiq, A., Baloch, A., Khan, S., Sezer, N., Mahmoud, S., Jama, M., Abdelaal, A., 2019. Towards modern sustainable cities: review of sustainability principles and trends. Journal of Cleaner Production 227, 972–1001.

Sousa, V., Dias, C., Vaz, J., Meireles, I., 2018. Life-cycle cost as basis to optimize waste collection in space and time: a methodology for obtaining a detailed cost breakdown structure. Waste Management & Research 36 (9), 788–799.

Toth, P., Vigo, D. (Eds.), 2002. The Vehicle Routing Problem. Monographs on Discrete Mathematics and Applications. Society for Industrial and Applied Mathematics (SIAM), Philadelphia.

Vázquez Brust, A., 2018. Ruteo de alta perfomance con OSRM. RPubs. Available in: https://rpubs.com/HAVB/osrm.

Vazquez, Y., Barragán, F., Castillo, L., Barbosa, S., 2020. Analysis of the relationship between the amount and type of MSW and population socioeconomic level: Bahía Blanca case study, Argentina. Heliyon 6 (6), e04343.

World Bank Group, 2016. Clean and Inclusive Cities in Argentina.

Zbib, H., Wøhlk, S., 2019. A comparison of the transport requirements of different curbside waste collection systems in Denmark. Waste Management 87, 21–32.

CHAPTER

11

A knowledge roadmap for digitally enabled sustainable construction and building supply chain management

Vidyasagar Gembali, Aalok Kumar and P.R.S. Sarma

Indian Institute of Management, Visakhapatnam, Andhra Pradesh, India

1. Introduction

Importantly, construction and building industry (CBI) takes high investments, generates huge revenues, and plays a key role in balancing the socioeconomic conditions of the global economy (Kozlovska et al., 2021; Nasir et al., 2017; Schönbeck et al., 2020). Construction and its related costs are occupying 13% of all global spending, and with an estimated revenue generation of $ 10 trillion, this revenue is expected to increase to $ 14 trillion by 2025. Additionally, CBI incorporates a significant economic spillover effect on the spending of every $ 1 with CBI, making an extra economic benefit of $ 2.86. Therefore, slight improvements in CBI can greatly improve a country's GDP (Barbosa et al., 2017; Leviäkangas et al., 2017). GDP and investments of a nation are highly influenced by CBI. Therefore, as a concern, improving innovation and productivity of construction operations is important because the average value addition from construction workers is $ 25/h compared with average of $37/h from all other sectors combined. To this, innovation and productivity are important factors to improve in construction fields (McKinsey&Company, 2017). CBI can generate high revenue but struggle with high construction costs, poor project practices, poor construction safety, low productivity, lack of innovation, and poor sustainability practices. The construction sector is improving significantly by creating huge amounts of built spaces for human activities. Built environment enhances the quality of life by fulfilling the socioeconomic requirements of people, societies, and nations. CBI contributes to 25% of the global GDP and 7% global job market (Norouzi et al., 2021); currently, this industry is struggling with severe sustainability

Evolution and Trends of Sustainable Approaches
https://doi.org/10.1016/B978-0-443-21651-0.00005-X

issues due to its stringent natural resource requirements and pollution generation (Ghaffar et al., 2020; Oluleye et al., 2023). In total, 40% of raw material consumption, 40% of waste generation, and 25% CO_2 emissions globally occur from CBI (Darko and Chan, 2016; Mahpour, 2018; Nasir et al., 2017). Industry 4.0 (I 4.0) technologies can enhance innovation and productivity with sustainability benefits into CBI (Balasubramanian et al., 2021; Hu et al., 2022) and presented in Tables 11.1 and 11.2.

TABLE 11.1 Sustainability benefits from digital transformation for the construction and building industry.

S.No.	Author (s)	Digital technology application	Sustainability benefit
1	Balasubramanian et al. (2021)	IoT + Bloch chain + Additive manufacturing (AM)	Positively affecting all sustainability pillars (economic, social, environmental) and surpassing negative effects imposed on organizations to take digitization.
2	Akbarieh et al. (2020)	BIM + EoL	Use of BIM with EoL scenarios easing deconstruction activities and promoting environmental sustainability and use of CE practices in construction.
3	Yevu et al. (2021)	Blockchain	Blockchain-enabled smart contracts are used toward digitization CBI, improving trust in sharing data in IoT environments. Lack of confidence is an important barrier that hamper many digital technologies use. Trust promotes a healthy work culture and improves social sustainability.
4	Chen et al. (2022), Liu (2015)	BIM	Integrating BIM at different stages of construction processes, from design to building construction to building operation, promotes sustainability by generating digital information about physical places and their functionalities. It eases access to information about the built environment and simplifies the reuse and recycling of building components.
5	Beltrami et al. (2021)	IoT	IoT improves economic and environmental sustainability by improving operational efficiency through the coordination of logistic operations and improves process flexibility.
6	Balasubramanian et al. (2021), Zabidin et al. (2020)	IoT + Blockchain	The combination of IoT + blockchain enhances social sustainability and stimulates the use of I 4.0 technologies in CBI, by improving security. The data collected from IoT devices can be securely stored in the cloud with blockchain and use for advanced analytics works.
7	Chen et al. (2022), López Martínez et al. (2021)	Big data	Big data enables an architecture for using meta tools to collect specifications and information from all elements in the digital ecosystem. This is helpful for the auto-generation of information required for the deployment of applications. Therefore, big data increases economic sustainability.

TABLE 11.1 Sustainability benefits from digital transformation for the construction and building industry.—cont'd

S.No.	Author (s)	Digital technology application	Sustainability benefit
8	Turner et al. (2021)	Industrial robot	Industrial robots have both positive and negative effects on social sustainability and positively affect operational efficiencies. Industrial robots are employed for highly production-intense tasks and at places where the human workforce is not safe to accompany in that location and with those materials. For example, cleaning window glasses at high altitudes
9	Darko et al. (2020), Maureira et al. (2021)	AI	AI improves economic sustainability by improving decision-making capability, AI enabling a complete platform for mining textual information from a corpus of information from literature studies to a firm database.
10	Ahmed (2019)	AR & VR	Improving economic and social sustainability. AR is successfully used in many construction projects for scheduling, human resource training, time management, cost management, and quality checking. VR used for visualization and improving safety and defects management.
11	Costin and Teizer (2015), Majrouhi Sardroud (2012)	Radio-frequency identification (RFID)	RFID has been used in construction operations for tracking machines, managing inventory, and theft control. RFID implies safety by providing onsite equipment maintenance providing data to engineers timely

Several practices and methods are available to mitigate the sustainability issues faced by CBI, and digital transformation is a prominent practical way of making CBI sustainable, productive, and resource-efficient (Fathalizadeh et al., 2022; Hilty and Aebischer, 2015). For digital transformation, different digital technologies (DTs) are used to enable industries to access real-time information, reduce waste generation, improve efficiency, reduce cost, improve quality, and finally help to achieve promised sustainability and productivity concerns. Digital transformation improves investments and promotes sustainable digital business models with better customer satisfaction. The sustainability benefits of DT are interconnected in the economic, social, and environmental facets of construction practices (Grubic and Jennions, 2018; Kaklauskas et al., 2020). Digital transformation in the construction sector is hampering due to traditional organizational practices for incorporating change for accompanying new practices and procedures into existing set of activities and fear of cost implications (Bazan and Estevez, 2022; Nagy et al., 2021).

Many of I 4.0–related DTs, from 3D printing to big data, deep learning, digital twins, augmented reality (AR), virtual reality (VR), internet of things (IoT), artificial intelligence (AI), blockchain (BC), and cloud computing, are part of digital transformation in the CBI (Chen et al., 2022; Karmakar and Delhi, 2021). From a sustainability perspective, DTs are impacting construction firms' ecology and social relations and affecting organizational work culture (Chen et al., 2022). For instance, some of the AI methods (genetic algorithms,

TABLE 11.2 Important studies on Construction 4.0 with their key findings and future scope.

S.No.	Author (s)	Key findings or contributions	Future scope
1	Karmakar and Delhi (2021)	To drive construction firms in C 4.0 ecosystem, firm's project management has to be driven by data management. Which includes data retrieving, data flow, data supply, and storage on entire project life cycle to support different stakeholders on different layers of hierarchy	Need to develop methodologies and frameworks to understand and implementation of complex technologies at different spheres of a firm.
2	Begić and Galić (2021)	Potential drivers for implementing C 4.0 are integrative big data, BIM, and IoT approaches. The integrative approaches of BIM with IoT or big data in construction projects show benefits in data management, real-time decision-making, and construction design, planning and building.	The benefits of C 4.0 are still mainly concentrated into the design stage i.e., preconstruction. Needed studies to explore the benefits in other stages of construction. It is important to consider other stages of construction because BIM offers many opportunities to optimize operations and data.
3	Saini et al. (2022)	Proposed a method for developing an intelligent decision engine using the Markov decision process (MDPs), and Monte Carlo tree search for making action planning (MCTS) for construction operations.	Similar further decision engines can be developed for other different construction operations, such as parameter optimization of drilling, cementing, well cleaning, site control etc.
4	Hu et al. (2022)	Developed a methodology based on 6Ms by incorporating Ishikawa's diagram and life cycle approach of a building for demonstrating the advantages of digital twins for efficient construction operations of a project	Digital technology applications in CBI have to conduct research in the integration of lean concepts into digital platforms, time analysis, economic analysis, multifunction sensors, AI tools for enhancing functionalities, multifunction digital systems, life cycle integration, servitization of assets, and sustainability.
5	Sepasgozar (2020)	Developed a module of digitally immersive multi player avatars using mixed reality and digital twins. These modules will be helpful for teaching purposes without the requirement of physical hands-on practice for complicated construction operations as well as for online students.	Required studies on different mixed reality modules, including all types of construction operations at different locations. This study is qualitative type limited by examining three modules; empirical studies are needed by including surveys from a large group of people who are familiar with digital apps and meta tools for evaluating virtual course modules.
6	Akanmu et al. (2020)	Developed a postural training system with attachments of Vive trackers and wearable sensors for tracking the kinetics of human body. From the sensors' feedback and managed with ML and VR, to understand human postures and associated risks related. Based on these feedbacks	For efficient training, future studies explore reinforcement learning models for proceeding with the state of learning and on evaluation methods of formative workload.
7	Shirowzhan et al. (2020)	With a systematic and critical analysis of the BIM-Compatibility literature, it is found that, compatibility issues of organizations and technical issues of interoperability hinder the adoption of BIM.	Significant knowledge gap exists on the compatibility of organizations for taking final decisions on adopting BIM, studies needed on compatibility as a key construct on BIM adoption

TABLE 11.2 Important studies on Construction 4.0 with their key findings and future scope.—cont'd

S.No.	Author (s)	Key findings or contributions	Future scope
8	AbouRizk (2010)	The simulation was successful in academia and research rather than in industry; important factors in industry adoption include collaboration and the development of programs. Which results of practical solutions and tool for adoption in real-world applications	For adopting simulations in complex construction projects with greater reliability, future studies has to concentrate on high-level architecture-based models, with construction synthetic environment framework for gaining modeling freedom for adopting simulation tools.
9	Louis and Dunston (2018)	Provided a sensor-agnostic operational implementation of an interoperable IoT network for taking real-time data from various parts of the large construction site	Future studies have to be conducted to test the methodology in real-world scenarios on different construction operations, including human presence with intense works of human workforce; other future studies include exploration on latency and data analysis techniques.
10	Danel et al. (2021)	For improving productivity of the construction site proposed a methodology with the input of tower crane data, data obtained from data logger for special and load information from structural movement when lifting weights	BIM integration is for future study, in this work used the data only with tower crane related coordinates which does not include every movement of object; constriction project data has to combine with crane data the integration of BIM data gives holistic perspective
11	Maskuriy et al. (2019)	Information and communication technology (ICT) is positively influencing growth of CBI, the construction industry can benefit from partnerships with ICT experts	I 4.0 technologies are affecting the organization's branding, marketing, operations, and social practices. I 4.0 has future research challenges on developing optimization models at firm level
12	Liu et al. (2017)	The collaborative nature of BIM affects the organization in process and procedures, people involved and technologies using. BIM Improves project performance, and BIM collaborations are affected by capabilities of information technologies, social behavior, roles of workers, technology management, trust between partners, learning experience, leadership, and communication	Country-specific studies has to conduct on collaboration and BIM themes

machine learning, fuzzy logic, and neural networks) are used in CBI to make it easy in taking better decisions and improving the efficiency of firm operations, but it necessitates the workforce to build capabilities on understanding data and cope up with data-oriented work system (Darko et al., 2020). I 4.0 is the basis for the concept of Construction 4.0 (C 4.0). C 4.0 is promoting the use of I 4.0 practices in CBI, through I 4.0 technologies, for digitally mapping the physical world and capturing data related to people, space, and different mechanical objects in the system, such as vehicles and machines. Using these data at a city scale can optimize building operations, city operations, city logistics, and many production-related problems in construction and able to make city a smart city (Kozlovska et al., 2021). In construction, DTs are incorporated for integrating sustainability with planning, modernization, and urban development at city-level operations for perceiving the concept of a smart city (Safiullin et al., 2019). As an extension to I 4.0, Industry 5.0 (I 5.0) is proposed, which is intended to solve additionally, the social problems along with production problems in CBI, and paving the way toward Construction 5.0 (C 5.0) (Martynov et al., 2019). The usefulness of DT in CBI is quite apparent for solving operational issues and social issues. Complexity and information challenges in construction systems are slowing down DTs' adoption in CBI (Karmakar and Delhi, 2021; Predescu et al., 2019).

Digitalization and use of DT can make a leap to efficient business models from existing traditional business models followed by firms to increase revenue generation and value creation opportunities. Here, digitalization means carrying business functions in a digital ecosystem where data are available at every stage of decision-making and promotes better decision-making capabilities to individuals. This is also perceived as keeping efforts on modernization in an entire organizational ecosystem with a complete digital set of processes and procedures (Lipsmeier et al., 2020; Plekhanov and Netland, 2019). In responding to sustainability concerns, it is required to meet the needs of the present without compromising future needs; with a practical conscience of three inclusive basic set points or pillars, "economic" outcomes have to be groomed in a way that it does not hamper future economic opportunities, "social" sustainability, emphasizing the need for "put the people" for development and empowering the people to make resilient societies, and "environmental" of considering organizations interactive physical surroundings in using energy, water, and other natural resources, so that current usage of natural resources does not hinder future natural resource requirements (Purvis et al., 2019). The digitalization helps construction firms to incorporate all three sustainability pillars in organizational decisions with a clear understanding and mapping of firm operations digitally (Klinc and Turk, 2019; Nagy et al., 2021).

This chapter further explains, in the next section, on extensive literature review on sustainability benefits from I 4.0 technologies, elements of I 4.0, and the concept of C 4.0. In the third section, management and operational challenges faced by construction firms with the advent of C 4.0 are emphasized; in the fourth section, the implications of I 4.0 technologies on organizational practices and a workforce of construction firms with a sustainability perspective are provided; and finally, in the fifth section, the conclusion of this chapter is provided.

2. Literature review

2.1 Construction sustainability with industry 4.0 technologies

The usual perception of sustainability is costly for many construction supply chain (CSC) stakeholders. Sustainability concerns the inhabitant's well-being and promotes quality of life in built environment. Therefore, the primary discussion is reducing the gap between sustainable housing and housing affordability. Sustainable housing does not harm the environment with use of raw materials efficiently and reduces waste generation (Kineber et al., 2020). Adopting DT can make the realization of sustainability with enhancement in process control, the ability to monitor construction practices, and transform a sustainable built environment

(Balasubramanian et al., 2021). It is important to understand the digitalization implications on sustainable construction because design teams and operation (construction) teams will be benefitted from the integration of DT with physical objects presented in construction. After all, this will lead to the accessibility of real-time data, improve communication, reduce the project delays, and reduce cost and raw material usage (Akanmu et al., 2021). In the digitalization of construction practices, building information modeling (BIM) is a prominent topic of discussion; BIM can support deconstruction by incorporating sustainability by following circular building stock 4.0. As per circular economy (CE), it reduces the use of raw materials and disposal of end-of-life (EoL) building components sustainably. CE requires design for deconstruction, design for material retrieval and reuse, and design for building flexibility. BIM can support these CE requirements by estimating the EoL properties of building components and improving the disassembly procedure (Akbarieh et al., 2020; Akinade et al., 2017).

With growing advancements in I 4.0 and related DT, digitalization is occurring in construction supply chain and procurement (CSCP) activities. Digitalization promotes a friendly environment for sustainability innovation in CSCP with many economic and social benefits (Yevu et al., 2021). CSCP mainly consists of four stages; 1. Planning, 2. Precontract, 3. Contract, and 4. Postcontract. DTs are efficient in promoting sustainability in all four stages of CSCP. Even further digitalization occurred at sourcing raw materials to check the materials for ecological compliance and to measure the sustainability performance of suppliers and contractors (Chen et al., 2022; Klinc and Turk, 2019). BIM has substantial sustainability benefits when incorporated with IoT, digital twin, and BC technologies into CSCs. However, sustainability and digitalization-related topics in the construction context are lagging attention from academic research (Li et al., 2019; Yevu et al., 2021). Procurement practices in CBI are unstructured in nature and hamper digital procurement usage; BIM can reduce ambiguity and improve structuredness in construction procurement by facilitating new ways of collaboration. Grilo and Jardim-Goncalves (2011) explained how BIM is challenging and promotes e-procurement in construction by combining BIM with cloud computing, model-driven architecture, and service-oriented architecture. BIM also combined with BC to provide information of value chains from workflows of buildings from the design stage (Li et al., 2019). Other combined uses of DT include AI with smart contracts and IoT with BC in supply chains for improving value chains (Rahman et al., 2019; Rejeb et al., 2019).

2.2 Elements of industry 4.0 and its effect on construction operations

Conceptually, I 4.0 consists of eight elements: 1. Collaboration of humans and machines, 2. The digitalization of production systems for optimizing and personalizing the production process, 3. Automation of procedures for adopting real-time changes, 4. Warehouse optimization and value-added optimization, 5. Communication with real-time data, 6. Digital integration of technical system and economic system, 7. Providing digital product offerings and digital services, 8. New market approaches with digital ecosystem and digital business models (Roblek et al., 2016; Zezulka et al., 2016). With these eight I 4.0 elements, a complete set of currently existing value chains and management practices are affected and changing, and made easy to incorporate sustainability consciousness in operations, deploy waste minimization, and manage resource efficiency. Sustainability is a necessity in CBI due to resource scarcity and existing biodiversity losses. From an established set of market practices, management procedures, organizational procedures, and workforce facing challenges for implementing I 4.0 technologies. Drivers for implementing I 4.0 include the benefits of DT on better customer service, data-enabled decision-making, improved flexibility, and improved control on production activities, especially in series production for a size of one (Horváth and Szabó, 2019; Szabo et al., 2020). I 4.0-related automation works effectively, especially on repetitive construction tasks. But many of the construction operations are different, typically scattered on a wide range of construction site areas and operated with a mix of remote and disparate resources and human workforce. These characteristics are making difficult to deploy automation into construction worksites (Kozlovska et al., 2021). Compared with construction, automation is rather an easy task in manufacturing, for example, in a product line operation, a product is moved to different workstations, and at every workstation, prescribed work performs on the product and moved to the next workstation, where materials, machines are standardized, but it is not the case in construction. IoT made it easy for construction operations by providing real-time decision support (Louis and Dunston, 2018).

Among different I 4.0 technologies, reports on IoT are promising to handle real-time information from a dispersive construction work environment. IoT has the potential for predictive maintenance of machinery in construction, mining, and oil and gas sectors with the ability to improve onsite operations and supply chain activities (Manyika et al., 2015). Many articles insist that infrastructure projects and built environment are prime focus for implementing IoT, because it has clear application on modular construction, offsite prefabricated construction equipment, and implementing lean principles. IoT promises CBI to support broadly improving methods and procedures in construction sector with its efficient communication abilities when dealing with real-time decisions (Gubbi et al., 2013; Miorandi et al., 2012). IoT application on construction simulation from physical construction practices, discrete event simulation (DES) is one of the efficient tools for collecting data from IoT devices from a broad and complex physical construction systems and embedding into a cybertwin for real-time decision support (AbouRizk, 2010; Louis and Dunston, 2018). Louis and Dunston (2018) used the DES tool to make a framework for realizing IoT benefits of real-time data in construction and conceded that IoT support for entirely all different operations in construction is a troublesome and highly unfeasible task, real-time support is practical and feasible for repetitive construction operations.

To make complete automation of construction operations, it is not possible to consider one single operation or one single technology such as IoT; it's a matter of different technologies and considering many operations with a thread of interoperability. Technology should connect to different machinery and heavy equipment such as excavators and bulldozers to digitally monitor for controlling operations (Shirowzhan et al., 2020). Apart from IoT, AR/VR technologies are used to train workers to improve posture and practice. Akanmu et al. (2020), Sepasgozar (2020) conducted a study on integrating AR/VR with digital twins to develop training modules for the design and execution of tunnel projects. Digital twin application has a wider scope with onsite relevance with increased functionalities and able to provide a holistic perspective. Digital twins can imply servitization with multidatasets from different resources for a quality asset management and benefit construction operations by leveraging data (Hu et al., 2022). Digital twinning also helps to eliminate and reduce biased human decisions, which helps to reduce problems with operational inefficiencies and improve safety, reduce cost, and maintain ecological compliance of materials at the same time with reinforcement learning for preparing action planning for operational objectives (Saini et al., 2022).

2.3 Concepts of construction 4.0

As a construction version of I 4.0, the concept of C 4.0 has emerged, in recent years, C 4.0 is gaining momentum in academic works with an exceptional scope. The aim of C 4.0 is to implement I 4.0 technologies in the construction sector to gain efficiency and minimize waste (Adepoju and Aigbavboa, 2021; Chen et al., 2022). C 4.0 is an embedment of I 4.0 principles and elements in construction practices for the digitalization of construction practices. Using DT, CBI can bring viable solutions to sustainability issues in CBI (Kozlovska et al., 2021). The primary emphasis of C4.0 is on applying new industrial methods, new business models, and related requirements on innovation, materials, and tools in the present construction sector. Implementing the C 4.0 concept can stimulate cross-industry collaboration and connect different professionals on a common platform for interaction (Chen et al., 2022; Karmakar and Delhi, 2021). CBI took more time to incorporate I 4.0 principles into practice than other industries such as manufacturing, automobile, and aviation. According to the Boston Consulting Group hypothesis in 2016, construction practices will soon be connected through communication systems, software applications, intelligent tools, and machinery. These are from I 4.0 group of technologies, and in CBI, all these are integrated into the core module as BIM (Gerbert et al., 2016). Some of the reasons for slow adoption C 4.0 are related to craft-based approaches, complex procedures, low-tech activities, low use of robotics, and the need of dynamic change to adapt for the continuous change of onsite construction practices from project to project (Begić and Galić, 2021; Craveiro et al., 2019). BIM is the key essence in the topic of C 4.0; it supports for keeping information regarding the physical and functional aspects of the buildings. It is believed that C 4.0 has a true potential for a required sustainability makeover to existing CBI through supporting modularized designs and offsite construction practices with AM. With AM, customized set of properties and functions can be printed along with the physical structure of the building. In AM, raw material waste is minimal compared with traditional building procedures (Chen et al., 2022; Norouzi et al., 2021).

Different hybrid methods from C 4.0 are taking practice in CBI, amalgamating BIM with AI, and lean thinking designing with digital twins application in construction, where physical world data on construction practices are mapped and modeled in virtual (Chen et al., 2022). Product life cycle management, which manages the products from inception (design to end of use), combines data systems to promote construction-specific customized tools, simplifying innovation. In addition, using RFID and industrial robots is an extra advantage to the construction industry for implementing C 4.0. Other technologies from I 4.0 such as IoT, AR, and VR are used in construction practices with the help of I 4.0 devices and tools (Lekan et al., 2020; Zabidin et al., 2020). The designed hybrid or customized technology adoption into construction organizations occurs through three methods. 1. Horizontal – establishing a network between different organizations, 2. Vertical – establishing a network between different technologies available in the same firm, and 3. End-to-end integration connects all steps from starting to final stage of a product with different technologies. The CBI activities are highly fragmented and interconnected to each other and involve many different sets of firms; it makes difficult to incorporate horizontal and vertical integration methods. For end-to-end integration, many firms have to agree, and it is quite difficult to accompany all the firms for one construction process improvement (Liao et al., 2017; Liu et al., 2017). As a topic C 4.0 is currently in the conceptual stage, very few studies are available. Minimal studies focus on applying C 4.0 and its realization with managerial and organizational perspectives (Maskuriy et al., 2019). Regarding the stakeholder perspective, comparatively, no study has been done that describes the details about what are the new stakeholders going to be part of the C 4.0 environment and how stakeholders' interests are managed compared with traditional construction practices (Maskuriy et al., 2019). C 4.0 is a collaboration between different construction professionals, construction firms, technology professionals, and technology firms with I 4.0 principles. Some literature of I 4.0 principles into construction shows that construction firms face sociotechnical barriers for implementing DT in CBI; these barriers cannot be handled alone by organizational economic policies or with interorganizational cooperation and support because new technological adoption demands a paradigm shift for viewing and solving problems through continuous learning. A few studies are available on the organizational perspective with I 4.0 implementation in construction compared with manufacturing, aviation, automobiles, and other sectors (Cimini et al., 2021; Danel et al., 2021; Müller et al., 2018).

3. Frameworks for digital transformation in CBI

Previous studies discussed different frameworks for taking digital transformation in construction practices. Grilo and Jardim-Goncalves (2011) explained the SOA4BIM framework, where SOA meant for service-oriented architecture is a generic framework based on model-drivenarchitecture (MDA), cloud computing, and SOA. SOA4BIM considers entire building life cycle from design, construction to operational maintenance of the building. SOA4BIM connects different stakeholders from all stages of building, including clients, designers, architects, specialist designer 1, specialist designer 2, ... specialist designer n, subcontractors, main contractor, and operational maintenance providers with BIM. SOA4BIM is built on integration of computational independent model (CIM), platform-independent model (PIM), platform-specific model (PSM). CIM is not dependent on ICT platforms for designing, building, and maintaining of built environment; this is advantage of this framework from business perspective; CIM is grounded on process protocol model (Kagioglou et al., 2000). According to design principles given by Hermann et al. (2016) for I 4.0, Hossain and Nadeem (2019) provided a framework for implementing C 4.0. The design principles include interconnection and interoperability, technical assistance, information transparency, and decentralized decisions. This C 4.0 framework consists of six consecutive steps; the first step is about mapping. In mapping, the existing state of a firm and plan for future are mapped; this step mainly considers the usage of BIM, automation of construction practices, digitization of construction supply chains and evaluates the management procedure of documents. After finishing of mapping stage, in the second stage, a pilot project is selected to implement learned lessons, incorporate end-to-end C 4.0 integration, and observe value chains and collaborations. In the pilot project, it is required to design strategies for implementation and decide required capacities, which is the third step. After deciding the capacity, data analytics is the fourth step. Data analytics play key role in making decisions. After making decisions, firm must push digital transformation with clear leadership; this is the fifth step. This transformation required a sustainable system to work; sustainable system can be built through collaborations, partnerships, and building culture on C 4.0 safety (Badri et al., 2018; Woodhead et al., 2018). Osunsanmi et al. (2020) provided a framework for C 4.0 by focusing on capability building on site-specific digitization activities, simulation, and virtualization of activities. Alade and Windapo (2020a,b), demonstrated a framework for leadership in I 4.0 scenarios. A framework is proposed in this chapter as shown in Fig. 11.1. This framework included an exhaustive extension capability building and leadership steps of framework given by Hossain and Nadeem (2019), and other major modifications taken place on data analytics part; in this framework, data analytics step is connected to mapping, capability building, and transformations stages of C 4.0 implementation, by accompanying benefit of data analytics at every decision-making situation in mapping, capability building, and transformation. As shown in Fig. 11.1, the framework proposes the implementation of C 4.0; this framework starts with mapping of existing digital maturity of firm and choosing of a pilot project for assessing capabilities needed for C 4.0 implementation and for understanding requirements. BIM or Revit is used to digitally map the construction processes in mapping. SOA4BIM is an appropriate framework for implementing BIM (Grilo and Jardim-Goncalves, 2011); along with BIM, supply chain automation and managing documentation are parts of mapping (Dallasega et al., 2018; Garrick and Chan, 2017; Li and Yang, 2017). After evaluating the mapped situation with the pilot project, capability building is

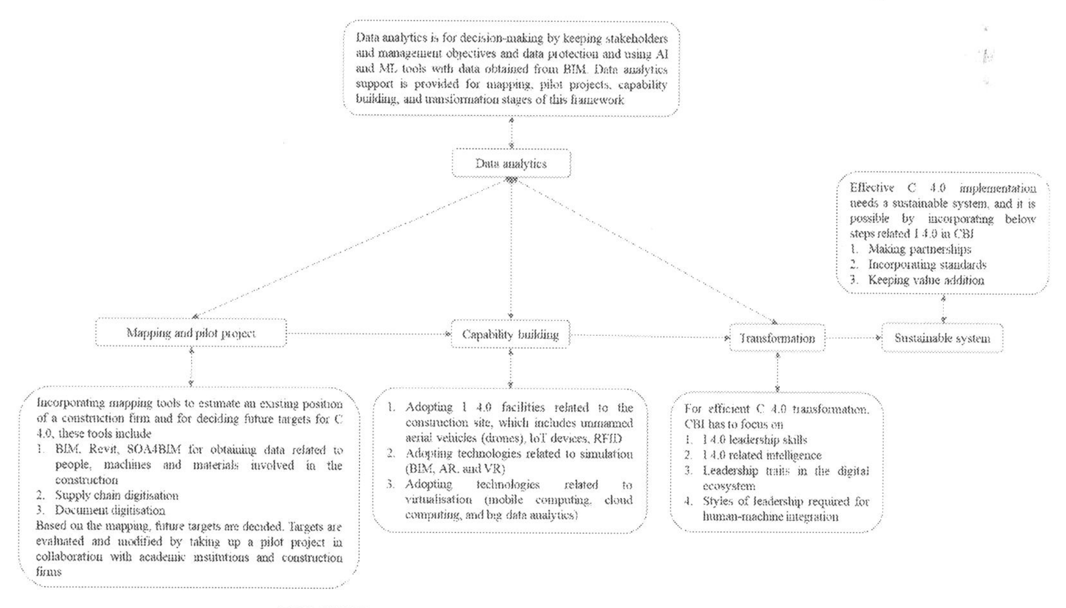

FIGURE 11.1 Framework for C 4.0 implementation.

the next step where one has to give priority to three aspects: 1. Digitizing construction site by using drones, IoT devices, and RFID; 2. Emphasizing on simulation; and 3. Incorporating virtualization tools and platforms (Osunsanmi et al., 2020). After acquiring capabilities required for C 4.0, leadership is pivotal on execution of framework. In C 4.0 situation, construction firms have to push fourth industrial revolution leadership skills and related intelligence, develop digital leadership traits, adopt different management styles according to the human, machine integration (Alade and Windapo, 2020b). The last step is establishing a sustainable system for practicing C 4.0 operations; to achieve a sustainable system, a firm has to take actions on making partnerships, which reduces risk and increases success, safety of C 4.0 practices, incorporates standards-related CBI and I 4.0, keeping value addition in C 4.0 operations (Badri et al., 2018; Woodhead et al., 2018). Digital analytics is interconnected to mapping, capability building, and simulation steps to make informed decisions.

4. Management and operational challenges with construction 4.0

Management faces challenges in dealing with cost and knowledge management for incorporating new technology into the existing practice. This is on providing expensive training from workers to dealing with new technology. No studies are available on supporting to provide detailing and improve confidence on investment ventures with clear cost–benefit trade-off analysis. This leads to lack of support for construction firms to proceed with C 4.0 technologies (Bazan and Estevez, 2022; Maskuriy et al., 2019; Osunsanmi et al., 2020). In the data world jobs will be different from routine; these jobs demand skill such as digital computation and data-related decision-making capabilities. Management needs to be decentralized to provide decision-making authority at many levels of organization and to many individuals. Strategically, management has to make modifications on hierarchy levels of the organization system (Bazan and Estevez, 2022; Maskuriy et al., 2019). Sharing of data plays a prominent role in C 4.0, but many research findings suggest that in accordance with problems of work force, culture within the firm regarding sharing of information is making a problem for organizations, and it is further intertwined and complicated with trust issues among different hierarchy levels and among different firms in CBI (Aghimien et al., 2022; Nagy et al., 2021).

As challenges for digital partnering, companies need to be partnering with other companies with complementary digital resources for competitive advantage, less risk of investment in DT, and improved performance of projects. Firms are facing challenges due to lack of clarity in partnering goal, lack of right selection of partner, lack of trust between parties (Aghimien et al., 2022). Institutional challenges are occurring with changes in sociotechnical system from digital transformation. Firms needs to understand its sociotechnical aspects for understanding challenges and to prepare a roadmap for digitization (Jacobsson et al., 2017; Lavikka et al., 2018). Fragmentation and complexity in the construction process from design to construction need different set of actors for different works, and they differed by their educational qualifications and styles of working. This fragmentation created a knowledge barrier and hampered a uniform ecosystem from digitalization for information communication (Lavikka et al., 2015, 2018; Miorandi et al., 2012). With uncertainty in technology developments and depreciation, construction professionals are not sure about which technology has to be adopted. However, construction experts can overcome these issues by incorporating suitable methods, which can predict the future. For example, scenario planning is an effective method for dealing with uncertainty holistically (Lavikka et al., 2018).

Organizations are facing a problem shifting from existing business models, which are material type of business models to information and provider of digital products type of business models. That is a data-oriented business model, where data play a key role. The construction industry faces severe competition from other engineering fields on integrating materials with data. In construction, technology is a digital layer upon what traditionally CBI is contributing over the years. From functional aspects, construction sector is serving to provide built environment to live and work, and other human activities are not changing. Therefore, construction professionals must adopt DT and make the layer on usual construction. This layering is done by construction firms rather than others on making digital layer over the physical real layer (Karmakar and Delhi, 2021; Klinc and Turk, 2019). BIM has clear construction feasibility for adopting DT on managing building material data, and BIM is an important element in C 4.0 for digital integration of building system. Organizations are facing operational challenges in integrating BIM with the project life cycle, because desired evolvements need to be taken place in BIM with cost management and platform-aiding design (Gerrish et al., 2017; Karmakar and Delhi, 2021).

Along with keeping the digital layer, facing safety issues from use of drones (unmanned aerial vehicles) is a mainstream use of robots. Robots and drones are used for inspection and coordination of construction activities, but due to the nature of construction facility, continuous mobilization and demobilization on a novel location happen, and usually robots are employed to those places where human accessibility is not possible. The complexity of locations and heavy moving loads on the construction site are operationally challenging to mainstreaming the use of robots to practice (Bryson and Winfield, 2017; Buchli et al., 2018).

Human resource problems for skilled workers, especially from youth workforce, are faced currently in construction firms. Young people are not interested in this sector due to lack of promising career paths and positive response on social change and technology adoption. Old workers are reluctant to change their mindset; it is causing challenges to make social change in the organization. Along these issues, organizations are skeptical on ethical issues, which may arise from use of robots because robots will reduce the need of humans in construction sites (Bazan and Estevez, 2022; Klinc and Turk, 2019; Maskuriy et al., 2019; Muñoz-La Rivera et al., 2020). C 4.0 promises on reducing spendings on wages and materials usage, but technology intervention into existing construction organizations causes complex sociotechnical problems; these problems are connected to supply chain, machines, and materials of built environment (Cimini et al., 2021; Danel et al., 2021). C 4.0 has to improve in time and cost-related aspects; for this time and cost dimensions need to add into common data environments (CDE). That is, time-related information (4D-BIM) and cost-related information (5D-BIM) of built materials. CDE is a data environment where data are collected from smart equipment, humans, and stored in shared space to provide accessibility of information to machines and humans for supporting decision-making capabilities. Even with these additions (4D-BIM, 5D-BIM), CBI is facing challenges from data exchange formats for complete realization of potential of C4.0 for construction sector (Karmakar and Delhi, 2021; Klinc and Turk, 2019; Radl and Kaiser, 2019).

5. Implications of construction 4.0 on construction firms, construction workforce, and organizational practices

5.1 Implications of construction 4.0 on construction firms

C 4.0 is going to change the way firms compete with each other in managing portfolios, managing professional skills, and technological capabilities. Therefore, implementation of C 4.0 technologies requires special attention on incorporating technologies into the organizational DNA. C 4.0 is redefining traditional construction practices with new efficiencies, occupational safety and health, and making more attractive to different stakeholders than ever before (García de Soto et al., 2019; Karmakar and Delhi, 2021). At the same time, continuous discussions are going on job creation and other disruptive effects from C 4.0 technologies on construction firms (García de Soto et al., 2019). Compared with other industries, CBI needs special care for incorporating I 4.0, because CBI is an important industry to balance socioeconomic conditions of community and nations. COVID-19 pandemic validated this by reconsidering the importance of economic well-being, balancing social conditions of workers (Lingard et al., 2021). C 4.0 injects I 4.0 technologies into construction practices, by means of transferring knowledge from construction ecosystem to robots and other I 4.0 devices; this infusion may promote analysis of social networks among workers for making knowledge transfer from a different group of workers (García de Soto et al., 2019; Lavikka et al., 2018; Maskuriy et al., 2019). In C 4.0, with ease of getting information, many players are equipped with better decision-making in the market; this enhanced competitiveness among all different types of construction firms made local firms to compete in global markets. In other words, local competitiveness is increasing with digital sharing and partnering (Aghimien et al., 2022; Lavikka et al., 2018).

Through digital partnering, SMEs become the beneficiaries with implementation of C 4.0 technologies. Digital partnering is becoming the new way of dealing with the implementation of I 4.0 technologies in construction organizations. Innovation becomes a necessity for firms and implying firms to keep efforts on research and development, promoting cross-functional collaboration among different stages of firm hierarchy, and keeping efforts for deploying innovation at every stage of construction strategy (Nagy et al., 2021; Schönbeck et al., 2020). C 4.0 made easy for firms to maintain sustainability in megaprojects with efficient collaborative learning and with inclusivity of firms for experimentation and innovation practices (Barata and da Cunha, 2019; Bonanomi et al., 2020). In the digital scenario, firms have to develop new approaches such as situational leadership, and construction firms have to train managers to understand data and decision-making capabilities on digital ecosystem (Aghimien et al., 2022).

5.2 Implications of construction 4.0 on construction workforce

CBI is often characterized by diversified set of workers, who are reluctant to change, due to high fragmentation and poor coordination between people. Implementation of C 4.0 necessitates understanding the job roles of construction workers to enhance clarity for efficiently demonstrating construction practices on digital platforms (García de Soto et al., 2019). The inherent rigidity of construction workforce makes it difficult to adopt new changes from technologies because it needs changes in social behavior of the workforce; in aligning to this, construction firms fear of availability of skilled workforce, especially in developing countries. With a note on skilled workforce, construction sector is providing jobs in large scales, but most of the workforce is not skilled intact with digitalization. Therefore, it is difficult to maintain most of the existing workforce in C 4.0 era; new skills are needed to implement C 4.0, and consideration of existing workforce on leaping into digital transformation is a crucial task (Adepoju and Aigbavboa, 2021). Layoff is not an easy task in CBI; it has been seen from COVID-19 pandemic that due to lack of work available, the burnout and deterioration of mental health occurred to construction workers. To implement DT, social change is the key for accepting technological change, by developing coordination between human and machines, because it implies new job roles into existing set of work culture. Along with social change, the high use of industrial robots creates ethical concerns and the unemployment rate for CBI (Adepoju and Aigbavboa, 2021; Lingard et al., 2021).

5.3 Implications of construction 4.0 on organizational practices of construction firms

The developments brought by fourth industrial revolution had left critical implications on competitiveness, society, and welfare and consequently influenced the organizational practices of many industries. Some of these implications were initially recognized in 2011 as a consequence of integrating information and communication systems into the industrial systems (Buhr, 2017; Schuh et al., 2014). Introduction of digitalization into construction sector is taken slowly compared with other sectors, but construction sector is also affected with DT when it comes to dealing with organizational practice. This is due to C 4.0 being equipped with a new cluster of workflows, difficulties, and stressing firms on organizational change to take technology adoption. However, with inertia, continuation from the traditional construction activities and from rigidly defined hierarchy levels in traditional firms is hindering the adoption of technology (Bazan and Estevez, 2022; Maskuriy et al., 2019). Peculiar construction organizational practices have poor innovation culture, poor learning ability, and low technology adoption capabilities. This makes a firm less efficient in making decisions in digital competitiveness (Maskuriy et al., 2019a,b). Along with the construction sector's rigid organizational behavior, markets keep on updating adversely with increased number of new technologies and knowledge management. In the data world without investing on new technology taking decisions makes firms less competitive, in this way difficulty for construction industry for catching benefits of C 4.0. Digital decision-making is even problematic in organizations, considering no value for technological adoption and facing practical issues in digitalization (Bazan and Estevez, 2022; Maskuriy et al., 2019b; Osunsanmi et al., 2020).

6. Conclusion

This book chapter presented current digital scenarios taking place in the construction sector, using I 4.0 technologies such as IoT, AI, digital twins, and BIM into CBI. After reviewing important academic literature related to sustainability with I 4.0 and C 4.0, elements of I

4.0, and concepts of C 4.0, it provided predominant challenges facing existing managerial and operational practices to take transformation for digital construction activities, given a detailed set of implications imposed by digital environment on competitiveness, workforce skills, and construction firms' organizational practices. It is evident that, from C 4.0 technological infusion into construction firms, it can improve quality of built environment, quality of delivery, and help attain better position in the market. Technology can transform firms from existing set of product providings that are isolated consideration of buildings as a product into considering buildings as a service to gain a bigger picture for construction ecosystem sustainability in the long run. Along with providing benefits, C 4.0 is cratering to many challenges on organization and work force, making trust issues in digital partnering, fear of facing ethical issues, investments for providing training to existing workforce, investments for capacity building, and required to balance sociotechnical changes taking place with advent of new technology and practice into existing sociotechnical equilibrium in the firm. Adopting C 4.0 into construction organizations is possible by emphasizing research and learning practices (Nagy et al., 2021). Many studies focused on the implications concerned with managerial and operational considerations. A few studies concentrated on the implications followed by organizational and stakeholders consideration (Aghimien et al., 2022; Bonanomi et al., 2020; Nagy et al., 2021).

Future research directions include deploying cyber-physical technology to improve training methods for reducing musculoskeletal problems for construction workers. Studies are needed on digital productivity assessment tools, no efficient mechanisms are available to assess work productivity in digital ecosystem. Cost–benefit analysis is needed; no studies have been found on providing cost–benefit analysis to improve confidence for investing in technology and capacity management.

References

AbouRizk, S., 2010. Role of simulation in construction engineering and management. Journal of Construction Engineering and Management 136 (10). https://doi.org/10.1061/(asce)co.1943-7862.0000220.

Adepoju, O.O., Aigbavboa, C.O., 2021. Assessing knowledge and skills gap for construction 4.0 in a developing economy. Journal of Public Affairs 21 (3), e2264. https://doi.org/10.1002/PA.2264.

Aghimien, D., Aigbavboa, C., Oke, A., Thwala, W., Moripe, P., 2022. Digitalisation of construction organisations—a case for digital partnering. International Journal of Construction Management 22 (10). https://doi.org/10.1080/15623599.2020.1745134.

Ahmed, S., 2019. A review on using opportunities of augmented reality and virtual reality in construction project management. Organization, Technology and Management in Construction. https://doi.org/10.2478/otmcj-2018-0012.

Akanmu, A.A., Anumba, C.J., Ogunseiju, O.O., 2021. Towards next generation cyber-physical systems and digital twins for construction. Journal of Information Technology in Construction 26. https://doi.org/10.36680/j.itcon.2021.027.

Akanmu, A.A., Olayiwola, J., Ogunseiju, O., McFeeters, D., 2020. Cyber-physical postural training system for construction workers. Automation in Construction 117. https://doi.org/10.1016/j.autcon.2020.103272.

Akbarieh, A., Jayasinghe, L.B., Waldmann, D., Teferle, F.N., 2020. BIM-based end-of-lifecycle decision making and digital deconstruction: literature review. Sustainability. https://doi.org/10.3390/su12072670.

Akinade, O.O., Oyedele, L.O., Ajayi, S.O., Bilal, M., Alaka, H.A., Owolabi, H.A., Bello, S.A., et al., 2017. Design for Deconstruction (DfD): critical success factors for diverting end-of-life waste from landfills. Waste Management 60. https://doi.org/10.1016/j.wasman.2016.08.017.

Alade, K., Windapo, A., 2020a. 4IR leadership effectiveness and practical implications for construction business organisations. The Construction Industry in the Fourth Industrial Revolution. https://doi.org/10.1007/978-3-030-26528-1_7.

Alade, K., Windapo, A.O., 2020b. Developing effective 4IR leadership framework for construction organisations. Engineering Construction and Architectural Management 28 (5). https://doi.org/10.1108/ECAM-07-2020-0576.

Badri, A., Boudreau-Trudel, B., Souissi, A.S., 2018. Occupational health and safety in the industry 4.0 era: a cause for major concern? Safety Science. https://doi.org/10.1016/j.ssci.2018.06.012.

Balasubramanian, S., Shukla, V., Islam, N., Manghat, S., 2021. Construction industry 4.0 and sustainability: an enabling framework. IEEE Transactions on Engineering Management. https://doi.org/10.1109/TEM.2021.3110427.

Barata, J., da Cunha, P.R., 2019. The viable smart product model: designing products that undergo disruptive transformations. Cybernetics & Systems 50 (7), 629–655. https://doi.org/10.1080/01969722.2019.1646021.

Barbosa, F., Woetzel, J., Mischke, J., Ribeirinho, M.J., Sridhar, M., Parsons, M., Bertram, N., et al., 2017. Reinventing Construction through a Productivity Revolution | McKinsey. McKinsey Global Institute. https://www.mckinsey.com/capabilities/operations/our-insights/reinventing-construction-through-a-productivity-revolution. (Accessed 1 July 2023).

Bazan, P., Estevez, E., 2022. Industry 4.0 and business process management: state of the art and new challenges. Business Process Management Journal 28 (1). https://doi.org/10.1108/BPMJ-04-2020-0163.

Begić, H., Galić, M., 2021. A systematic review of construction 4.0 in the context of the BIM 4.0 premise. Buildings. https://doi.org/10.3390/BUILDINGS11080337.

Beltrami, M., Orzes, G., Sarkis, J., Sartor, M., 2021. Industry 4.0 and sustainability: towards conceptualisation and theory. Journal of Cleaner Production 312, 127733. https://doi.org/10.1016/J.JCLEPRO.2021.127733.

Bonanomi, M.M., Hall, D.M., Staub-French, S., Tucker, A., Talamo, C.M.L., 2020. The impact of digital transformation on formal and informal organisational structures of large architecture and engineering firms. Engineering Construction and Architectural Management 27 (4), 872–892. https://doi.org/10.1108/ECAM-03-2019-0119/FULL/PDF.

Bryson, J., Winfield, A., 2017. Standardising ethical design for artificial intelligence and autonomous systems. Computer 50 (5). https://doi.org/10.1109/MC.2017.154.

Buchli, J., Giftthaler, M., Kumar, N., Lussi, M., Sandy, T., Dörfler, K., Hack, N., 2018. Digital in situ fabrication - challenges and opportunities for robotic in situ fabrication in architecture, construction, and beyond. Cement and Concrete Research. https://doi.org/10.1016/j.cemconres.2018.05.013.

Buhr, D., 2017. Social Innovation Policy for Industry 4.0. Friedrich Ebert Stiftung.

Chen, Y., Huang, D., Liu, Z., Osmani, M., Demian, P., 2022. Construction 4.0, industry 4.0, and building information modeling (BIM) for sustainable building development within the smart city. Sustainability 14 (16). https://doi.org/10.3390/su141610028.

Cimini, C., Boffelli, A., Lagorio, A., Kalchschmidt, M., Pinto, R., 2021. How do industry 4.0 technologies influence organisational change? An empirical analysis of Italian SMEs. Journal of Manufacturing Technology Management 32 (3). https://doi.org/10.1108/JMTM-04-2019-0135.

Costin, A.M., Teizer, J., 2015. Fusing passive RFID and BIM for increased accuracy in indoor localisation. Visualization in Engineering 3 (1). https://doi.org/10.1186/s40327-015-0030-6.

Craveiro, F., Duarte, J.P., Bartolo, H., Bartolo, P.J., 2019. Additive manufacturing as an enabling technology for digital construction: a perspective on Construction 4.0. Automation in Construction. https://doi.org/10.1016/j.autcon.2019.03.011.

Dallasega, P., Rauch, E., Linder, C., 2018. Industry 4.0 as an enabler of proximity for construction supply chains: a systematic literature review. Computers in Industry. https://doi.org/10.1016/j.compind.2018.03.039.

Danel, T., Lafhaj, Z., Puppala, A., Lienard, S., Richard, P., 2021. Proposal for tower crane productivity indicators based on data analysis in the era of construction 4.0. Buildings 11 (1), 21. https://doi.org/10.3390/BUILDINGS11010021.

Darko, A., Chan, A.P.C., 2016. Critical analysis of green building research trend in construction journals. Habitat International. https://doi.org/10.1016/j.habitatint.2016.07.001.

Darko, A., Chan, A.P.C., Adabre, M.A., Edwards, D.J., Hosseini, M.R., Ameyaw, E.E., 2020. Artificial intelligence in the AEC industry: scientometric analysis and visualisation of research activities. Automation in Construction. https://doi.org/10.1016/j.autcon.2020.103081.

Fathalizadeh, A., Hosseini, M.R., Vaezzadeh, S.S., Edwards, D.J., Martek, I., Shooshtarian, S., 2022. Barriers to sustainable construction project management: the case of Iran. Smart and Sustainable Built Environment 11 (3). https://doi.org/10.1108/SASBE-09-2020-0132.

García de Soto, B., Agustí-Juan, I., Joss, S., Hunhevicz, J., 2019. Implications of construction 4.0 to the workforce and organisational structures. International Journal of Construction Management 22 (2), 205–217. https://doi.org/10.1080/15623599.2019.1616414.

Garrick, J., Chan, A., 2017. Knowledge management and professional experience: the uneasy dynamics between tacit knowledge and performativity in organizations. Journal of Knowledge Management 21 (21).

Gerbert, P., Castagnino, S., Rothballer, C., Renz, A., Filitz, R., 2016. The Transformative Power of Building Information Modeling. BCG. https://www.bcg.com/publications/2016/engineered-products-infrastructure-digital-transformative-power-building-information-modeling. (Accessed 27 July 2023).

Gerrish, T., Ruikar, K., Cook, M., Johnson, M., Phillip, M., Lowry, C., 2017. BIM application to building energy performance visualisation and management: challenges and potential. Energy and Buildings 144, 218–228. https://doi.org/10.1016/J.ENBUILD.2017.03.032.

Ghaffar, S.H., Burman, M., Braimah, N., 2020. Pathways to circular construction: an integrated management of construction and demolition waste for resource recovery. Journal of Cleaner Production 244. https://doi.org/10.1016/j.jclepro.2019.118710.

Grilo, A., Jardim-Goncalves, R., 2011. Challenging electronic procurement in the AEC sector: a BIM-based integrated perspective. Automation in Construction 20 (2). https://doi.org/10.1016/j.autcon.2010.09.008.

Grubic, T., Jennions, I., 2018. Remote monitoring technology and servitised strategies—factors characterising the organisational application. International Journal of Production Research 56 (6). https://doi.org/10.1080/00207543.2017.1332791.

Gubbi, J., Buyya, R., Marusic, S., Palaniswami, M., 2013. Internet of Things (IoT): a vision, architectural elements, and future directions. Future Generation Computer Systems 29 (7). https://doi.org/10.1016/j.future.2013.01.010.

Hermann, M., Pentek, T., Otto, B., 2016. Design principles for industrie 4.0 scenarios. In: Proceedings of the Annual Hawaii International Conference on System Sciences. https://doi.org/10.1109/HICSS.2016.488.

Hilty, L.M., Aebischer, B., 2015. Ict for sustainability: an emerging research field. Advances in Intelligent Systems and Computing 310. https://doi.org/10.1007/978-3-319-09228-7_1.

Horváth, D., Szabó, R.Z., 2019. Driving forces and barriers of Industry 4.0: do multinational and small and medium-sized companies have equal opportunities? Technological Forecasting and Social Change 146, 119–132. https://doi.org/10.1016/J.TECHFORE.2019.05.021.

Hossain, M.A., Nadeem, A., 2019. Towards digitising the construction industry: state of the art of construction 4.0. In: ISEC 2019 - 10th International Structural Engineering and Construction Conference. https://doi.org/10.14455/isec.res.2019.184.

Hu, W., Lim, K.Y.H., Cai, Y., 2022. Digital twin and industry 4.0 enablers in building and construction: a survey. Buildings. https://doi.org/10.3390/buildings12112004.

Jacobsson, M., Linderoth, H.C.J., Rowlinson, S., 2017. The role of industry: an analytical framework to understand ICT transformation within the AEC industry. Construction Management & Economics 35 (10). https://doi.org/10.1080/01446193.2017.1315148.

Kagioglou, M., Cooper, R., Aouad, G., Sexton, M., 2000. Rethinking construction: the generic design and construction process protocol. Engineering Construction and Architectural Management. https://doi.org/10.1108/eb021139.

Kaklauskas, A., Zavadskas, E.K., Binkyte-Veliene, A., Kuzminske, A., Cerkauskas, J., Cerkauskiene, A., Valaitiene, R., 2020. Multiple criteria evaluation of the EU country sustainable construction industry lifecycles. Applied Sciences 10 (11). https://doi.org/10.3390/app10113733.

Karmakar, A., Delhi, V.S.K., 2021. Construction 4.0: what we know and where we are headed? Journal of Information Technology in Construction. https://doi.org/10.36680/j.itcon.2021.028.

Kineber, A.F., Othman, I., Oke, A.E., Chileshe, N., Buniya, M.K., 2020. Identifying and assessing sustainable value management implementation activities in developing countries: the case of Egypt. Sustainability 12 (21). https://doi.org/10.3390/su12219143.

Klinc, R., Turk, Ž., 2019. Construction 4.0 – digital transformation of one of the oldest industries. Economic and Business Review 21 (3). https://doi.org/10.15458/ebr.92.

Kozlovska, M., Klosova, D., Strukova, Z., 2021. Impact of industry 4.0 platform on the formation of construction 4.0 concept: a literature review. Sustainability. https://doi.org/10.3390/su13052683.

Lavikka, R., Kallio, J., Casey, T., Airaksinen, M., 2018. Digital disruption of the AEC industry: technology-oriented scenarios for possible future development paths. Construction Management & Economics 36 (11), 635–650. https://doi.org/10.1080/01446193.2018.1476729.

Lavikka, R.H., Smeds, R., Jaatinen, M., 2015. Coordinating collaboration in contractually different complex construction projects. Supply Chain Management 20 (2). https://doi.org/10.1108/SCM-10-2014-0331.

Lekan, A., Clinton, A., Fayomi, O.S.I., James, O., 2020. Lean thinking and industrial 4.0 approach to achieving construction 4.0 for industrialisation and technological development. Buildings 10 (12). https://doi.org/10.3390/buildings10120221.

Leviäkangas, P., Mok Paik, S., Moon, S., 2017. Keeping up with the pace of digitisation: the case of the Australian construction industry. Technology in Society 50. https://doi.org/10.1016/j.techsoc.2017.04.003.

Li, J., Greenwood, D., Kassem, M., 2019. Blockchain in the built environment and construction industry: a systematic review, conceptual models and practical use cases. Automation in Construction 102. https://doi.org/10.1016/j.autcon.2019.02.005.

Li, J., Yang, H., 2017. A research on development of construction industrialisation based on BIM technology under the background of industry 4.0. MATEC Web of Conferences 100. https://doi.org/10.1051/matecconf/201710002046.

Liao, Y., Deschamps, F., Loures, E. de FR., Ramos, L.F.P., 2017. Past, present and future of Industry 4.0 - a systematic literature review and research agenda proposal. International Journal of Production Research. https://doi.org/10.1080/00207543.2017.1308576.

Lingard, H., Peihua Zhang, R., Räisänen, C., Miang Goh, Y., Bowen, P., Bhandari, S., 2021. Special issue: what have we learnt from the COVID-19 global pandemic: improving the construction industry's abilities to foresee, respond to and recover from future endemic catastrophes. Construction Management & Economics 39 (2). https://doi.org/10.1080/01446193.2020.1869480.

Lipsmeier, A., Kühn, A., Joppen, R., Dumitrescu, R., 2020. Process for the development of a digital strategy. Procedia CIRP 88. https://doi.org/10.1016/j.procir.2020.05.031.

Liu, S., 2015. Sustainable building design optimisation using building information modeling. In: ICCREM 2015 - Environment and the Sustainable Building - Proceedings of the 2015 International Conference on Construction and Real Estate Management. https://doi.org/10.1061/9780784479377.038.

Liu, Y., van Nederveen, S., Hertogh, M., 2017. Understanding effects of BIM on collaborative design and constructionAn empirical study in China. International Journal of Project Management 35 (4). https://doi.org/10.1016/j.ijproman.2016.06.007.

López Martínez, P., Dintén, R., Drake, J.M., Zorrilla, M., 2021. A big data-centric architecture metamodel for Industry 4.0. Future Generation Computer Systems 125. https://doi.org/10.1016/j.future.2021.06.020.

Louis, J., Dunston, P.S., 2018. Integrating IoT into operational workflows for real-time and automated decision-making in repetitive construction operations. Automation in Construction 94. https://doi.org/10.1016/j.autcon.2018.07.005.

Mahpour, A., 2018. Prioritising barriers to adopt circular economy in construction and demolition waste management. Resources, Conservation and Recycling 134 (November 2017), 216–227. https://doi.org/10.1016/j.resconrec.2018.01.026.

Majrouhi Sardroud, J., 2012. Influence of RFID technology on automated management of construction materials and components. Scientia Iranica 19 (3). https://doi.org/10.1016/j.scient.2012.02.023.

Manyika, J., Chui, M., Bisson, P., Woetzel, J., Dobbs, R., Bughin, J., Aharon, D., 2015. The Internet of Things: Mapping the Value beyond the Hype. McKinsey Global Institute.

Martynov, V.V., Shavaleeva, D.N., Zaytseva, A.A., 2019. Information technology as the basis for transformation into a digital society and industry 5.0. In: Proceedings of the 2019 IEEE International Conference Quality Management, Transport and Information Security, Information Technologies IT and QM and IS 2019. https://doi.org/10.1109/ITQMIS.2019.8928305.

Maskuriy, R., Selamat, A., Ali, K.N., Maresova, P., Krejcar, O., 2019a. Industry 4.0 for the construction industry-How ready is the industry? Applied Sciences 9 (14). https://doi.org/10.3390/app9142819.

Maskuriy, R., Selamat, A., Maresova, P., Krejcar, O., David, O.O., 2019b. Industry 4.0 for the Construction Industry: Review of Management Perspective. Economies. https://doi.org/10.3390/economies7030068.

Maureira, C., Pinto, H., Yepes, V., Garcia, J., 2021. Towards an AEC-AI industry optimisation algorithmic knowledge mapping: an adaptive methodology for macroscopic conceptual analysis. IEEE Access 9. https://doi.org/10.1109/ACCESS.2021.3102215.

McKinsey and Company, 2017. McKinsey-Studie: Produktivität der Baubranche in Deutschland stagniert. Pressemitteilung.

Miorandi, D., Sicari, S., De Pellegrini, F., Chlamtac, I., 2012. Internet of things: vision, applications and research challenges. Ad Hoc Networks. https://doi.org/10.1016/j.adhoc.2012.02.016.

Müller, J.M., Kiel, D., Voigt, K.I., 2018. What drives the implementation of Industry 4.0? The role of opportunities and challenges in the context of sustainability. Sustainability 10 (1). https://doi.org/10.3390/su10010247.

Muñoz-La Rivera, F., Mora-Serrano, J., Valero, I., Oñate, E., 2020. Methodological-technological framework for construction 4.0. Archives of Computational Methods in Engineering 28 (2), 689–711. https://doi.org/10.1007/S11831-020-09455-9.

Nagy, O., Papp, I., Szabó, R.Z., 2021. Construction 4.0 organisational level challenges and solutions. Sustainability 13 (21). https://doi.org/10.3390/su132112321.

Nasir, M.H.A., Genovese, A., Acquaye, A.A., Koh, S.C.L., Yamoah, F., 2017. Comparing linear and circular supply chains: a case study from the construction industry. International Journal of Production Economics 183. https://doi.org/10.1016/j.ijpe.2016.06.008.

Norouzi, M., Chàfer, M., Cabeza, L.F., Jiménez, L., Boer, D., 2021. Circular economy in the building and construction sector: a scientific evolution analysis. Journal of Building Engineering 44, 102704. https://doi.org/10.1016/J.JOBE.2021.102704.

Oluleye, B.I., Chan, D.W.M., Antwi-Afari, P., Olawumi, T.O., 2023. Modeling the principal success factors for attaining systemic circularity in the building construction industry: an international survey of circular economy experts. Sustainable Production and Consumption 37, 268–283. https://doi.org/10.1016/J.SPC.2023.03.008.

Osunsanmi, T.O., Aigbavboa, C.O., Emmanuel Oke, A., Liphadzi, M., 2020. Appraisal of stakeholders' willingness to adopt construction 4.0 technologies for construction projects. Built Environment Project and Asset Management 10 (4). https://doi.org/10.1108/BEPAM-12-2018-0159.

Plekhanov, D., Netland, T., 2019. Digitalisation stages in firms: towards a framework. In: 26th EurOMA Conference, No. June.

Predescu, A., Mocanu, M., Lupu, C., 2019. ARMAX: a mobile geospatial augmented reality platform for serious gaming. In: Proceedings - 2019 IEEE 15th International Conference on Intelligent Computer Communication and Processing, ICCP 2019. https://doi.org/10.1109/ICCP48234.2019.8959671.

Purvis, B., Mao, Y., Robinson, D., 2019. Three pillars of sustainability: in search of conceptual origins. Sustainability Science 14 (3). https://doi.org/10.1007/s11625-018-0627-5.

Radl, J., Kaiser, J., 2019. Benefits of implementation of common data environment (CDE) into construction projects. IOP Conference Series: Materials Science and Engineering 471. https://doi.org/10.1088/1757-899X/471/2/022021.

Rahman, M.A., Rashid, M.M., Shamim Hossain, M., Hassanain, E., Alhamid, M.F., Guizani, M., 2019. Blockchain and IoT-based cognitive edge framework for sharing economy services in a smart city. IEEE Access 7. https://doi.org/10.1109/ACCESS.2019.2896065.

Rejeb, A., Keogh, J.G., Treiblmaier, H., 2019. Leveraging the internet of things and blockchain technology in supply chain management. Future Internet 11 (7). https://doi.org/10.3390/fi11070161.

Roblek, V., Meško, M., Krapež, A., 2016. A complex view of industry 4.0. Sage Open 6 (2). https://doi.org/10.1177/2158244016653987.

Safiullin, A., Krasnyuk, L., Kapelyuk, Z., 2019. Integration of Industry 4.0 technologies for 'smart cities' development. IOP Conference Series: Materials Science and Engineering 497. https://doi.org/10.1088/1757-899X/497/1/012089.

Saini, G.S., Pournazari, P., Ashok, P., van Oort, E., 2022. Intelligent action planning for well construction operations demonstrated for hole cleaning optimization and automation. Energies 15 (15). https://doi.org/10.3390/en15155749.

Schönbeck, P., Löfsjögård, M., Ansell, A., 2020. Quantitative review of construction 4.0 technology presence in construction project research. Buildings. https://doi.org/10.3390/buildings10100173.

Schuh, G., Potente, T., Wesch-Potente, C., Weber, A.R., Prote, J.P., 2014. Collaboration mechanisms to increase productivity in the context of industrie 4.0. Procedia CIRP 19. https://doi.org/10.1016/j.procir.2014.05.016.

Sepasgozar, S.M.E., 2020. Digital twin and web-based virtual gaming technologies for online education: a case of construction management and engineering. Applied Sciences 10 (13). https://doi.org/10.3390/app10134678.

Shirowzhan, S., Sepasgozar, S.M.E., Edwards, D.J., Li, H., Wang, C., 2020. BIM compatibility and its differentiation with interoperability challenges as an innovation factor. Automation in Construction. https://doi.org/10.1016/j.autcon.2020.103086.

Szabo, R.Z., Herceg, I.V., Hanák, R., Hortovanyi, L., Romanová, A., Mocan, M., Djuričin, D., 2020. Industry 4.0 implementation in b2b companies: cross-country empirical evidence on digital transformation in the cee region. Sustainability 12 (22). https://doi.org/10.3390/su12229538.

Turner, C.J., Oyekan, J., Stergioulas, L., Griffin, D., 2021. Utilising industry 4.0 on the construction site: challenges and opportunities. IEEE Transactions on Industrial Informatics 17 (2). https://doi.org/10.1109/TII.2020.3002197.

Woodhead, R., Stephenson, P., Morrey, D., 2018. Digital construction: from point solutions to IoT ecosystem. Automation in Construction 93. https://doi.org/10.1016/j.autcon.2018.05.004.

Yevu, S.K., Yu, A.T.W., Darko, A., 2021. Digitalisation of construction supply chain and procurement in the built environment: emerging technologies and opportunities for sustainable processes. Journal of Cleaner Production. https://doi.org/10.1016/j.jclepro.2021.129093.

Zabidin, N.S., Belayutham, S., Ibrahim, C.K.I.C., 2020. A bibliometric and scientometric mapping of Industry 4.0 in construction. Journal of Information Technology in Construction 25. https://doi.org/10.36680/j.itcon.2020.017.

Zezulka, F., Marcon, P., Vesely, I., Sajdl, O., 2016. Industry 4.0 – an introduction in the phenomenon. IFAC-PapersOnLine 49. https://doi.org/10.1016/j.ifacol.2016.12.002.

PART 3

Sustainability and technology

CHAPTER

12

A cyber-physical system for improving the sustainability of freight logistics industry: A case of developing nation

Umabharati Rawat[1] *and Aalok Kumar*[2]

[1]Indian Institute of Technology Roorkee, Roorkee, Uttarakhand, India; [2]Indian Institute of Management, Visakhapatnam, Andhra Pradesh, India

1. Introduction

The logistics industry, known for its reliance on manual labor, is moving toward digital transformation (Sun et al., 2022). However, the complex and dynamic logistics market requirements create various challenges in the transformation: for instance, dynamic customer demands, sustainability concerns, technology investments, security considerations, workforce requirements, and information sharing with business partners (Manavalan and Jayakrishna, 2019). However, to be competitive and sustainable in the logistics market, it is crucial to address various challenges such as high costs, low efficiency, limited transparency, and inadequate coordination among stakeholders (Facchini et al., 2020). Specifically, communication failures among users may lead to data retention risk for transaction history and contribute to significant energy loss in the logistics industry (Esmaeilian et al., 2020). Moreover, the industry must contend with rising customer demands, global competition, external influences, errors, and costs (Issaoui et al., 2021). Additionally, the escalating environmental issues, rising fuel prices, and mounting congestion on road networks necessitate innovative solutions for freight transport operations (Harris et al., 2015). Achieving a harmonious balance among economic growth, social progress, and environmental protection is of utmost importance (Roblek et al., 2020). This balance guarantees the fulfillment of the present generation's needs while safeguarding the capacity of future generations to fulfill their own needs.

Evolution and Trends of Sustainable Approaches
https://doi.org/10.1016/B978-0-443-21651-0.00006-1

Logistics activities are substantial energy consumers and recognized sources of greenhouse gas (GHG) emissions; the optimization of energy demand in logistics becomes crucially important (Munsamy et al., 2020). An enterprise's logistics system's effectiveness and efficiency heavily impact its cost, customer satisfaction, and profitability (Sun et al., 2022). To address these challenges, it is essential to adopt Industry 4.0 technologies and formulate a strategic plan for developing an Industry 4.0 vision (Facchini et al., 2020). Industry 4.0 presents an opportunity to overcome these challenges in the logistics industry by leveraging cost-effective IoT platforms with data collection and analysis capabilities (Esmaeilian et al., 2020). For example, IoT enables real-time tracking of goods, waste reduction, optimized inventory management, and overall efficiency improvement. By investing in IoT, freight logistics companies can shift their focus toward sustainability and improve their logistics practices (Manavalan and Jayakrishna, 2019). Big Data analytics plays a pivotal role in identifying areas for improved sustainability, including waste reduction, enhanced energy efficiency, and optimized transportation routes. It provides valuable insights and supports decision-making to achieve sustainability goals in freight logistics (Chalmeta and Santos-deLeón, 2020). To mitigate risks associated with economic, social, and technical aspects, Logistics 4.0 can leverage blockchain technology, which ensures a secure and transparent method of recording and sharing data (Kodym et al., 2020). Thus, logistics companies can enhance their sustainability performance and minimize their environmental impact by embracing Industry 4.0 (Esmaeilian et al., 2020). Employing Industry 4.0 strategies in freight logistics such as IoT, big data analysis, cloud computing, cyber-physical systems, autonomous vehicles, and logistics clusters can sustain a competitive edge over the long term (Gružauskas et al., 2018).

The logistics industry recognizes the essentiality of sustainability, driven by diverse customer demands, stringent environmental regulations, and intense global competition (Sun et al., 2022). Given the significant role played by the industry in contributing to greenhouse gas emissions, air pollution, and noise pollution, sustainability has become increasingly critical in freight logistics (Luthra and Mangla, 2018). Moreover, sustainability holds paramount importance within the context of Industry 4.0, as it ensures that newly established business models and structures are environmentally and socially responsible and capable of long-term sustenance (Prause, 2016). Consequently, there is a growing demand for sustainable logistics practices to mitigate the environmental impact of logistics activities. Embracing these practices offers potential cost savings and enhanced efficiency and customer satisfaction (Luthra and Mangla, 2018).

Industry 4.0 technologies present promising opportunities to improve logistics' economic effectiveness, environmental sustainability, and societal influence (Sun et al., 2022). By optimizing production processes and organizational structures, these technologies enable companies to reduce energy consumption, minimize waste, and enhance overall efficiency. This optimization facilitates better decision-making, reducing environmental impact (Facchini et al., 2020). Integrating Industry 4.0 technologies with sustainability initiatives empowers companies to achieve their sustainability objectives effectively and significantly contribute to a more sustainable future (Facchini et al., 2020).

The logistics industry is currently experiencing a transition toward intelligent systems, where cyber-physical systems (CPSs) have emerged as the leading technology within the framework of Industry 4.0. This adoption addresses the considerable challenges related to intelligence and coordination between logistics operation systems and information systems

(Lee et al., 2018). By leveraging ambient intelligence and real-time information sharing, CPS enables effective decision-making and adaptation to evolving customer demands (Lee et al., 2018). The advancement of smart logistics is closely tied to the integration of CPS, which combines deep computing, control, and communication technologies (Zhang, 2018). Smart logistics offers notable benefits to the transportation sector, such as optimizing road usage, reducing transportation time, cutting costs, and enhancing inventory management (Issaoui et al., 2021). Smart Logistics-based cyber-physical systems (SLCPSs) are vital in ensuring steady IoT device coverage and connectivity (Abbas and Marwat, 2020).

The focus on sustainable development has intensified due to the escalating concerns surrounding environmental and social challenges. Many companies have recognized the importance of incorporating sustainability into their logistics operations to improve their social reputation and gain a competitive advantage (Luthra and Mangla, 2018). The advent of Industry 4.0 and the increasing urgency to implement the 17 Sustainable Development Goals (SDGs) outlined in Agenda 2030 have prompted a critical examination of technological advancements within the context of the fourth industrial revolution (Roblek et al., 2020). Additionally, the complex and dynamic logistics market requirements and the vast nature of the logistics network pose various challenges. There is a pressing need for freight logistics to evolve and transition from traditional approaches to more sustainable solutions (Gružauskas et al., 2018). CPS has emerged as a promising solution to tackle these challenges, particularly in the transportation sector. CPS finds various applications in the logistics and transportation industry, including enhancing operational efficiency, reducing costs, and promoting sustainability.

This analysis aims to explore the impact of CPS on the evolution of sustainable development policies by addressing the following research questions:

RQ1: What are the capabilities of CPS in advancing freight logistics operations and processes?

RQ2: How can assessing CPS potential significantly enhance sustainability in freight logistics operations in an uncertain decision-making environment?

To address these research questions, the following objectives were formed.

RO1: To identify the CPS potential that can significantly enhance sustainable freight logistics operations.

RO2: To evaluate the CPS potentials through the fuzzy best worst method to investigate their influence on sustainability within freight logistics.

This study seeks to provide valuable insights by exploring the CPS potential and their impact on sustainable freight logistics operations.

2. Literature review

In this section, we delve into how CPS impacts the sustainability of the freight logistics industry.

Sun et al. (2022) noted the absence of comprehensive analyses that connect sustainable logistics practices with specific Industry 4.0 technologies. Among the technologies within Industry 4.0, CPS is the predominant one. Considering this, we have examined the key advantages of CPS in achieving sustainable freight logistics, as outlined in Table 12.1.

TABLE 12.1 Sustainability impact of CPS potentials on freight logistics.

S. No.	CPS potentials	Sustainability impact	References
1	Real-time data analytics (Cp_1)	Real-time data analytics provide numerous advantages in optimizing route planning, improving inventory management, and enabling proactive maintenance. These advantages encompass reduced energy and fuel consumption, minimized emissions, efficient resource utilization, decreased product waste, mitigated stockouts, and early detection of equipment failure or malfunction.	(Luthra and Mangla, 2018; Sun et al., 2022; Winkelhaus and Grosse, 2020; Yavas and Ozkan-Ozen, 2020)
2	Autonomous vehicles OR Autonomous logistics systems (Cp_2)	Autonomous logistics systems help in resource utilization, improving energy efficiency, and enhancing security and working efficiency in logistics operations. It can enable end-to-end visibility and transparency in logistics processes. CPS has revolutionized goods delivery methods, emphasizing intelligent and autonomous vehicles such as self-driving trucks and lorries. These innovations promise to reduce costs, lower accident rates, and decrease CO_2 emissions. Additionally, automation minimizes detours and idling time, reducing fuel consumption and transportation-related greenhouse gas (GHG) emissions. Furthermore, autonomous logistics systems digitize paperwork, contributing to paperless operations.	(Albrecht et al., 2023; Gružauskas et al., 2018; Sun et al., 2022; Winkelhaus and Grosse, 2020; Yavas and Ozkan-Ozen, 2020)
3	Predictive maintenance (Cp_3)	Predictive maintenance (PM) minimizes maintenance costs and downtime while adopting a zero-failure manufacturing approach. PM is achieved by continuously monitoring the equipment's operational state and forecasting potential failures. Predictive maintenance is crucial in logistics operations for identifying unexpected failures and improving vehicle reliability and the lifespan of logistics assets. It can reduce the environmental impact of emergency repairs, transportation delays, and excessive energy consumption. This CPS potential reduces the ecological impact of resource-intensive repairs, replacements, and associated activities.	(Abbas and Marwat, 2020; Li et al., 2017; Winkelhaus and Grosse, 2020)
4	Smart traffic management OR Efficient route planning (Cp_4)	CPS provides operational scheduling support for route optimization, such as resource allocation and utilization. For instance, real-time data processing can be useful in analyzing traffic conditions, vehicle information, better transportation planning, and timely decision-making to minimize transport delays. It is crucial to achieve sustainability in freight logistics operations through their ability to optimize route planning, reduce traffic congestion, and provide real-time traffic updates. This potential reduces fuel consumption, lowers emissions, improves operational efficiency, and creates a greener and more sustainable freight transportation ecosystem.	(Albrecht et al., 2023; Sun et al., 2022)

TABLE 12.1 Sustainability impact of CPS potentials on freight logistics.—cont'd

S. No.	CPS potentials	Sustainability impact	References
5	Smart material handling OR Intelligent cargo management (Cp_5)	CPS can automate the manual tasks and assist with the manual tasks. In logistics, cargo management or material handling can be automated, which is crucial in promoting sustainability in freight logistics. It includes smart loading/unloading systems and associated activities such as cargo reception, unloading, inspection, and storage preparation. Intelligent material handling facilitates efficient resource utilization, improves energy efficiency, reduces waste, optimizes inventory management, ensures safety, and enables data-driven optimization. Smart material handling optimizes various operational aspects and creates a more sustainable and efficient freight logistics ecosystem.	(Albrecht et al., 2023; Munsamy et al., 2020; Sun et al., 2022)
6	Advanced collaborative platforms for collaboration and coordination (Cp_6)	CPS has the potential to develop advanced collaborative platforms to enable seamless collaboration and information sharing among stakeholders in the freight logistics ecosystem, including shippers, carriers, suppliers, and regulatory bodies. It can promote collaborative decision-making, improve coordination, and reduce inefficiencies like empty truck miles using real-time data updates. It can support multimodal freight transport operations and minimizes the overall environmental impact.	(Abbas and Marwat, 2020; Albrecht et al., 2023; Strandhagen et al., 2017; Zhang, 2018)
7	Advanced safety and security (Cp_7)	CPS can improve logistics operations' security and safety, which may reduce the risk of accidents. This potential can be helpful in enhancing worker safety, protecting assets, complying with regulations, mitigating risks, and creating trust among stakeholders. Additionally, CPS improves safety measures, enables proactive risk mitigation, and enhances freight logistics' overall safety and sustainability.	(Luthra and Mangla, 2018; Sun et al., 2022)
8	Intelligent cargo tracking and monitoring (Cp_8)	Intelligent cargo tracking and monitoring is the significant potential of CPS. It can enhance end-to-end visibility in freight logistics, which may be useful in smart decision-making. In pursuing sustainability, these decisions enable efficient resource utilization, optimize freight logistics operations, prevent loss and damage, facilitate optimal route planning and delivery, and enhance the customer experience. In the end, waste can be minimized, and operation efficiency increased, which positively impacts the environment.	(Albrecht et al., 2023; Manavalan and Jayakrishna, 2019; Munsamy et al., 2020; Sun et al., 2022; Winkelhaus and Grosse, 2020)
9	System integration (Cp_9)	System integration is the critical potential of CPS. System integration can control the overall logistics value chain. It can integrate various logistics activities using sensors, actuators, and logistics systems at the different hierarchy levels and impact sustainability. Integration facilitates efficient data exchange at different levels and can be useful in optimizing route planning, load consolidation, collaboration, resource utilization, and performance monitoring. Logistics organizations with system integration facilities enhance operational efficiency, reduce waste, and minimize environmental impact.	(Beier et al., 2020)

Albrecht et al. (2023) identified various archetypes of digital technologies, such as platform utilization, connectivity, sensor-based data collection, and analytical insight generation. Sun et al. (2022) emphasized the significance of industry 4.0 enabled potentials for sustainable transportation. Real-time data analytics, autonomous vehicles, smart traffic management, intelligent cargo management, advanced safety and security, and intelligent cargo tracking and monitoring were identified as crucial factors in achieving sustainable logistics operations. Yavas and Ozkan-Ozen (2020) proposed a framework for logistics centers in the context of Industry 4.0. They highlighted the importance of real-time data analytics capabilities and autonomous logistics systems in optimizing logistics processes and improving overall efficiency. Winkelhaus and Grosse (2020) emphasized the distinguishing characteristics of Logistics 4.0, which include automated systems, predictive maintenance, enhanced visibility, and advanced cargo tracking and monitoring capabilities.

Abbas and Marwat (2020) underscored the significance of predictive maintenance and advanced collaboration as CPS potentials that enable intelligent and efficient logistics operations. Munsamy et al. (2020) highlighted the potential of Logistics 4.0 in reducing greenhouse gas (GHG) emissions. Monitoring and tracking capabilities and smart material handling practices are vital in energy modeling and sustainability efforts. Beier et al. (2020) examined the sociotechnical perspective of Industry 4.0 and emphasized the importance of system integration potential as a key factor in achieving sustainability goals within the context of Industry 4.0. Manavalan and Jayakrishna (2019) examined the possibility of embedded systems, particularly their ability to track physical objects in real time, which enables supply chain partners to have accurate information regarding the availability time of goods.

Luthra and Mangla (2018) identified critical factors, including real-time data analytics and advanced security and safety, as CPS potentials that contribute to sustainable freight transportation. Gružauskas et al. (2018) highlighted the long-term competitive advantage that can be achieved by deploying autonomous vehicles equipped with CPS and IoT capabilities. Zhang (2018) emphasized the potential of CPS in facilitating coordination among various logistics processes. Strandhagen et al. (2017) emphasized the transformative role of logistics 4.0 features, such as instant information exchange, automated solutions, and real-time big data analysis, in driving the emergence of new business models within the logistics industry.

3. Research methodology

Fuzzy best worst method (Fuzzy-BWM) is used to analyze the CPS potentials. The BWM is used in multi-criteria decision-making (MCDM) that compares the best and worst criteria/alternatives against other criteria/alternatives through pairwise comparisons (Rezaei, 2015). Compared with other MCDM techniques, this method requires fewer comparison data points and yields more consistent comparisons, giving more reliable results (Rezaei, 2015). Integrating fuzzy concepts into MCDM processes addresses the uncertainty and vagueness in decision-makers judgments, which gives more reliable decision outcomes (Guo and Zhao, 2017). Integrating the BWM with fuzzy theory can effectively address practical problems in uncertain environments. In BWM, the nonlinearly constrained optimization problem determines the fuzzy weights of criteria and alternatives based on different criteria (Guo and Zhao, 2017). The following steps outline the computational methodology.

Step 1: Develop a robust decision criteria system for assessing the sustainability impacts of CPS.

The decision criteria system encompasses nine CPS potentials, which are pivotal in conducting a thorough evaluation of sustainability impact. The values assigned to the decision criteria can effectively indicate the performance levels of various CPS potentials. Suppose there are n potentials in decision criteria $\{ Cp_{1,} Cp_{2,} \dots Cp_n\}$.

Step 2: Formation of fuzzy linguistic scale

A five-point fuzzy linguistic scale is employed to handle the ambiguities of expert assessments. Five linguistics terms (Equally, Low, Moderately, Highly, Utterly) are assigned to the linguistic variable "significant." The positive triangular fuzzy numbers (l_{ij}, c_{ij}, u_{ij}) represent the linguistic terms, which have been shown in Table 12.2.

Step 3: Determine the most and least significant criterion.

During this stage, decision-makers identify the most significant (best) potential and the least significant (worst) potential following the established decision criteria framework. The most significant potential is denoted as Cp_B, while the least significant potential is labeled as WCp.

Step 4: Performing the fuzzy reference comparisons for the best criterion.

During this step, the fuzzy preferences of the best criterion compared with all other criteria are determined using the fuzzy linguistic scale. These fuzzy preferences are then transformed into TFNs based on the predefined transformation rules outlined in Table 12.2. The resulting fuzzy best-to-others vector is as follows:

$$\widetilde{CP}_B = \left(\widetilde{Cp}_{B1,} \widetilde{Cp}_{B1,} \dots \widetilde{Cp}_{B1}\right) \tag{12.1}$$

$\widetilde{CP}_B$ denotes the fuzzy vector indicating preference toward the best alternative compared with others. The vector $\widetilde{CP}_{Bj} - \left(\widetilde{Cp}^{l}_{Bj}, \widetilde{Cp}^{m}_{Bj}, \widetilde{Cp}^{u}_{Bj}\right)$, symbolizes the fuzzy preference of the best criterion over criterion j, where j ranges from 1 to n ($j = 1, 2, \cdots, n$). It is evident that $\widetilde{CP}_{BB} = (1, 1, 1)$.

Step 5: Conduct fuzzy reference comparisons for the worst criterion.

The fuzzy preference of each criterion over the worst criterion is determined using a fuzzy linguistic scale. These fuzzy preferences are then transformed into TFNs based on the

TABLE 12.2 Linguistic terms and associated triangular fuzzy numbers (TFNs).

Linguistic terms	Equally significant (EqS)	Low significant (LoS)	Moderately significant (MoS)	Highly significant (HiS)	Utterly significant (UtS)
TFNs	(1, 1, 1)	(2/3, 1, 3/2)	(3/2, 2, 5/2)	(5/2, 3, 7/2)	(7/2, 4, 9/2)

transformation rules specified in Table 12.2. The resulting others-to-worst vector can be obtained as follows:

$$\widetilde{CP}_W = \left(\widetilde{Cp}_{1W,}\ \widetilde{Cp}_{2W,} \cdots\ \widetilde{Cp}_{nW}\right) \tag{12.2}$$

where $\widetilde{P}_W$ represents the fuzzy others-to-worst vector; $\widetilde{CP}_{iW} = \left(\widetilde{Cp}^{l}_{iW}, \widetilde{Cp}^{m}_{iW}, \widetilde{Cp}^{u}_{iW}\right)$, represents the fuzzy preference of the worst criterion over criterion $i, i = 1, 2, \ldots, n$. It can be known that $\widetilde{CP}_{WW} = (1, 1, 1)$.

Step 6: Determine the optimal fuzzy weights $(\widetilde{w}^*_1, \widetilde{w}^*_2, \ldots, \widetilde{w}^*_n)$.

The optimal weight for each criterion should satisfy the conditions $\frac{\widetilde{w}_B}{\widetilde{w}_j} = \widetilde{Cp}_{Bj}$ and $\frac{\widetilde{w}_j}{\widetilde{w}_W} = \widetilde{Cp}_{iW}$ for each pair of $\widetilde{w}_B/\widetilde{w}_j$ and $\widetilde{w}_j/\widetilde{w}_W$. To fulfill these requirements involves determining a solution where the maximum absolute gaps $\left|\frac{\widetilde{w}_B}{\widetilde{w}_j} - \widetilde{Cp}_{Bj}\right|$ and $\left|\frac{\widetilde{w}_j}{\widetilde{w}_W} - \widetilde{Cp}_{iW}\right|$ for all j are minimized. Here $\widetilde{w}_B$, $\widetilde{w}_j$, and $\widetilde{w}_W$ are triangular fuzzy numbers. Subsequently, the optimum fuzzy weights can be obtained by employing constrained optimization programming.

$$\min_j \max \left\{ \left|\frac{\widetilde{w}_B}{\widetilde{w}_j} - \widetilde{Cp}_{Bj}\right|, \left|\frac{\widetilde{w}_j}{\widetilde{w}_W} - \widetilde{Cp}_{iW}\right| \right\}$$

$$s.t. \begin{cases} \sum_{j=1}^{n} R(\widetilde{w}_j) = 1 \\ \widetilde{w}^l_j, \widetilde{w}^m_j, \widetilde{w}^u_j \\ \widetilde{w}^l_j \geq 0 \\ j = 1, 2, 3 \ldots, n \end{cases} \tag{12.3}$$

In Eq. (12.3), $\widetilde{w}_B = \widetilde{w}^l_B, \widetilde{w}^m_B, \widetilde{w}^u_B$, $\widetilde{w}_j = \widetilde{w}^l_j, \widetilde{w}^m_j, \widetilde{w}^u_j$, $\widetilde{w}_W = \widetilde{w}^l_W, \widetilde{w}^m_W, \widetilde{w}^u_W$, $\widetilde{Cp}_{Bj} = \left(\widetilde{Cp}^l_{Bj}, \widetilde{Cp}^m_{Bj}, \widetilde{Cp}^u_{Bj}\right)$, and $\widetilde{Cp}_{iW} = \left(\widetilde{Cp}^l_{iW}, \widetilde{Cp}^m_{iW}, \widetilde{Cp}^u_{iW}\right)$.

Furthermore, $R(\widetilde{w}_j)$ signifies the graded mean integration of $\widetilde{w}_j$, and its computation can be performed as follows using the subsequent equation.

$$R(\widetilde{w}_j) = \widetilde{w}^l_j + 4 * \widetilde{w}^m_j + \widetilde{w}^u_j\ /\ 6 \tag{12.4}$$

The constraint problem stated in Eq. (12.3) can be transformed into the following nonlinear constraint optimization problem, as indicated in Eq. (12.5).

$$\min \delta$$

$$
s.t.\begin{cases}
\left|\frac{\widetilde{w}_B}{\widetilde{w}_j}-\widetilde{Cp}_{Bj}\right|\leq\delta \\
\left|\frac{\widetilde{w}_j}{\widetilde{w}_W}-\widetilde{Cp}_{iW}\right|\leq\delta \\
\sum_{j=1}^{n}R(\widetilde{w}_j)=1 \\
\widetilde{w}_j^l,\widetilde{w}_j^m,\widetilde{w}_j^u \\
\widetilde{w}_j^l\geq 0 \\
j=1,2,3\ldots,n
\end{cases}
\tag{12.5}
$$

Where $\delta=(\delta^l,\delta^m,\delta^u)$, with the consideration that $\delta^l\leq\delta^m\leq\delta^u$, it is supposed that $\delta=(k^*,k^*,k^*)$, $k^*\leq\delta^l$. Therefore, Eq. (12.5) can be transformed into Eq. (12.6)

$$
s.t.\begin{cases}
\left|\frac{\widetilde{w}_B^l,\widetilde{w}_B^m,\widetilde{w}_B^u}{\widetilde{w}_j^l,\widetilde{w}_j^m,\widetilde{w}_j^u}-\widetilde{Cp}_{Bj}^l,\widetilde{Cp}_{Bj}^m,\widetilde{Cp}_{Bj}^u\right|\leq(k^*,k^*,k^*),j=1,2,\ldots,n \\
\left|\frac{\widetilde{w}_j^l,\widetilde{w}_j^m,\widetilde{w}_j^u}{\widetilde{w}_W^l,\widetilde{w}_W^m,\widetilde{w}_W^u}-\widetilde{Cp}_{iW}^l,\widetilde{Cp}_{iW}^m,\widetilde{Cp}_{iW}^u\right|\leq(k^*,k^*,k^*),i,j=1,2,\ldots,n \\
\sum_{j=1}^{n}R(\widetilde{w}_j)=1 \\
\widetilde{w}_j^l,\widetilde{w}_j^m,\widetilde{w}_j^u \\
\widetilde{w}_j^l\geq 0 \\
j=1,2,3\ldots,n
\end{cases}
\tag{12.6}
$$

The optimal fuzzy weights $(\widetilde{w}_1,\widetilde{w}_2,\ldots,\widetilde{w}_n)$ can be obtained by solving Eq. (12.6).

Step 7: Transforming fuzzy optimal weights of the n CPS potentials through defuzzification

The process of transforming the fuzzy optimal weights $(\widetilde{w}_1,\widetilde{w}_2,\ldots,\widetilde{w}_n)$ of the n potentials into crisp numbers is achieved using Eq. (12.4).

Step 8: Check the consistency.

The consistency ratio (CR) serves as a crucial indicator for assessing the consistency of the pairwise decision-making process. It is calculated using Eq. (12.7).

$$CR = k^{*}/CI \tag{12.7}$$

4. Illustration of proposed model

4.1 Data collection and analysis procedure

Initially, nine CPS potentials are identified through literature review for the assessment of their impact on sustainability within the freight transport industry. To illustrate the case, five hypothetical experts' responses are considered to assess the sustainability impact of CPS. This assessment focuses on the nine potential aspects of CPS outlined in Table 12.1. To address the experts' vagueness, a fuzzy linguistic scale is created in step 2 and presented in Table 12.2. Based on experts' opinions, the best and the worst potentials were identified using step 3. For example, Expert 1 designated "autonomous logistics systems" (Cp_2) as the best CPS potential, while considering "advanced safety and security" (Cp_7) as the worst for promoting sustainable logistics operations. After performing fuzzy reference comparisons in step 4, Table 12.3 presents the linguistic terms used by decision-makers to express their fuzzy preferences for the best criterion over all other criteria.

The fuzzy best-to-others vector ($\widetilde{CP}_B$) can then be derived by referring to Table 12.2 and applying Eq. (12.1) in the subsequent manner:

$$\widetilde{CP}_{B1} = \left(\left(\frac{3}{2},2,\frac{5}{2}\right),(1,1,1),\left(\frac{3}{2},2,\frac{5}{2}\right),\left(\frac{3}{2},2,\frac{5}{2}\right),\left(\frac{2}{3},1,\frac{3}{2}\right),\left(\frac{3}{2},2,\frac{5}{2}\right),\left(\frac{7}{2},4,\frac{9}{2}\right),\left(\frac{3}{2},2,\frac{5}{2}\right),\left(\frac{2}{3},1,\frac{3}{2}\right)\right)$$

$$\widetilde{CP}_{B2} = \left((1,1,1),\left(\frac{3}{2},2,\frac{5}{2}\right),\left(\frac{3}{2},2,\frac{5}{2}\right),\left(\frac{2}{3},1,\frac{3}{2}\right),\left(\frac{7}{2},4,\frac{9}{2}\right),\left(\frac{3}{2},2,\frac{5}{2}\right),\left(\frac{5}{2},3,\frac{7}{2}\right),\left(\frac{3}{2},2,\frac{5}{2}\right),\left(\frac{3}{2},2,\frac{5}{2}\right)\right)$$

TABLE 12.3 Best-to-other CPS potential rating.

Experts	CPS potentials	Cp_1	Cp_2	Cp_3	Cp_4	Cp_5	Cp_6	Cp_7	Cp_8	Cp_9
Expert 1	Best (Cp_2)	MoS	EqS	MoS	MoS	LoS	MoS	UtS	MoS	LoS
Expert 2	Best (Cp_1)	EqS	MoS	MoS	LoS	UtS	MoS	HiS	MoS	MoS
Expert 3	Best (Cp_4)	MoS	MoS	HiS	EqS	UtS	LoS	HiS	MoS	MoS
Expert 4	Best (Cp_4)	LoS	LoS	HiS	EqS	MoS	MoS	UtS	HiS	MoS
Expert 5	Best (Cp_2)	MoS	EqS	MoS	LoS	UtS	LoS	HiS	MoS	LoS

$$\widetilde{CP}_{B3} = \left(\left(\frac{3}{2},2,\frac{5}{2}\right),\left(\frac{3}{2},2,\frac{5}{2}\right),\left(\frac{5}{2},3,\frac{7}{2}\right),(1,1,1),\left(\frac{7}{2},4,\frac{9}{2}\right),\left(\frac{2}{3},1,\frac{3}{2}\right),\left(\frac{5}{2},3,\frac{7}{2}\right),\left(\frac{3}{2},2,\frac{5}{2}\right),\left(\frac{3}{2},2,\frac{5}{2}\right)\right)$$

$$\widetilde{CP}_{B4} = \left(\left(\frac{2}{3},1,\frac{3}{2}\right),\left(\frac{2}{3},1,\frac{3}{2}\right),\left(\frac{5}{2},3,\frac{7}{2}\right),(1,1,1),\left(\frac{3}{2},2,\frac{5}{2}\right),\left(\frac{3}{2},2,\frac{5}{2}\right),\left(\frac{7}{2},4,\frac{9}{2}\right),\left(\frac{5}{2},3,\frac{7}{2}\right),\left(\frac{5}{2},3,\frac{7}{2}\right)\right)$$

$$\widetilde{CP}_{B5} = \left(\left(\frac{3}{2},2,\frac{5}{2}\right),(1,1,1),\left(\frac{3}{2},2,\frac{5}{2}\right),\left(\frac{2}{3},1,\frac{3}{2}\right),\left(\frac{7}{2},4,\frac{9}{2}\right),\left(\frac{2}{3},1,\frac{3}{2}\right),\left(\frac{5}{2},3,\frac{7}{2}\right),\left(\frac{3}{2},2,\frac{5}{2}\right),\left(\frac{2}{3},1,\frac{3}{2}\right)\right)$$

Table 12.4 lists the linguistic terms of decision-makers for fuzzy preferences of all the criteria over the worst criterion, following the execution of fuzzy reference comparisons in step 5. Subsequently, the fuzzy others-to-worst vector ($\widetilde{CP}_W$) can be obtained by referring to Table 12.2 and employing Eq. (12.2) as follows:

$$\widetilde{CP}_{W1} = \left(\left(\frac{5}{2},3,\frac{7}{2}\right),\left(\frac{7}{2},4,\frac{9}{2}\right),\left(\frac{5}{2},3,\frac{7}{2}\right),\left(\frac{3}{2},2,\frac{5}{2}\right),\left(\frac{3}{2},2,\frac{5}{2}\right),\left(\frac{3}{2},2,\frac{5}{2}\right),(1,1,1),\left(\frac{3}{2},2,\frac{5}{2}\right),\left(\frac{2}{3},1,\frac{3}{2}\right)\right)$$

$$\widetilde{CP}_{W2} = \left(\left(\frac{7}{2},4,\frac{9}{2}\right),\left(\frac{5}{2},3,\frac{7}{2}\right),\left(\frac{3}{2},2,\frac{5}{2}\right),\left(\frac{5}{2},3,\frac{7}{2}\right),(1,1,1),\left(\frac{3}{2},2,\frac{5}{2}\right),\left(\frac{2}{3},1,\frac{3}{2}\right),\left(\frac{2}{3},1,\frac{3}{2}\right),\left(\frac{3}{2},2,\frac{5}{2}\right)\right)$$

$$\widetilde{CP}_{W3} = \left(\left(\frac{5}{2},3,\frac{7}{2}\right),\left(\frac{3}{2},2,\frac{5}{2}\right),\left(\frac{3}{2},2,\frac{5}{2}\right),\left(\frac{7}{2},4,\frac{9}{2}\right),(1,1,1),\left(\frac{3}{2},2,\frac{5}{2}\right),\left(\frac{2}{3},1,\frac{3}{2}\right),\left(\frac{2}{3},1,\frac{3}{2}\right),\left(\frac{3}{2},2,\frac{5}{2}\right)\right)$$

TABLE 12.4 Other-to-worst CPS potential rating.

Experts	**Expert 1**	**Expert 2**	**Expert 3**	**Expert 4**	**Expert 5**
CPS potentials	**Worst (Cp_7)**	**Worst (Cp_5)**	**Worst (Cp_5)**	**Worst (Cp_7)**	**Worst (Cp_5)**
Cp_1	HiS	UtS	HiS	HiS	HiS
Cp_2	UtS	HiS	MoS	HiS	UtS
Cp_3	HiS	MoS	MoS	MoS	MoS
Cp_4	MoS	HiS	UtS	UtS	HiS
Cp_5	MoS	EqS	EqS	LoS	EqS
Cp_6	MoS	MoS	MoS	MoS	MoS
Cp_7	EqS	LoS	LoS	EqS	LoS
Cp_8	MoS	LoS	LoS	MoS	LoS
Cp_9	LoS	MoS	MoS	LoS	MoS

$$\widetilde{CP}_{W4} = \left(\left(\frac{5}{2},3,\frac{7}{2}\right),\left(\frac{5}{2},3,\frac{7}{2}\right),\left(\frac{3}{2},2,\frac{5}{2}\right),\left(\frac{7}{2},4,\frac{9}{2}\right),\left(\frac{2}{3},1,\frac{3}{2}\right),\left(\frac{2}{3},1,\frac{3}{2}\right),(1,1,1),\left(\frac{3}{2},2,\frac{5}{2}\right),\left(\frac{2}{3},1,\frac{3}{2}\right)\right)$$

$$\widetilde{CP}_{W5} = \left(\left(\frac{5}{2},3,\frac{7}{2}\right),\left(\frac{7}{2},4,\frac{9}{2}\right),\left(\frac{3}{2},2,\frac{5}{2}\right),\left(\frac{5}{2},3,\frac{7}{2}\right),(1,1,1),\left(\frac{3}{2},2,\frac{5}{2}\right),\left(\frac{2}{3},1,\frac{3}{2}\right),\left(\frac{2}{3},1,\frac{3}{2}\right),\left(\frac{3}{2},2,\frac{5}{2}\right)\right)$$

The subsequent step involves evaluating the optimal fuzzy weights for each expert for the nine CPS potentials by utilizing Eq. (12.6) as shown in Table 12.5.

In the subsequent step, the crisp values of the fuzzy optimal weights for each expert are obtained using Eq. (12.4). These values are presented in Table 12.6.

Furthermore, average priority weights are calculated and incorporated into the last column of Tables 12.5 and 12.6.

In the final step, the evaluation of consistency is performed. For instance, in the case of expert 1, the consistency index yields 0.098, a value less than 1 that signifies result consistency. The calculated average consistency value is 0.079, as determined by Eq. (12.7). The consistency index (CI) for the fuzzy BWM is obtained from the study conducted by (Guo and Zhao, 2017).

TABLE 12.5 Optimal fuzzy weights.

	Expert 1	Expert 2	Expert 3	Expert 4	Expert 5	Aggregate
δ	(0.7913, 0.7913, 0.7913)	(0.6471, 0.6471, 0.6471)	(0.6719, 0.6719, 0.6719)	(0.5000, 0.5000, 0.5000)	(0.5600, 0.5600, 0.5600)	(0.5000, 0.6340, 0.7913)
$\widetilde{w}_1$	(0.1011, 0.1157, 0.1870)	(0.1912, 0.1912, 0.1912)	(0.1102, 0.1102, 0.1102)	(0.1592, 0.1592, 0.1592)	(0.1028, 0.1269, 0.1515)	(0.1011, 0.1406, 0.1912)
$\widetilde{w}_2$	(0.1398, 0.1398, 0.1728)	(0.1032, 0.1032, 0.1032)	(0.1102, 0.1265, 0.1502)	(0.1739, 0.1739, 0.1863)	(0.1831, 0.1995, 0.1995)	(0.1032, 0.1486, 0.1995)
$\widetilde{w}_3$	(0.1011, 0.1157, 0.1870)	(0.1032, 0.1161, 0.1305)	(0.0713, 0.0866, 0.1102)	(0.0621, 0.0745, 0.0931)	(0.0807, 0.0865, 0.1051)	(0.0621, 0.0959, 0.1870)
$\widetilde{w}_4$	(0.1011, 0.1157, 0.1434)	(0.1599, 0.1599, 0.1720)	(0.2015, 0.2015, 0.2015)	(0.1863, 0.1863, 0.1863)	(0.1221, 0.1279, 0.1489)	(0.1011, 0.1583, 0.2015)
$\widetilde{w}_5$	(0.1183, 0.1216, 0.1434)	(0.0415, 0.0439, 0.0557)	(0.0473, 0.0473, 0.0603)	(0.0727, 0.0745, 0.0931)	(0.0506, 0.0520, 0.0530)	(0.0415, 0.0679, 0.1434)
$\widetilde{w}_6$	(0.1011, 0.1157, 0.1434)	(0.1032, 0.1161, 0.1305)	(0.1265, 0.1265, 0.1502)	(0.0931, 0.1242, 0.1397)	(0.1030, 0.1331, 0.1511)	(0.0931, 0.1231, 0.1511)
$\widetilde{w}_7$	(0.0436, 0.0436, 0.0516)	(0.0670, 0.0722, 0.0891)	(0.0713, 0.0792, 0.1028)	(0.0466, 0.0497, 0.0621)	(0.0568, 0.0627, 0.0671)	(0.0436, 0.0615, 0.1028)
$\widetilde{w}_8$	(0.1011, 0.1157, 0.1434)	(0.0722, 0.0722, 0.0891)	(0.0792, 0.0792, 0.1028)	(0.0621, 0.0745, 0.0931)	(0.0652, 0.0811, 0.0959)	(0.0621, 0.0845, 0.1434)
$\widetilde{w}_9$	(0.0754, 0.0780, 0.0998)	(0.1032, 0.1161, 0.1305)	(0.1102, 0.1265, 0.1502)	(0.0727, 0.0745, 0.0931)	(0.1035, 0.1331, 0.1489)	(0.0727, 0.1057, 0.1502)

TABLE 12.6 Optimal crisp weights.

	Expert 1	Expert 2	Expert 3	Expert 4	Expert 5	Average
δ	0.7913	0.6471	0.6719	0.5000	0.5600	0.6379
$\widetilde{w}_1$	0.1251	0.1912	0.1102	0.1592	0.1270	0.1425
$\widetilde{w}_2$	0.1453	0.1032	0.1277	0.1759	0.1968	0.1495
$\widetilde{w}_3$	0.1251	0.1163	0.0880	0.0756	0.0887	0.1054
$\widetilde{w}_4$	0.1179	0.1620	0.2015	0.1863	0.1304	0.1559
$\widetilde{w}_5$	0.1247	0.0454	0.0495	0.0773	0.0519	0.0761
$\widetilde{w}_6$	0.1179	0.1163	0.1304	0.1216	0.1311	0.1228
$\widetilde{w}_7$	0.0449	0.0742	0.0818	0.0512	0.0624	0.0654
$\widetilde{w}_8$	0.1179	0.0750	0.0831	0.0756	0.0809	0.0906
$\widetilde{w}_9$	0.0812	0.1163	0.1277	0.0773	0.1308	0.1076

5. Results and discussion

Nine CPS potentials were identified through a comprehensive literature review to evaluate the sustainability impact on freight logistics. To address the experts' vagueness and ensure the validity of the findings, the fuzzy best worst method (FBWM) approach was employed. This approach, detailed in Section 4, encompasses an eight-step process that includes consistency analysis. By considering the experts' opinions and accounting for their uncertainties, the FBWM approach enables a robust assessment of the sustainability impact on freight logistics.

The analysis reveals the significance of different CPS potentials in enhancing sustainability in freight logistics operations. The most significant CPS potential, as demonstrated in both Figs. 12.1 and 12.2, is smart traffic management or efficient route planning (Cp_4), as pointed out by Sun et al. (2022).

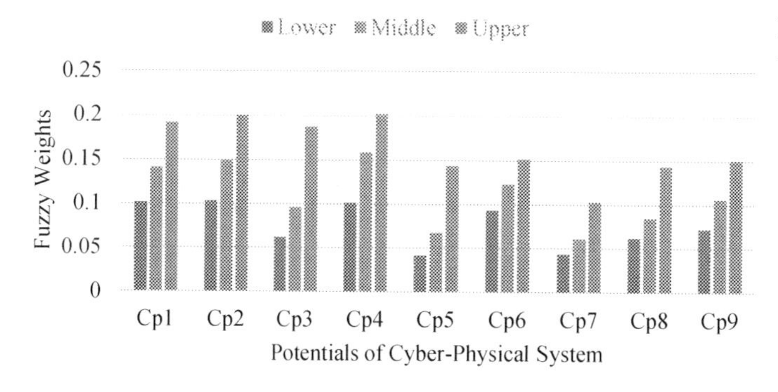

FIGURE 12.1 Average priority fuzzy weights of CPS potentials.

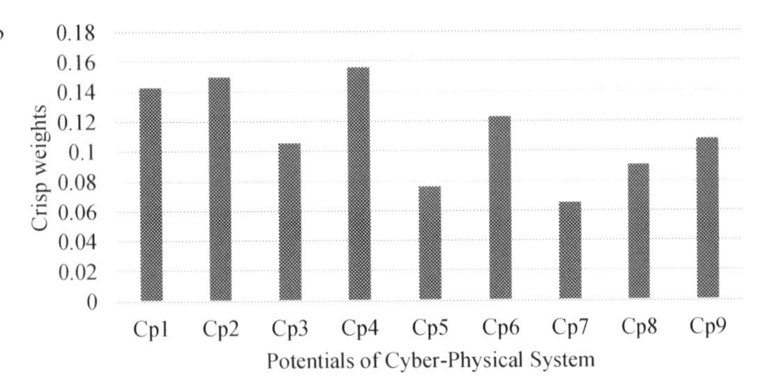

FIGURE 12.2 Average priority crisp weights of CPS potentials.

Autonomous logistics system (Cp_2) emerges as the second most significant CPS potential. This finding aligns with the emphasis placed by Yavas and Ozkan-Ozen (2020) on the crucial sustainability impact of automated vehicles in freight logistics processes. Real-time data analytics (Cp_1) ranks as the third most crucial CPS potential, effectively driving logistics sustainability. Sun et al. (2022) also highlighted the importance of real-time data analytics capabilities as enablers of sustainability in logistics operations. Advanced collaborative platform (Cp_6) emerges as the fourth most crucial CPS potential, as emphasized by both Sun et al. (2022) and Strandhagen et al. (2017) in their respective studies. Strandhagen et al. (2017) suggest that real-time information sharing of predictive maintenance can enhance product life and durability.

System integration (Cp_9) capabilities rank as the fifth top CPS potential, as pointed out by Beier et al. (2020). Moreover, Table 12.7 consists of each CPS's potential and its total sustainability impact (both positive and negative) in freight logistics.

TABLE 12.7 Total sustainability effects of CPS in freight logistics.

CPS potentials	Sustainability effect (positive)	Sustainability effect (negative)
Real-time data analytics	**Economic:** Enhanced efficiency, detection of equipment failure or malfunction **Environmental:** Minimized emissions, efficient resource utilization, reduced energy consumption **Social:** Risk reduction of accidents, visualization, and informed decision-making	**Economic:** Overreliance on data-driven decisions may decrease overall performance **Environmental:** Excessive data usage can increase energy consumption and carbon emissions, E-waste generation **Social:** Risk of compromising sensitive information
Autonomous vehicles OR Autonomous logistics systems	**Economic:** Optimization of operations **Environmental:** Resource consumption reduction, lower CO_2 emissions, reduced fuel consumption, lower energy consumption **Social:** Minimization of delays, prevention of inefficiencies	**Economic:** Disruptions in supply chain operations, maintenance of automated systems, security and privacy risks **Environmental:** E-waste generation **Social:** Potential displacement of jobs, reduce the need for human intervention

TABLE 12.7 Total sustainability effects of CPS in freight logistics.—cont'd

CPS potentials	Sustainability effect (positive)	Sustainability effect (negative)
Predictive maintenance	**Economic:** Improving vehicle reliability, mitigating the risk of unexpected failures and transportation delays, minimizing downtime **Environmental:** Reducing excessive energy consumption **Social:** Decreasing the likelihood of accidents or disruptions	**Economic:** Disruption to operations **Environmental:** Increased energy consumption, resource consumption and waste generation, transportation, and emissions **Social:** Job displacement, workforce skill and training requirement
Smart traffic management OR efficient route planning	**Economic:** Improving operational efficiency **Environmental:** Reducing fuel consumption, lowering emissions **Social:** Creating a greener freight logistics ecosystem	**Economic:** Infrastructure requirements **Environmental:** Increased energy consumption, **Social:** Data privacy and security risks
Smart material handling OR Intelligent cargo management	**Economic:** Optimizing inventory management, data-driven optimization, efficient freight logistics ecosystem, optimizing operations **Environmental:** Efficient resource utilization, improving energy efficiency, reducing waste **Social:** Ensuring safety and ergonomics, Minimizing cargo loss or damage	**Economic:** Technological obsolescence **Environmental:** Energy consumption, electronic waste generation **Social:** Workforce displacement
Advanced collaborative platforms for collaboration and coordination	**Economic:** Freight route optimization, shipment consolidation, enhanced efficiency **Environmental:** Efficient resource utilization, minimize carbon footprint **Social:** Improved collaboration and coordination, eco-friendly modal shift, stakeholder engagement	**Economic:** Technological complexity and dependence **Environmental:** Energy consumption, electronic waste generation, **Social:** Data security and privacy risks
Advanced safety and security	**Economic:** Mitigating risks, proactive risk mitigation **Environmental:** Reducing environmental impact **Social:** Prioritizing worker safety, building stakeholder confidence	**Economic:** Technological complexity and dependence **Environmental:** Energy consumption, electronic waste generation **Social:** Data security and privacy risks
Intelligent cargo tracking and monitoring	**Economic:** Optimizing freight logistics operations, optimal route planning and delivery **Environmental:** Efficient resource utilization, reduce waste **Social:** Enhance the customer experience, drive data-driven decision making	**Economic:** Technological complexity and dependence **Environmental:** Energy consumption, electronic waste generation **Social:** Data security and privacy risks
System integration	**Economic:** Process automation, enhanced operational efficiency, optimized route planning **Environmental:** Reduce waste, minimize environmental impact **Social:** Collaboration among supply chain partners	**Economic:** Technological complexity and obsolescence **Environmental:** Energy consumption, electronic waste generation **Social:** Data security and privacy risks

6. Conclusion

This chapter presents a roadmap for achieving sustainability in freight logistics operations using cutting-edge digital disruption such as CPS. This study aims to identify and analyze key CPS potential and to explore their sustainability impact on logistics. Initially, CPS potentials were identified through a literature review, and nine CPS potentials were considered for logistics operations to measure the sustainability impact after experts' opinions. After that, an illustrative example is discussed to evaluate the CPS potentials using a MCDM method, i.e., fuzzy BWM on logistics processes, using linguistic responses of experts.

The evaluation also considers uncertainty and vagueness in decision-making by integrating fuzzy logic with the best-worst method. This method provides a structured approach for analyzing the CPS potential, which can empower logistics companies to make informed decisions toward CPS adoption in achieving sustainable practices. According to the analysis findings, the most significant CPS potential is "efficient route planning or smart traffic management" in achieving sustainability goals in freight logistics. Conversely, the least significant CPS potential is identified as "advanced safety and security" for attaining sustainability in logistics.

The study identified the most significant potentials based on their higher sustainability impact on freight transportation. For instance, "efficient route planning or smart traffic management," "autonomous logistics systems," "real-time data analytics," "collaborative platforms," and "system integration" capabilities are key CPS potentials for achieving higher sustainability impact in the industry. These potentials impact operational performance and are crucial for optimizing resources and reducing environmental externalities. Moreover, the study emphasizes the role of CPS in attaining collaboration and coordination among stakeholders and integrating CPS technologies into existing infrastructure and processes. This study explores how effective implementation of CPS can influence the total sustainability of the logistics industry. It can enable emerging transport companies to remain competitive in the rapidly evolving freight logistics industry.

In summary, this study provides awareness about CPS for transforming the freight logistics industry to attain operational efficiency and sustainable growth. It contributes to the body of knowledge by identifying the CPS's potential and its sustainability impact on freight logistics. CPS implementation in the logistics industry can achieve long-term success by improving sustainable practices and positively impacting the society.

References

Abbas, A.W., Marwat, S.N.K., 2020. Scalable emulated framework for IoT devices in smart logistics based cyber-physical systems: bonded coverage and connectivity analysis. IEEE Access 8, 138350–138372. https://doi.org/10.1109/ACCESS.2020.3012458.

Albrecht, T., Baier, M.S., Gimpel, H., Meierhöfer, S., Röglinger, M., Schlüchtermann, J., Will, L., 2023. Leveraging digital technologies in logistics 4.0: insights on affordances from intralogistics processes. Information Systems Frontiers. https://doi.org/10.1007/s10796-023-10394-6.

Beier, G., Ullrich, A., Niehoff, S., Reißig, M., Habich, M., 2020. Industry 4.0: how it is defined from a sociotechnical perspective and how much sustainability it includes – a literature review. Journal of Cleaner Production 259. https://doi.org/10.1016/j.jclepro.2020.120856.

Chalmeta, R., Santos-deLeón, N.J., 2020. Sustainable supply chain in the era of industry 4.0 and big data: a systematic analysis of literature and research. Sustainable Times 12. https://doi.org/10.3390/su12104108.

Esmaeilian, B., Sarkis, J., Lewis, K., Behdad, S., 2020. Blockchain for the future of sustainable supply chain management in Industry 4.0. Resources, Conservation and Recycling. https://doi.org/10.1016/j.resconrec.2020.105064.

Facchini, F., Oleśków-Szłapka, J., Ranieri, L., Urbinati, A., 2020. A maturity model for logistics 4.0: an empirical analysis and a roadmap for future research. Sustainable Times 12, 1–18. https://doi.org/10.3390/SU12010086.

Gružauskas, V., Baskutis, S., Navickas, V., 2018. Minimizing the trade-off between sustainability and cost effective performance by using autonomous vehicles. Journal of Cleaner Production 184, 709–717. https://doi.org/10.1016/j.jclepro.2018.02.302.

Guo, S., Zhao, H., 2017. Fuzzy best-worst multi-criteria decision-making method and its applications. Knowledge-Based Systems 121, 23–31. https://doi.org/10.1016/j.knosys.2017.01.010.

Harris, I., Wang, Y., Wang, H., 2015. ICT in multimodal transport and technological trends: unleashing potential for the future. International Journal of Production Economics 159, 88–103. https://doi.org/10.1016/j.ijpe.2014.09.005.

Issaoui, Y., Khiat, A., Bahnasse, A., Ouajji, H., 2021. Toward smart logistics: engineering insights and emerging trends. Archives of Computational Methods in Engineering 28, 3183–3210. https://doi.org/10.1007/s11831-020-09494-2.

Kodym, O., Kubáč, L., Kavka, L., 2020. Risks associated with Logistics 4.0 and their minimization using Blockchain. Open Engineering 10, 74–85. https://doi.org/10.1515/eng-2020-0017.

Lee, C.K.M., Lv, Y., Ng, K.K.H., Ho, W., Choy, K.L., 2018. Design and application of internet of things-based warehouse management system for smart logistics. International Journal of Production Research 56, 2753–2768. https://doi.org/10.1080/00207543.2017.1394592.

Li, Z., Wang, Y., Wang, K.S., 2017. Intelligent predictive maintenance for fault diagnosis and prognosis in machine centers: industry 4.0 scenario. Advanced Manufacturing 5, 377–387. https://doi.org/10.1007/s40436-017-0203-8.

Luthra, S., Mangla, S.K., 2018. Evaluating challenges to Industry 4.0 initiatives for supply chain sustainability in emerging economies. Process Safety and Environmental Protection 117, 168–179. https://doi.org/10.1016/j.psep.2018.04.018.

Manavalan, E., Jayakrishna, K., 2019. A review of Internet of Things (IoT) embedded sustainable supply chain for industry 4.0 requirements. Computers & Industrial Engineering 127, 925–953. https://doi.org/10.1016/j.cie.2018.11.030.

Munsamy, M., Telukdarie, A., Dhamija, P., 2020. Logistics 4.0 energy modelling. International Journal of Business Analytics 7, 98–121. https://doi.org/10.4018/IJBAN.2020010106.

Prause, G., 2016. Sustainable business models and structures for industry 4. Journal of Security and Sustainability Issues 158–169. https://doi.org/10.9770/jssi.2015.5.2(3)CITATIONS. December 2015. 5 (2), 2.

Rezaei, J., 2015. Best-worst multi-criteria decision-making method. Omega 53, 49–57. https://doi.org/10.1016/j.omega.2014.11.009.

Roblek, V., Thorpe, O., Bach, M.P., Jerman, A., Meško, M., 2020. The fourth industrial revolution and the sustainability practices: a comparative automated content analysis approach of theory and practice. Sustainable Times 12, 1–28. https://doi.org/10.3390/su12208497.

Strandhagen, J.O., Vallandingham, L.R., Fragapane, G., Strandhagen, J.W., Stangeland, A.B.H., Sharma, N., 2017. Logistics 4.0 and emerging sustainable business models. Advanced Manufacturing 5, 359–369. https://doi.org/10.1007/s40436-017-0198-1.

Sun, X., Yu, H., Solvang, W.D., Wang, Y., Wang, K., 2022. The application of Industry 4.0 technologies in sustainable logistics: a systematic literature review (2012–2020) to explore future research opportunities. Environmental Science & Pollution Research 29, 9560–9591. https://doi.org/10.1007/s11356-021-17693-y.

Winkelhaus, S., Grosse, E.H., 2020. Logistics 4.0: a systematic review towards a new logistics system. International Journal of Production Research 58, 18–43. https://doi.org/10.1080/00207543.2019.1612964.

Yavas, V., Ozkan-Ozen, Y.D., 2020. Logistics centers in the new industrial era: a proposed framework for logistics center 4.0. Transplantation Research Part E Logist. Transp. Rev. 135, 101864. https://doi.org/10.1016/j.tre.2020.101864.

Zhang, N., 2018. Smart logistics path for cyber-physical systems with internet of things. IEEE Access 6, 70808–70819. https://doi.org/10.1109/ACCESS.2018.2879966.

CHAPTER

13

Analysis of decision support systems for the green hydrogen value chain design: A comparison of literature and real-world tools

Natalia Yulissa Velastín Osorio[1], *Andrea Teresa Espinoza Pérez*[1,2], *Pavlo Santander-Tapia*[1,2], *Yunesky Masip Macía*[3], *René André Garrido Lazo*[2,4], *Daniel Serafini*[5], *Ángel Alexander Rodríguez Soto*[3], *Marcelo Mena Carrasco*[6] and *Pablo Rodríguez Machuca*[4]

[1]Industrial Engineering Department, Faculty of Engineering, University of Santiago of Chile (USACH), Santiago, Chile; [2]Program for the Development of Sustainable Production Systems (PDSPS), Industrial Engineering Department, Faculty of Engineering, University of Santiago of Chile (USACH), Santiago, Chile; [3]Escuela de Ingeniería Mecánica, Pontificia Universidad Católica de Valparaíso, Quilpué, Chile; [4]Facultad de Ingeniería, Departamento de Geografía, Universidad de Santiago de Chile, Santiago, Chile; [5]Departamento de Física, Universidad de Santiago de Chile, Santiago, Chile; [6]Escuela de Ingeniería Bioquímica, Centro de Acción Climática, Quilpué, Chile

1. Introduction

Globally, climate change has been proven to bring diverse environmental and social problems (Sepúlveda et al., 2021), and the increasing greenhouse gases due to human activities have triggered it. Moreover, approximately 73% of greenhouse gas (GHG) emissions worldwide originate from the energy sector (Grant et al., 2021). Consequently, the demand for environmentally friendly energy has increased, bringing the need for innovative and sustainable

Evolution and Trends of Sustainable Approaches
https://doi.org/10.1016/B978-0-443-21651-0.00013-9

energy production, storage, and distribution solutions. In this context, green hydrogen has emerged as an attractive proposal as a pure energy source because its production produces no greenhouse gas pollution and can be used in various ways, from power generation to transportation (Tashie-Lewis and Nnabuife, 2021). However, several challenges remain for green hydrogen to be a viable and practical option. For instance, decision-making in the green hydrogen supply chain is critical to reaching sustainability. A sustainable strategy for green hydrogen production and distribution is critical for addressing economic and technological issues and environmental and societal consequences. Sustainable green hydrogen production, storage, and distribution can help reduce greenhouse gas emissions and achieve climate objectives while providing economic and social opportunities for local communities and minimizing adverse environmental effects. Different decision-making tools and platforms in the green hydrogen value chain have been created to address these concerns by improving the planning, design, and administration of green hydrogen production, transport, and use, as described throughout this study. Therefore, this research aims to assess if literature developments have evolved to be implemented as decision support tools for designing the hydrogen value chain. This comparison aims to provide relevant stakeholders in the hydrogen industry and academics interested in this area with helpful information to make accurate decisions and enhance the sustainability and effectiveness of the green hydrogen supply chain. To achieve this purpose, this research systematically analyzes the literature on decision-making in the green hydrogen value chain and the tools or platforms created for the same purpose. Characteristics and functionalities are recognized, followed by a critical and comparative analysis.

2. Methodology

A search was conducted for scientific articles in which tools or methodologies for decision-making are developed. The details of each search are specified further. It is essential to mention that the terms "supply chain" and "value chain" are used interchangeably in the reviewed literature. Therefore, throughout this chapter, reference is made to both the value and supply chains, depending on how each author has defined them. However, the value chain of green hydrogen considers an additional phase: hydrogen conditioning (Masip Mací et al., 2021).

The methodology implemented for the literature review consists of three principal stages. The first stage is to execute a strategy search in which (1) keywords associated with the research subject, (2) the search equation, (3) the search period, and (4) the database used to obtain the research documents are defined. The keywords used were "hydrogen," "green hydrogen," "model," "tool," "platform," "decision making," and "decision-making." These keywords were selected since articles are being sought that address models, tools, or platforms supporting decision-making regarding hydrogen, whether green or not. As for the database, Web of Science and Scopus were used. In the final stage, studies are selected according to selection criteria related to the research topic and using review procedures such as reading abstracts and full articles to choose the most relevant articles for the topic of interest. Those dealing with implementing models, methods, or systems to assist decision-making for hydrogen as an energy vector were chosen.

3. Results

As a result of the methodology application, we selected 32 articles according to the selection criteria.

A cooccurrence analysis was conducted using the 32 selected articles according to the selection criteria. In Fig. 13.1, the cooccurrence analysis of keywords from the documents obtained from Scopus is presented. By employing the VOSviewer software, two clusters of keywords were identified, all of which are related to the concept of decision-making.

Upon examining Figs. 13.1–13.3, it is evident that most of the articles primarily focus on decision-making, hydrogen production, and storage, which are pivotal phases in the hydrogen supply chain. Additionally, topics such as multiobjective optimization, multicriteria decision-making, and location selection are addressed.

Fig. 13.2 illustrates the temporal evolution of keywords found in the selected articles. It can be observed that terms such as "integer programming" and "multiobjective optimization" were more prevalent in years prior to 2019. Starting in 2019, there was a noticeable increase in the number of articles related to decision-making and location selection. In the year 2022, articles related to multicriteria decision-making were introduced.

It is worth noticing that sensitivity analysis is a topic that has a connection with the different terms presented in these 32 articles, indicating its relevance.

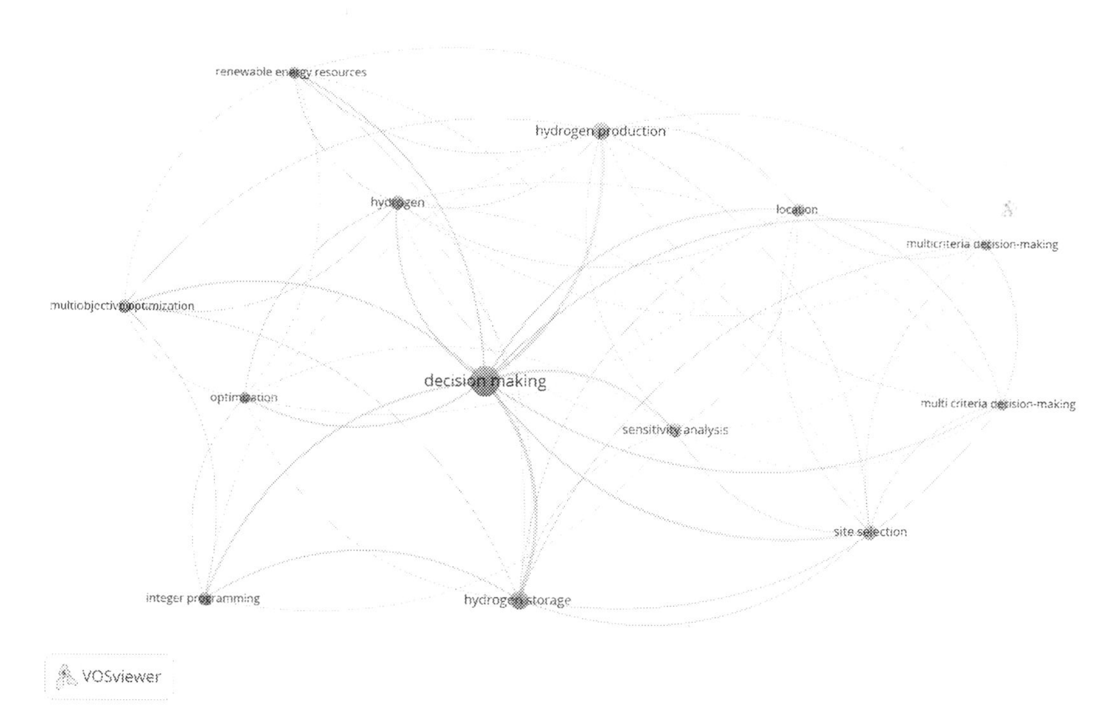

FIGURE 13.1 Keyword co-occurrence network of the 32 articles selected.

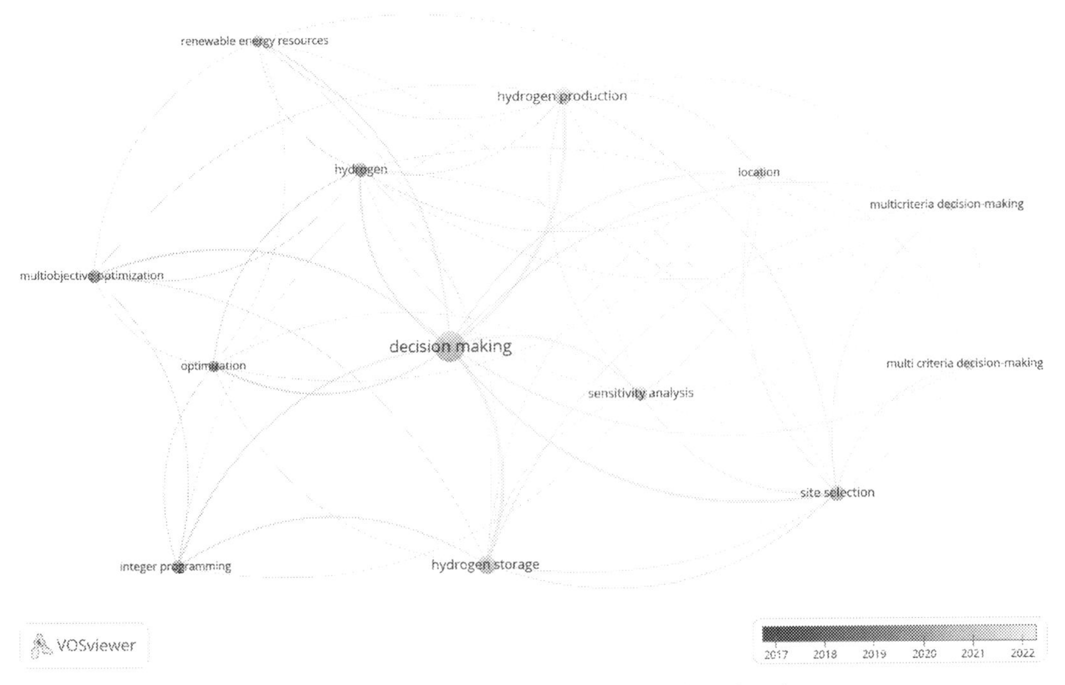

FIGURE 13.2 Density map of the 32 articles selected.

The 32 selected articles were examined in depth, and their analysis can be found in Table 13.1. Thus, Table 13.1 summarizes the 32 articles obtained from the literature review, each briefly describing its purpose, sustainable criteria, and country of each article. These articles were analyzed and classified according to the sustainability dimension studied (economic, social, political, environmental, and technological), where the economic dimension involves economic variables that determine the performance of the organization and its economic viability; the social dimension is referred to the effects of the development of green hydrogen on society, both positive and bad, including societal acceptance and governmental support; the political dimension is understood as the legal and governmental policies, such as taxes and subsidies (Espinoza Pérez et al., 2017), and building restrictions in specific locations, for example, in areas inhabited by native people or protected areas; the environmental dimension refers to the reduction in the emission level of carbon dioxide, the environmental impact, global warming potential, and wastewater management; and the technological dimension is understood as the production, transport, and storage technologies available for the green hydrogen industry. Furthermore, it is also interesting to categorize the countries where these investigations and case studies have been developed.

Regarding sustainability dimensions, the results show that the articles assessed several sustainable aspects simultaneously. For example (Alexopoulos et al., 2022) used geographic information systems to select the optimal site for storing hydrogen produced by wind energy

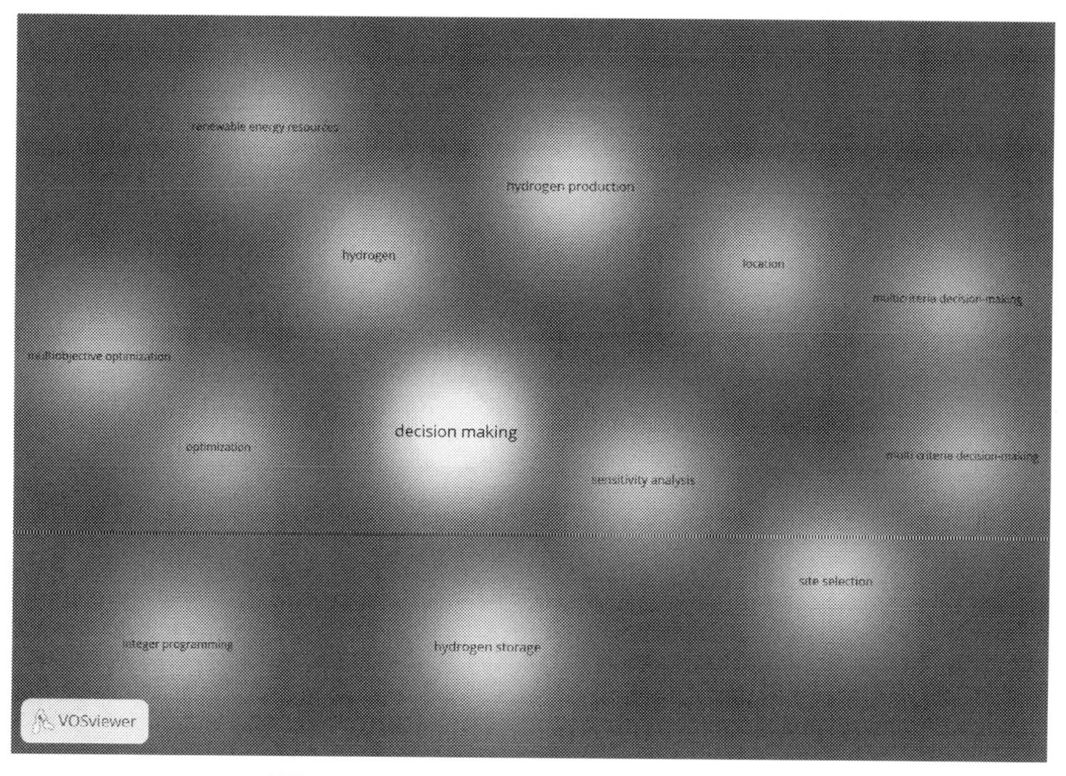

FIGURE 13.3 Cluster map of the 32 articles selected.

using three sustainability criteria: environmental, economic, and social. These criteria include ecological impact, noise impact, investment cost, payback, resident acceptance, and government support. On the other hand, in terms of societal elements, it incorporates social and political acceptance and the creation of new employment, among others.

Fig. 13.4 depicts the sustainability combination elements, with 87.5% accounting for economic considerations. For a better understanding, out of the 28 articles that consider economic factors in their evaluation, 21 include environmental aspects, and similarly, 14 of them consider social factors.

The models found in the research papers serve various purposes, which means they have a precise aim regarding how they aid decision-making. These decisions were divided into four categories: (1) design decisions, (2) location decisions, (3) sustainable production decisions, and (4) other decisions. For example (Oner and Khalilpour, 2022) suggest evaluating 11 hydrogen transporters using a multicriteria decision-making support tool.

It is important to note that, regarding sustainability criteria, it is common for research articles to incorporate more than one sustainability criterion. However, the research articles concentrate on economic and environmental elements.

TABLE 13.1 Summary of literature evaluated.

References	Purpose	Sustainability evaluated criteria					Country
		Eco.	Env.	Soc.	Pol.	Tec.	
Alexopoulos et al. (2022)	Evaluate 23 options for flexibility in terms of energy management.	x				x	No applied
Oner and Khalilpour (2022)	Develop an MCDM model to evaluate multiple hydrogen carriers.	x	x				AU & JP
Tebibel (2022)	Develop a decision-making tool to solve multi-objective optimization problems with constraints to minimize cost and energy losses and maximize the reliability of a small-scale wind to hydrogen (WindtH2) system subject to various constraints.	x				x	DZ
Ao Xuan et al. (2022)	Determine the appropriate location for solar hydrogen production plants in 13 provinces of Uzbekistan.	x					UZ
Hosseini Dehshiri and Hosseini Dehshiri (2022)	Determine suitable areas for building wind farms to produce hydrogen.	x		x			IR
Almutairi (2022)	Identify the best location for hydrogen generation from wind energy using multiple criteria decision-making (MCDM) techniques.			x		x	SA
Gao et al. (2021)	Develop a scientific method for the location of photovoltaic coupled hydrogen storage (PVPCHS) projects.	x	x	x			MN
Wu et al. (2021a)	Provide a practical model for the decision-making of the site selection of the photovoltaic hydrogen production project (PVHPP) based on the multi-criteria decision-making method (MCDM).	x	x	x		x	CN
Zhou et al. (2021a)	Optimize the configuration of the integrated natural gas-wind-photovoltaic-hydrogen energy system and analyze the optimization results to serve decision-making.	x	x				CN
Zhou et al. (2021b)	Build a two-stage location optimization model based on big data of geographic information to provide a special guide for the HRS site decision.	x		x			CN
Wu et al. (2021b)	Optimal site selection for hydrogen storage with wind energy using GIS and MCDM approach.	x	x	x			CN
Lin et al. (2021)	Develop a systematic prioritization framework for hydrogen production pathways from a sustainability perspective, addressing selection problems with hybrid information and vagueness in determining criteria weights.	x	x	x			No applied

Zhao et al. (2021)	Identify optimal energy alternatives for investment in hydrogen production.	x	x	x	x	x	No applied
Abdel-Basset et al. (2021)	Evaluate the sustainability of different options for green hydrogen production.	x	x	x		x	No applied
Robles et al. (2020)	Develop a methodological framework for quantifying the potential social benefits of hydrogen fuel cell vehicles.	x	x	x	x		FR
Guleria and Bajaj (2020)	Design an approach to decide the site of the hydrogen energy plant.	x	x	x		x	No applied
Seker and Aydin (2020)	Propose a fuzzy MCDM technique for selecting the location of hydrogen production facilities from hydrogen sulfide decomposition on the north coast of Turkey (Black Sea) in an uncertain environment.	x	x	x		x	TR
Messaoudi et al. (2019)	Develop a decision support system (DSS) that can assist authorities and decision-makers in identifying priority sites for solar resources to produce hydrogen energy for the transportation sector in Algeria, using multiple criteria along with geographic information systems.	x				x	DZ
Matani et al. (2019)	Developing a dynamic and flexible technology roadmap using a combination of the classical roadmap development method with a novel multi-criteria decision-making approach and fuzzy multiple aspects (F-MaMcDm).	x	x	x	x	x	No applied
Galan et al. (2019)	Implement a real-time large-scale optimization decision support tool for hydrogen under uncertainty.						No applied
Bique et al. (2018)	Identify the best hydrogen infrastructure routes taking into account local factors such as the location of hydrogen supply and demand, and the distribution between the location of hydrogen production and hydrogen demand points.	x	x				DE
Wang et al. (2019)	Propose a fuzzy MCDM model for selecting the site of the hydrogen power plant in Vietnam under fuzzy environment conditions.	x	x	x		x	VN
Lewandowska-Śmierzchalska et al. (2018)	Identify the most suitable geologic structure for underground hydrogen storage.						PL
Kim and Kim (2017)	Develop a new optimization-based approach (minimizing the daily total cost of the WPHS system) for strategic planning of a renewable hydrogen supply system using land-based and marine wind energy.	x				x	KR
Almansoori and Betancourt-Torcat (2016)	Optimize the hydrogen supply chain under emission constraints.	x	x		x	x	DE

(*Continued*)

TABLE 13.1 Summary of literature evaluated.—cont'd

References	Purpose	Sustainability evaluated criteria					Country
		Eco.	Env.	Soc.	Pol.	Tec.	
De-León Almaraz et al. (2015)	Design a five-step HSC (energy source, production, transportation, storage and refueling station) using multi-objective optimization, minimizing cost, environmental impact, and security risk for a multi-period problem. (2) Examine the robustness of the developed methodology and the impact of territory discretization and economies of scale for two HSCs. (3) Identify the main challenges in adapting the model when studying different geographical scales for the HSC. (4) Use a geographical tool before and after optimization to have a more precise snapshot of the HSC taking into account geographical limitations.	x	x				FR
Yu (2014)	Evaluate hydrogen production technologies in China, based on the intuitional theory of fuzzy sets with interval values.	x	x	x	x		CN
De-León Almaraz et al. (2013)	Design a three-level HSC (production, storage, and transportation) taking into account the minimum cost, minimum environmental impact, and minimum security risk.	x	x				GB
Han et al. (2013)	Develop a mathematical model to design hydrogen infrastructure with a profitable H_2 supply, safety, and CO_2 mitigation.	x	x				KR
Hong et al. (2022)	Find a set of optimal solutions by performing a multi-objective optimization that simultaneously satisfies the maximization of the thermal efficiency of the SMR process and the minimization of CO_2 emissions from the process.		x	x			No applied
Almutairi et al. (2021)	Determine the best locations in the fars province for hydrogen production from energy produced by wind and solar hybrid stations.	x	x	x	x		IR
Chang et al. (2012)	Develop an evaluation model to assess hydrogen fuel cell applications and thus provide a decision-making tool for decision-makers.	x	x	x			TW

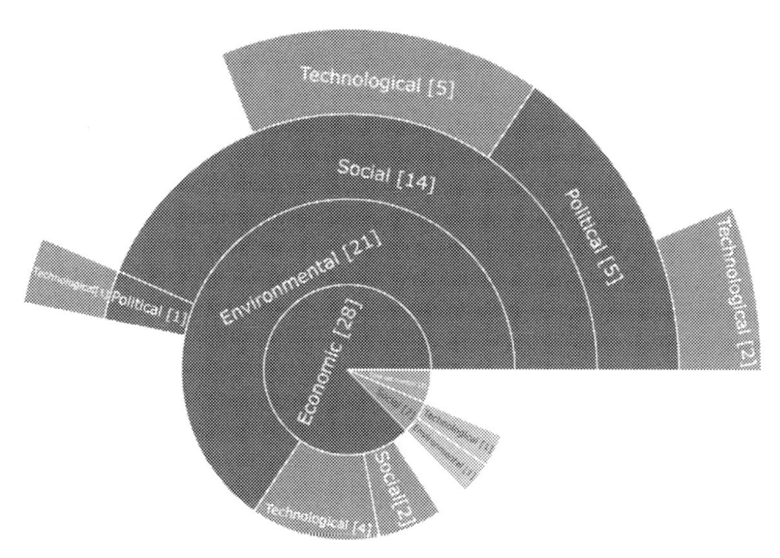

FIGURE 13.4 Sustainability criteria results in the literature.

4. Discussion

A search for decision support tools was conducted to contrast the literature findings with those currently available in the real world. Since indexed scientific journals do not explicitly mention platforms that assist in the decision-making process involved in designing the green hydrogen value chain, the Google platform served for a second search. We selected keywords associated with green hydrogen tools and the decision-making process, such as "green hydrogen tools," "tool for decision making in green hydrogen," and "platform for decision making in green hydrogen." The selection criteria focused on platforms that effectively support the hydrogen value or supply chain decision-making, which corresponds to remove from the scope information platforms. For example, (H2Valleys—Mission Innovation Hydrogen Valley Platform) only shows the location of hydrogen projects worldwide (cofinanced by the European Union).

Seventeen tools were found that effectively support decision-making in the hydrogen value chain or supply chain. These platforms cover simulation software, models, websites, and Excel spreadsheets; even some platforms use Geographic Information Systems (GIS) to visualize factors such as the cost of producing hydrogen in a specific area. These tools were analyzed and classified by (1) the evaluated aspects, (2) the platform purpose, and (3) the country. The evaluated aspects consider sustainability dimensions (economic, environmental, social, technological, and political). Table 13.2 shows the results, where Ec. means Economic, Env. means environmental, Soc. means social, Pol. means political, and Tec. means Technological.

Regarding the evaluated aspects, Fig. 13.5 shows that most platforms consider multiple criteria.

TABLE 13.2 Summary of real-world tools evaluated.

References	Purpose	Sustainable criteria evaluated					Country
		Ec.	Env.	Soc.	Pol.	Tec.	
Geoscience Australia	Data portal allows access to free geo-scientific data and tools (including the previous one) to map and understand the potential of hydrogen pro- duction in Australia subject to spatial constraints.	x	x	x		x	AU
NRELa	Provides access to geospatial data on hydrogen demand, re- sources, infrastructure, costs, production, and distribution.	x	x				US
NRELb	Provides hydrogen production models that deliver transparent reports on the design assumptions and a consistent cost analysis methodology for hydrogen production in both central and distributed facilities.	x					US
Kuroki et al. (2021)	Provides thermodynamic modeling to track and report changes in temperature, pressure, and mass flows of hydrogen when fueling a vehicle with fuel cells.	x				x	US
NREL (2017)	It provides users with detailed annual financial projections in the form of income statements, cash flow statements, and balance sheets; graphical representation of financial performance parameters.	x					US
Bush et al. (2019)	Provides information that can guide the development of infrastructure and investment decisions in transportation and accelerate the adoption of hydrogen at scale.	x				x	US
HYPOS	Simulation software for evaluating the value chains of hydrogen.	x					DE
HYPOS	Allows for a dynamic view of the value chains that are particularly relevant for green hydro- gen production.	x			x	x	DE
Argonne National Laboratorya	Allows researchers and analysts to evaluate energy per mile and emissions effects of various vehicle and fuel combination based on a full fuel cycle.	x	x			x	US
Argonne National Laboratoryb	Used to analyze the technical economic aspects of the pro- duction, supply and replenishment of hydrogen.	x				x	US
Mathieu (2021)	Simulates the entire hydrogen production process, predicting subsystem interactions and performance.	x	x			x	DE

TABLE 13.2 Summary of real-world tools evaluated.—cont'd

References	Purpose	Sustainable criteria evaluated					Country
		Ec.	Env.	Soc.	Pol.	Tec.	
GHD	The green hydrogen supply chain model was used to: evaluate trade-offs between the cost of renewable energy generation and the utilization of the production plant, compare market routes of those purchasing the hydrogen, and find the optimal product transportation method based on distance.	x				x	RU
Ministry of the Environment and Japan (2021)	Simulate the production of green hydrogen, with the goal of promoting medium and long-term measures against climate change in the local community by analyzing the life cycle of the project.		x				JP
H2 Go Power	Provide predictive algorithms to assist the energy management system by considering weather, prices, and network management factors.		x	x	x		UK
National Technology and Engineering Solutions of Sandia and LLC	Software that integrates data And models to evaluate hydrogen's safety, transportation, and infrastructure.			x	x	x	US
Ministry of Business et al. (2022)	The model identifies plausible scenarios for the future hydrogen economy and identifies key variables that determine how those scenarios could develop.	x	x		x	x	NZ

However, (NRELb), (HYPOS), (NREL, 2017), and (HYPOS) exclusively evaluate economic aspects mainly represented by the net present value. Instead, the platform proposed by the Japanese government only considers environmental factors using a life cycle analysis of hydrogen production (Ministry of the Environment and Japan, 2021). The remaining platforms consider economic (including the interest rate, different costs, and net present value), environmental (including carbon dioxide emission level control and wastewater management), social (including restrictions such as physical spaces where indigenous peoples exist and therefore projects cannot be developed), technological such as type of production technology, political (including regulations), and geospatial aspects. Among those platforms, (Geoscience Australia) highlights integrating five factors: economic, environmental, social, technical, and geospatial. This Australian government platform has a wide range of options for hydrogen production projects, including geographical information of the country to show different characteristics of hydrogen production through a map. For example, the levelized

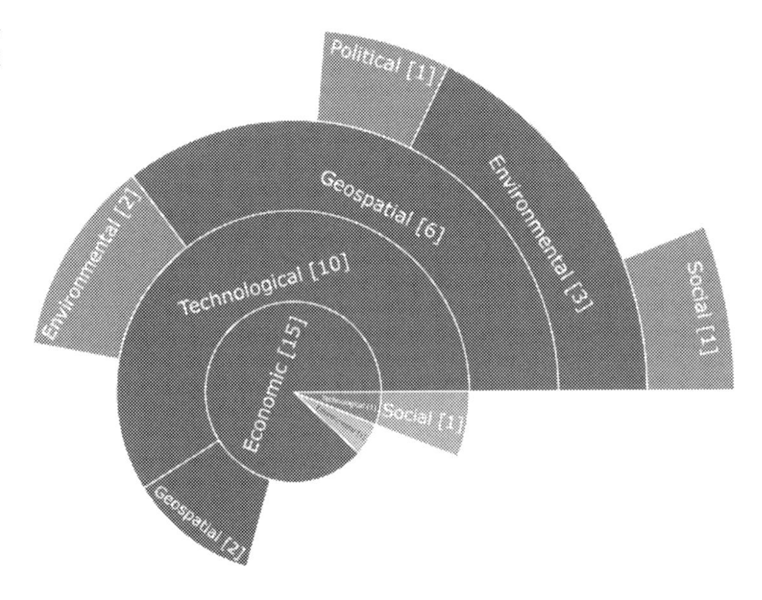

FIGURE 13.5 Sustainability criteria results in the real-world tools.

cost of producing hydrogen (LCOH) in a particular area of Australia depends on the restrictions selected by the user, which generates layers of geographical information where projects cannot be developed: proximity to airports, prohibition of activities in protected areas, sectors where indigenous peoples live.

Considering the purpose of the platforms, 76.5% allow users to simulate processes related to the hydrogen value chain, such as production, providing them with data without the need to incur the risk of tangible value chain implementation failure. In particular, (NRELa) developed a model that allows a private company to design the production plant by simulation, analyze any associated risks, and optimize the return on investment. In addition, this platform evaluates trade-offs between renewable energy generation costs and plant utilization, compares market routes of those buying hydrogen, finds the optimal method of transportation based on distance, compares supplier technologies and evaluates their effectiveness, understands critical cost drivers and how to minimize them, calculates required storage capacity, tests changes, and assesses how they affect the overall green hydrogen life cycle and different parts of the supply chain.

Considering who can decide on the platforms and their functionalities, 94.1% of the support systems provide the information, but the user makes the final decision (16 out of 17 tools). Moreover, (Kuroki et al., 2021) is the only platform in which artificial intelligence does decision-making dynamically, which uses optimization and learning algorithms.

Finally, considering the type of platform, most platforms are freely accessible and developed by different government actors, such as NREL. However, three systems correspond to private companies to help other companies in decision-making, using specific modeling software (Geoscience Australia), (NRELa) (Kuroki et al., 2021).

Comparing the results from the literature and the real-world applications of decision-making tools, they behave similarly when considering the assessed sustainability aspects.

This is depicted in Figs. 13.4 and 13.5, where it is evident that both articles and tools evaluate multiple sustainability criteria simultaneously. However, the research articles concentrate on economic and environmental elements. In contrast, tools and platforms primarily focus on economic and technological aspects. In addition, these platforms include Geographical Information System (GIS) tools.

Another point of contrast between literature and real-world applications of decision-making tools for designing the green hydrogen value chain is that literature primarily focuses on developing mathematical models, frameworks, and tools (self-defined by their authors).

The entire literature article reviewed developed models, as shown in Fig. 13.6. In contrast, tools outside the literature encompass various tool types, including web platforms, software, and tools developed using Excel, as shown in Fig. 13.7.

Considering the countries where the tools and research are developed, Fig. 13.8 shows that only five countries have developed decision-making platforms or tools accessible by Google search. Additionally, it is possible to notice that, despite a large amount of related literature, China lacks decision-making platforms or tools. However, this may be because the platforms are not found when searching in English, or because such public platforms have yet not been developed. In contrast, the United States has a significant presence in platform development, but no scientific articles have been related to those developments. Furthermore, the development of these tools has been concentrated in the United States, China, Russia, and Europe, with scarce progress in Latin America and Africa.

Therefore, we can say that the gap between the literature and real-world uses for decision-making tools lies in the purposes of the tool and the aspects evaluated. On the one hand, the literature has primarily concentrated on creating multicriteria and multiobjective optimization models to acquire an optimum arrangement among several options, such as those intended for selecting green hydrogen production locations and considering several sustainability dimensions at the same time. In distinction, the tools or platforms discovered have a more centralized goal, such as performing a life cycle analysis that primarily focuses on

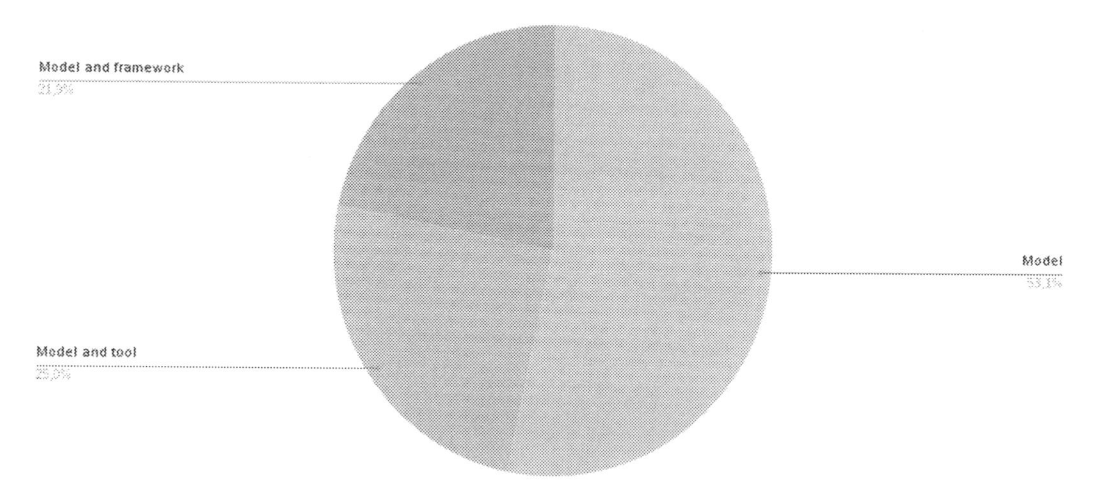

FIGURE 13.6 Type of article developed in the literature.

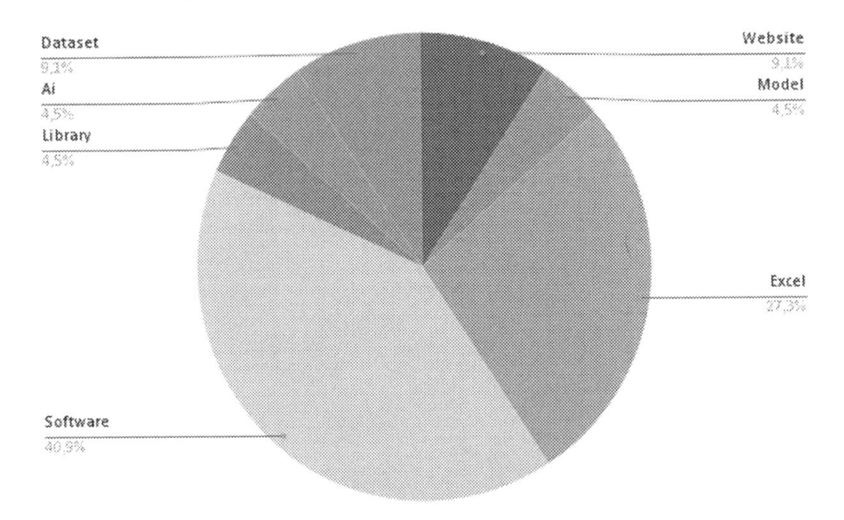

FIGURE 13.7 Type of tool developed in real-world tools.

FIGURE 13.8 Worldwide comparison of literature and real-world tools/platforms adoption by countries.

environmental and social factors or in tools that allow calculating the equilibrium price of the kg of hydrogen produced according to certain variables through a series of data provided by the user, which in no case allows the user to have a vision of how to improve or expand their alternatives, but only calculates.

5. Conclusions

Green hydrogen is a clean energy vector capable of transporting energy cleanly, contributing to the mitigation goals of climate change proposed by various countries to the United Nations (UN). However, its study has only been recently proposed, indicating that further research is still needed to ensure energy security when using it. Therefore, literature authors and tool/platform developers must collaborate in a complementary manner. Furthermore, the literature should broaden its modeling options by accounting for the stochastic behavior of nonrenewable energy sources to be ready to make decisions when these energy sources are unavailable, either by anticipating such a situation and modeling an energy storage system or by supplementing the production plant with other types of energy. In the case of tools, on the other hand, they focus on specific applications in which the user must provide all the data and attempt again and again until they discover a favorable outcome. However, that result is not necessarily optimal. Developing mathematical models based on multiobjective optimization will allow a wider range of energy types, technologies, storage, and distribution options to assist decision-makers in design and operational decisions in determining the optimal configuration while always meeting the organization's objectives. Existing theoretical models can be adapted in the future. Thus, instruments that aid decision-making and encourage the development of new clean-energy initiatives, such as green hydrogen, can help mitigate climate change's negative impacts. Countries with high technological development, such as the United Kingdom, Australia, China, the United States, Japan, and Germany, have developed decision-making support systems and models using hydrogen as an energy vector. However, at the same time, Africa and Latin America lag despite their hydrogen potential. Therefore, a significant challenge is to develop platforms adapted to these territories to support hydrogen development as an energy vector.

In summary, decision support tools for the design of green hydrogen supply chains or value chains play a crucial role in achieving sustainable development goals and moving toward a future that relies less on fossil fuels, thereby contributing to the reduction of negative impacts of climate change. It is also important to note that there may be other tools currently available but were not found in the search conducted. On the other hand, it is worth highlighting those certain technologically advanced countries such as Australia, the United States, and Germany have been able to develop systems, tools, and models that can significantly assist decision-makers in the context of using hydrogen as an energy carrier. However, there are regions across the globe, such as Africa and Latin America, which, despite their renewable energy potential for harnessing this energy vector, have not seen a wide range of studies related to green hydrogen as an energy carrier, missing out on its benefits.

Acknowledgments

The researchers gratefully acknowledge the Agencia Nacional de Investigación y Desarrollo: Project ACT210050 fund for financial support of this study. In addition, the authors gratefully acknowledge the support and constructive suggestions made for this chapter by members of the Program for the Development of Sustainable Production Systems (PDSPS).

References

Abdel-Basset, M., Gamal, A., Chakrabortty, R.K., Ryan, M.J., 2021. Evaluation of sustainable hydrogen production options using an advanced hybrid MCDM approach: a case study. International Journal of Hydrogen Energy 46 (5), 4567–4591. https://doi.org/10.1016/j.ijhydene.2020.10.232.

Alexopoulos, D.K., Anastasiadis, A.G., Kaminaris, S.D., Vokas, G.A., Psomopoulos, C.S., 2022. Assessing flexibility options in power systems using Fuzzy Logic and AHP methods. Energy Reports 8, 776–791. https://doi.org/10.1016/j.egyr.2022.07.079.

Almansoori, A., Betancourt-Torcat, A., 2016. Design of optimization model for a hydrogen supply chain under emission constraints - a case study of Germany. Energy (Oxford, England) 111, 414–429. https://doi.org/10.1016/j.energy.2016.05.123.

Almutairi, K., Mostafaeipour, A., Jahanshahi, E., Jooyandeh, E., Himri, Y., Jahangiri, M., Issakhov, A., Chowdhury, S., Hosseini Dehshiri, S., Hosseini Dehshiri, S., Techato, K., 2021. Ranking locations for hydrogen production using hybrid wind-solar: a case study. Sustainability 13 (8), 4524. https://doi.org/10.3390/su13084524.

Almutairi, K., 2022. Determining the appropriate location for renewable hydrogen development using multi-criteria decision-making approaches. International Journal of Energy Research 46 (5), 5876–5895. https://doi.org/10.1002/er.7528.

Ao Xuan, H., Vu Trinh, V., Techato, K., Phoungthong, K., 2022. Use of hybrid MCDM methods for site location of solar-powered hydrogen production plants in Uzbekistan. Sustainable Energy Technologies and Assessments 52 (101979), 101979. https://doi.org/10.1016/j.seta.2022.101979.

Argonne National Laboratory, (n.d.). The Greenhouse Gases, Regulated Emissions, and Energy Use in Technologies Model. https://greet.es.anl.gov/ (Accessed 29 July 2022).

Argonne National Laboratory, (n.d.). Hydrogen Delivery Infrastructure Analysis. https://hdsam.es.anl.gov/ (Accessed 29 July 2022).

Bique, A.O., Maia, L.K.K., La Mantia, F., Manca, D., Zondervan, E., 2018. Balancing costs, safety and CO_2 emissions in the design of hydrogen supply chains. In: Friedl, A., Klemeš, J.J., Radl, S., Varbanov, P.S., Wallek, T. (Eds.), Computer Aided Chemical Engineering, vol 43. Elsevier, pp. 603–608. https://doi.org/10.1016/B978-0-444-64235-6.50108-X.

Bush, B., Muratori, M., Hunter, C., Zuboy, J., Melaina, M., 2019. Scenario Evaluation and Regionalization Analysis (SERA) Model: Demand Side and Refueling Infrastructure Buildout. National Renewable Energy Laboratory, Golden, CO. NREL/TP-5400- 70090.

Chang, P.-L., Hsu, C.-W., Lin, C.-Y., 2012. Assessment of hydrogen fuel cell applications using fuzzy multiple-criteria decision making method. Applied Energy 100, 93–99. https://doi.org/10.1016/j.apenergy.2012.03.051.

De-León Almaraz, S., Azzaro-Pantel, C., Montastruc, L., Pibouleau, L., Senties, O.B., 2013. Assessment of mono and multi-objective optimization to design a hydrogen supply chain. International Journal of Hydrogen Energy 38 (33), 14121–14145. https://doi.org/10.1016/j.ijhydene.2013.07.059.

De-León Almaraz, S., Azzaro-Pantel, C., Montastruc, L., Boix, M., 2015. Deployment of a hydrogen supply chain by multi-objective/multi-period optimisation at regional and national scales. Chemical Engineering Research and Design: Transactions of the Institution of Chemical Engineers 104, 11–31. https://doi.org/10.1016/j.cherd.2015.07.005.

Espinoza Pérez, A.T., Camargo, M., Narváez Rincón, P.C., Alfaro Marchant, M., 2017. Key challenges and requirements for sustainable and industrialized biorefinery supply chain design and management: a bibliographic analysis. Renewable and Sustainable Energy Reviews 69, 350–359. https://doi.org/10.1016/j.rser.2016.11.084.

Galan, A., de Prada, C., Gutierrez, G., Sarabia, D., Grossmann, I.E., Gonzalez, R., 2019. Implementation of RTO in a large hydrogen network considering uncertainty. Optimization and Engineering 20 (4), 1161–1190. https://doi.org/10.1007/s11081-019-09444-3.

Gao, J., Men, H., Guo, F., Liang, P., Fan, Y., 2021. A multi-criteria decision-making framework for the location of photovoltaic power coupling hydrogen storage projects. Journal of Energy Storage 44 (103469), 103469. https://doi.org/10.1016/j.est.2021.103469.

Geoscience Australia, (n.d.). AusH2 - Australia's Hydrogen Opportunities Tool. https://portal.ga.gov.au/persona/hydrogen (Accessed 29 July 2022).

Geoscience Australia, 2021. Hydrogen Economic Fairways Tool. https://portal.ga.gov.au/persona/heft.

GHD, (n.d.). Planning Green Hydrogen Production and Transportation. https://www.anylogic.com/resources/case-studies/planning-green-hydrogen-production-and-transportation/ (Accessed 29 July 2022).

Grant, D., Zelinka, D., Mitova, S., 2021. Reducing CO_2 emissions by targeting the world's hyper-polluting power plants. Environmental Research Letters 16 (9), 094022. https://doi.org/10.1088/1748-9326/ac13f1.

Guleria, A., Bajaj, R.K., 2020. A robust decision making approach for hydrogen power plant site selection utilizing (R, S)-Norm Pythagorean Fuzzy information measures based on VIKOR and TOPSIS method. International Journal of Hydrogen Energy 45 (38), 18802–18816. https://doi.org/10.1016/j.ijhydene.2020.05.091.

H2 Go Power, (n.d.). HyAi. https://www.h2gopower.com/hyai (Accessed 29 July 2022).

H2Valleys — Mission Innovation Hydrogen Valley Platform, (n.d.). https://h2v.eu/ (Accecessed 29 July 2022).

Han, J.-H., Ryu, J.-H., Lee, I.-B., 2013. Multi-objective optimization design of hydrogen infrastructures simultaneously considering economic cost, safety and CO_2 emission. Chemical Engineering Research and Design: Transactions of the Institution of Chemical Engineers 91 (8), 1427–1439. https://doi.org/10.1016/j.cherd.2013.04.026.

Hong, S., Lee, J., Cho, H., Kim, M., Moon, I., Kim, J., 2022. Multi-objective optimization of CO_2 emission and thermal efficiency for on-site steam methane reforming hydrogen production process using machine learning. Journal of Cleaner Production 359 (132133), 132133. https://doi.org/10.1016/j.jclepro.2022.132133.

Hosseini Dehshiri, S.S., Hosseini Dehshiri, S.J., 2022. Locating wind farm for power and hydrogen production based on Geographic information system and multi-criteria decision making method: an application. International Journal of Hydrogen Energy 47 (58), 24569–24583. https://doi.org/10.1016/j.ijhydene.2022.03.083.

Kim, M., Kim, J., 2017. An integrated decision support model for design and operation of a wind-based hydrogen supply system. International Journal of Hydrogen Energy 42 (7), 3899–3915. https://doi.org/10.1016/j.ijhydene.2016.10.129.

Kuroki, T., Nagasawa, K., Peters, M., Leighton, D., Kurtz, J., Sakoda, N., Monde, M., Takata, Y., 2021. Thermodynamic modeling of hydrogen fueling process from high pressure storage tanks to vehicle tank. International Journal of Hydrogen Energy 46 (42), 22004–22017.

Lewandowska-Śmierzchalska, J., Tarkowski, R., Uliasz-Misiak, B., 2018. Screening and ranking framework for underground hydrogen storage site selection in Poland. International Journal of Hydrogen Energy 43 (9), 4401–4414. https://doi.org/10.1016/j.ijhydene.2018.01.089.

Lin, R., Lu, S., Yang, A., Shen, W., Ren, J., 2021. Multi-criteria sustainability assessment and decision-making framework for hydrogen pathways prioritization: an extended ELECTRE method under hybrid information. International Journal of Hydrogen Energy 46 (24), 13430–13445. https://doi.org/10.1016/j.ijhydene.2021.01.018.

Masip Macía, Y., Rodríguez Machuca, P., Rodríguez Soto, A.A., Carmona Campos, R., 2021. Green hydrogen value chain in the sustainability for port operations: case study in the region of Valparaiso, Chile. Sustainability 13 (24), 13681. https://doi.org/10.3390/su132413681.

Matani, B., Shirazi, B., Soltanzadeh, J., 2019. F-MaMcDm: sustainable green-based hydrogen production technology roadmap using fuzzy multi-aspect multi-criteria decision-making. International Journal of Innovation and Technology Management 16 (08). https://doi.org/10.1142/s0219877019500573.

Mathieu, C., 2021. Green Hydrogen Production Simulation within Simcenter Amesim. https://blogs.sw.siemens.com/simcenter/green-hydrogen-production-simulation-within-simcenter-amesim/.

Messaoudi, D., Settou, N., Negrou, B., Settou, B., 2019. GIS based multi-criteria decision making for solar hydrogen production sites selection in Algeria. International Journal of Hydrogen Energy 44 (60), 31808–31831. https://doi.org/10.1016/j.ijhydene.2019.10.099.

Ministry of Business, Innovation, and Employment, 2022. New Zealand Hydrogen Online Modelling Tool. https://www.mbie.govt.nz/assets/Data-Files/Energy/hydrogen-supply-and-demand-dashboard.xlsm.

Ministry of the Environment, Japan, 2021. Hydrogen Supply Chain Platform. https://www.env.go.jp/seisaku/list/ondanka saisei/lowcarbon- h2-sc/en/index.html.

HYPOS, (n.d.). H2-Index-I. https://www.hypos-eastgermany.de/en/the-goals-of-the-project/hypos-projects/strategy/h2-index-i/ (Accessed 29 July 2022).

HYPOS, (n.d.) H2-Index-II. https://www.hypos-eastgermany.de/en/the-goals-of-the-project/hypos-projects/strategy/h2-index-ii/ (Accessed 29 July 2022).

National Technology and Engineering Solutions of Sandia, LLC. (n.d.). HyRAM+. https://energy.sandia.gov/programs/sustainable-transportation/hydrogen/hydrogen-safety-codes-and-standards/hyram/(Accessed 29 July 2022).

NREL, 2017. H2FAST: Hydrogen Financial Analysis Scenario Tool. https://www.nrel.gov/hydrogen/h2fast.html.

NREL, (n.d.). Hydra. https://gds-files.nrelcloud.org/archive/hydra.zip (Accessed 29 July 2022).

NREL, (n.d.). H2A: Hydrogen Analysis Production Models. https://www.nrel.gov/hydrogen/h2a-production-models.html (Accessed 29 July 2022).

Oner, O., Khalilpour, K., 2022. Evaluation of green hydrogen carriers: a multi-criteria decision analysis tool. Renewable and Sustainable Energy Reviews 168 (112764), 112764. https://doi.org/10.1016/j.rser.2022.112764.

Robles, J.O., Azzaro-Pantel, C., Garcia, G.M., Lasserre, A.A., 2020. Social cost-benefit assessment as a post-optimal analysis for hydrogen supply chain design and deployment: application to Occitania (France). Sustainable Production and Consumption 24, 105–120. https://doi.org/10.1016/j.spc.2020.06.010.

Seker, S., Aydin, N., 2020. Hydrogen production facility location selection for Black Sea using entropy based TOPSIS under IVPF environment. International Journal of Hydrogen Energy 45 (32), 15855–15868. https://doi.org/10.1016/j.ijhydene.2019.12.183.

Sepúlveda, M., Baeza, E., Carrillo, M., Risso, N., Rohten, J., 2021. Hydrogen technologies and PEM fuel cells: a sustainable alternative for Chile. In: En 2021 IEEE International Conference on Automation/XXIV Congress of the Chilean Association of Automatic Control (ICA-ACCA). Valparaíso, Chile, pp. 1–6. https://doi.org/10.1109/ICAACCA51523.2021.9465181.

Tashie-Lewis, B.C., Nnabuife, S.G., 2021. Hydrogen production, distribution, storage and power conversion in a hydrogen economy - a technology review. Chemical Engineering Journal Advances 8 (100172), 100172. https://doi.org/10.1016/j.ceja.2021.100172.

Tebibel, H., 2022. Battery energy storage system for enhancing the electrolyzer capacity factor in small-scale WindtH2 system with a smoothing control strategy: constrained multi-objective Pareto optimization and case study in Algeria. Journal of Energy Storage 52 (105017), 105017. https://doi.org/10.1016/j.est.2022.105017.

Wang, C.-N., Hsueh, M.-H., Lin, D.-F., 2019. Hydrogen power plant site selection under fuzzy multicriteria decision-making (FMCDM) environment conditions. Symmetry 11 (4), 596. https://doi.org/10.3390/sym11040596.

Wu, Y., Deng, Z., Tao, Y., Wang, L., Liu, F., Zhou, J., 2021a. Site selection decision framework for photovoltaic hydrogen production project using BWM-CRITIC-MABAC: a case study in Zhangjiakou. Journal of Cleaner Production 324 (129233), 129233. https://doi.org/10.1016/j.jclepro.2021.129233.

Wu, Y., He, F., Zhou, J., Wu, C., Liu, F., Tao, Y., Xu, C., 2021b. Optimal site selection for distributed wind power coupled hydrogen storage project using a geographical information system based multi-criteria decision-making approach: a case in China. Journal of Cleaner Production 299 (126905), 126905. https://doi.org/10.1016/j.jclepro.2021.126905.

Yu, D., 2014. Hydrogen production technologies evaluation based on interval-valued intuitionistic fuzzy multi-attribute decision making method. Journal of Applied Mathematics 1–10. https://doi.org/10.1155/2014/751249, 2014.

Zhao, Y., Xu, Y., Yüksel, S., Dinçer, H., Ubay, G.G., 2021. Hybrid IT2 fuzzy modelling with alpha cuts for hydrogen energy investments. International Journal of Hydrogen Energy 46 (13), 8835–8851. https://doi.org/10.1016/j.ijhydene.2020.12.211.

Zhou, J., Wu, Y., Zhong, Z., Xu, C., Ke, Y., Gao, J., 2021a. Modeling and configuration optimization of the natural gas-wind-photovoltaic-hydrogen integrated energy system: a novel deviation satisfaction strategy. Energy Conversion and Management 243 (114340), 114340. https://doi.org/10.1016/j.enconman.2021.114340.

Zhou, J., Wu, Y., Tao, Y., Gao, J., Zhong, Z., Xu, C., 2021b. Geographic information big data-driven two-stage optimization model for location decision of hydrogen refueling stations: an empirical study in China. Energy (Oxford, England) 225 (120330), 120330. https://doi.org/10.1016/j.energy.2021.120330.

CHAPTER

14

Digital technologies for fostering sustainability in Industry 4.0

Harpreet Kaur Channi[1] *and Raman Kumar*[2]

[1]Department of Electrical Engineering, Chandigarh University, Mohali, Punjab, India; [2]Department of Mechanical and Production Engineering, Guru Nanak Dev Engineering College, Ludhiana, Punjab, India

1. Introduction

"Industry 4.0," often known as the "fourth industrial revolution," refers to unprecedented technical innovation in the manufacturing sector. It's defined by a new degree of automation, data sharing, and real-time connection between equipment, goods, and systems thanks to incorporating digital technology and the Internet into conventional industrial processes. There are many different technologies included in Industry 4.0. Some examples include the Internet of Things (IoT), Artificial Intelligence (AI), Machine Learning (ML), Robotics, Cloud Computing, and Big Data Analytics. These developments allow businesses to set up "smart factories" and "smart supply chains," where computers and other automated technologies work to analyze data and improve manufacturing methods. Companies may reduce costs, boost product quality, reduce manufacturing times, and increase their worldwide competitiveness by using the potential of Industry 4.0 (Ortega-Gras et al., 2021). While technological progress has many benefits, it raises new challenges, such as training a more capable workforce and handling privacy and security issues. However, the revolutionary potential of Industry 4.0 is vast, allowing for a more connected, flexible, and digitally driven production world.

The fourth industrial revolution, or Industry 4.0, is defined by the widespread use of digital technology in production. Opportunities for economic development, innovation, and efficiency are higher than ever, thanks to the advent of the Fourth Industrial Revolution. While these innovations are exciting, the environmental concerns that come along with Industry 4.0 are not. Sustainable development objectives are threatened by the rapid expansion in output, energy use, resource depletion, and carbon emissions. Therefore, it is crucial to investigate and adopt methods that balance economic development with ecological consciousness (Thiede, 2021).

Evolution and Trends of Sustainable Approaches
https://doi.org/10.1016/B978-0-443-21651-0.00016-4

Within the framework of Industry 4.0, digital technologies are emphasized here as playing a critical role in promoting sustainability. The Internet of Things (IoT), AI, ML, blockchain, and integrating renewable energy sources can significantly improve environmental results and encourage more sustainable business practices globally. Industries may improve resource utilization, process efficiency, waste management, and environmental impact by using these technologies. This chapter delves further into the potential for Industry 4.0 to leverage digital technologies to enhance environmental friendliness. We'll investigate how Internet of Things (IoT) devices can provide timely data insights to help organizations reduce their environmental impact. In this article, we'll examine how AI and ML are reshaping several sectors by enhancing processes such as predictive maintenance and energy efficiency. We will also discuss how blockchain has the potential to promote environmentally responsible buying by increasing supply chain transparency (Kurniawan et al., 2022a).

This study sheds light on how businesses might use state-of-the-art technology to improve the environment and sustain economic development and competitiveness by examining the interplay between sustainability and digitalization in the context of Industry 4.0. Stakeholders, politicians, and industry participants must work together to embrace and execute these sustainable digital solutions to create a future in which technology advancement and environmental responsibility are not in conflict. This study aims to aid in comprehending and practical use of digital technologies as catalysts for supporting sustainability in the industrial environment as the path toward sustainable Industry 4.0 develops (Hassoun et al., 2022a; Mubarak et al., 2021; Kunkel et al., 2022). Nine digital technologies that support Industry 4.0 are discussed further, as shown in Fig. 14.1.

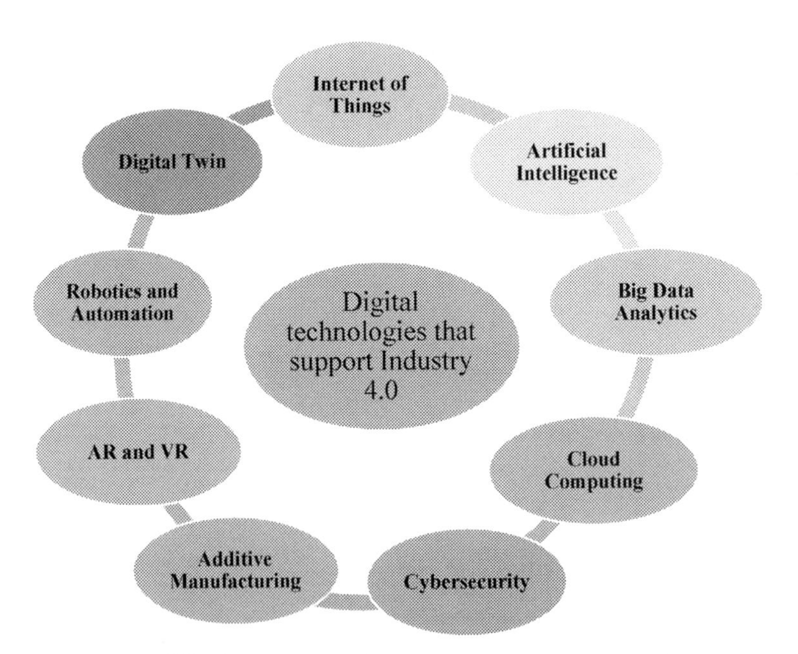

FIGURE 14.1 Digital technologies that support Industry 4.0.

- **Internet of Things (IoT):** IoT involves connecting physical devices and sensors to the Internet, enabling data exchange and real-time communication between machines, products, and people. This connectivity allows for better monitoring, analysis, and optimization of industrial processes.
- **Artificial Intelligence (AI):** AI technologies, such as machine learning and deep learning, play a crucial role in Industry 4.0 by enabling autonomous decision-making, predictive maintenance, quality control, and optimizing production processes.
- **Big Data Analytics:** The vast amounts of data generated by IoT devices and various sensors in the manufacturing environment require advanced analytics. Big data analytics helps identify patterns, trends, and insights to improve efficiency and productivity.
- **Cloud Computing:** Cloud services provide scalable and flexible computing resources, enabling industries to store and process large datasets, run complex simulations, and collaborate efficiently across geographies.
- **Cybersecurity:** With increased connectivity and data exchange, Industry 4.0 presents new cybersecurity challenges. Robust cybersecurity measures are essential to protect sensitive data, intellectual property, and critical infrastructure from cyber threats.
- **Additive Manufacturing (3D Printing):** 3D Printing allows for the creation complex, customized parts and products directly from digital designs. It reduces lead times, lowers production costs, and supports rapid prototyping.
- **Augmented Reality (AR) and Virtual Reality (VR):** AR and VR technologies enhance worker training, maintenance, and troubleshooting processes by overlaying digital information in the real-world environment.
- **Robotics and Automation:** Robotics and advanced automation can handle repetitive tasks with precision and speed. Collaborative robots (cobots) work alongside human workers, enhancing productivity and safety.
- **Digital Twin:** A digital twin is a virtual representation of a physical product, process, or system. It allows real-time monitoring, analysis, and optimization of the physical counterpart, enabling predictive maintenance and continuous improvement. These digital technologies collectively transform traditional industries into intelligent, efficient, interconnected systems defining Industry 4.0 (Cricelli and Strazzullo, 2021).

1.1 Importance of sustainability in Industry 4.0

As a new wave of industrial revolution, Industry 4.0 will have far-reaching effects on the natural world, human culture, and the economy, making sustainability a top priority. Sustainability is essential in the context of Industry 4.0 for several reasons (Lepore et al., 2021).

- **Environmental Conservation:** The environmental concerns of rising energy consumption, resource depletion, and pollution may be exacerbated by the adoption of Industry 4.0 technologies, even though they provide significant advantages in terms of productivity and efficiency. Sustainable practices are crucial for reducing these adverse effects and protecting the planet's natural resources for future generations.
- **Climate Change Mitigation:** Greenhouse gas emissions are a leading cause of global warming, and industrial activities are a vital source. Industry 4.0 can be crucial in

international efforts to minimize global warming and its repercussions by adopting sustainable practices and lowering carbon footprints.

- **Resource Efficiency:** Regarding raw materials, water, and energy, sustainable practices in Industry 4.0 may lead to optimization. Industries may save money and lessen their environmental impact by cutting down on trash and increasing recycling efforts.
- **Social Responsibility:** Disruptions in the labor market caused by Industry 4.0's technological advances have the potential to have far-reaching effects on people's livelihoods and communities. Promoting sustainability guarantees a responsible attitude to change, promoting equitable working conditions, ethical supply chains, and inclusive development for all parties involved.
- **Resilient Supply Chains:** By encouraging diversity, minimizing reliance on scarce resources, and planning for unexpected disruptions brought on by climate-related disasters or geopolitical concerns, sustainable practices may improve supply chain resilience.
- **Enhanced Competitiveness:** These days, sustainability is not just about doing the right thing; it's a significant consideration for customers and investors. Businesses prioritizing ecological considerations in their operations benefit from an influx of environmentally aware consumers and financiers.
- **Regulatory Compliance:** Environmental rules are being enacted by governments throughout the globe as a means of combating climate change and environmental deterioration. Adopting a sustainable approach in Industry 4.0 guarantees conformity with existing and new rules, protecting businesses from possible liabilities.
- **Long-Term Viability:** Long-term expenditures, supply chain disruptions, and brand harm may await enterprises that do not incorporate sustainability into their operations as resources grow scarcer and environmental concerns escalate.
- **Innovation and R&D Opportunities:** Industry 4.0's pursuit of sustainability encourages creativity and exploration. Innovations in green technology, renewable energy sources, and circular economies have potential uses outside the industrial sector.
- **Global Collaboration:** Collaboration between governments, businesses, academic institutions, and civil society is essential for Industry 4.0 to be sustainable. A focus on sustainability facilitates international collaboration toward common environmental objectives.

Finally, sustainability is important to Industry 4.0 since it affects the world and the bottom line. Sustainable practices and technology reduce negative environmental effects and provide businesses with a leg up in the fast-paced, competitive, and uncertain world of the fourth industrial revolution. Industry 4.0 has the potential to be a driving force for a more sustainable and inclusive future if it adopts sustainability as a guiding principle (Tavera et al., 2021; Di Maria et al., 2022; da Silva and Sehnem, 2022).

2. Sustainable development and Industry 4.0

Global economies and society have a lot riding on the intersection of sustainable development and Industry 4.0. Sustainable development aims to provide for the needs of the present without sacrificing the capacity of future generations to provide for their requirements regarding the environment, social justice, and economic well-being. In contrast, "Industry 4.0" refers to incorporating digital technology into traditional industrial methods to boost output, productivity, and creativity. These two ideas must coincide to combat climate change, resource depletion, and pollution (Sharma et al., 2021). If implemented properly, Industry 4.0 can greatly accelerate the adoption of renewable energy sources and improve resource efficiency. Sustainable practices, ethical sourcing, and environmentally friendly goods may all be driven by data-driven choices made possible by the Internet of Things (IoT), Artificial Intelligence (AI), and other digital technologies. In addition, sustainable development offers the overarching framework to ensure that the social responsibility, inclusive growth, and fair distribution of benefits brought about by Industry 4.0 are properly used. Sustainability in the context of Industry 4.0 paves the path for a sustainable and prosperous future for generations by helping society establish resilient economies and promote a more harmonious connection with the environment (Mukhuty et al., 2022; Zhang et al., 2021; Mourtzis et al., 2022).

2.1 Integration of sustainability in Industry 4.0

For sophisticated digital technologies to be used in manufacturing and production responsibly and ethically, sustainability must be included in the Industry 4.0 framework. Sustainability principles and the disruptive power of Industry 4.0 may work together to benefit the environment, society, and the economy (Kumar et al., 2021a). Fig. 14.2 depicts some prominent methods that emphasize incorporating sustainability in Industry 4.0.

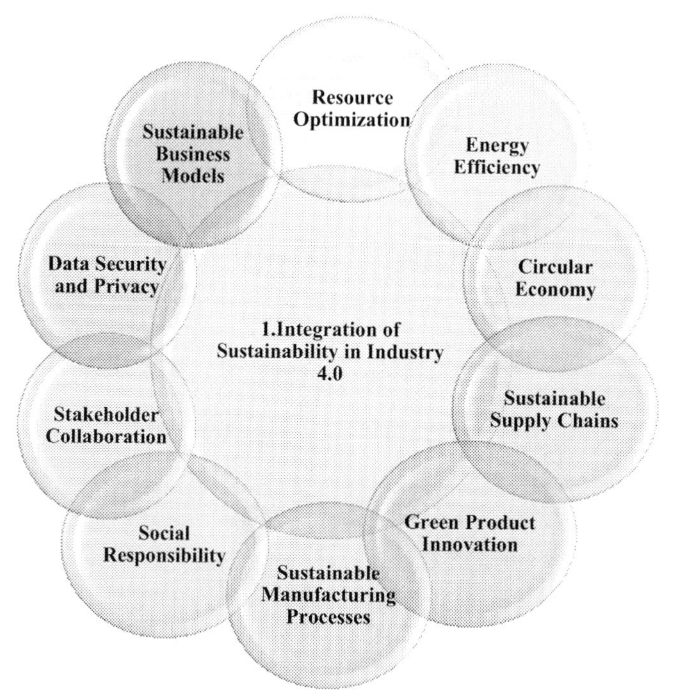

FIGURE 14.2 Integration of sustainability in Industry 4.0.

- **Resource Optimization:** By collecting and analyzing data in real time, Industry 4.0 tools such as the Internet of Things and artificial intelligence allow for more efficient use of resources. Industries may reduce waste and environmental effects by monitoring their energy use, water consumption, and raw material inputs.
- **Energy Efficiency:** Reducing carbon emissions and dependency on fossil fuels may be accomplished by implementing sustainable practices, such as using renewable energy sources in the production process. Energy efficiency may be improved by using smart energy management systems driven by Industry 4.0 technology.
- **Circular Economy:** The concepts of the circular economy, which include reducing waste, reusing and recycling resources, and maximizing the lifespan of products, may be further advanced by implementing Industry 4.0. The industry may take a more sustainable approach to resource usage with the help of digital technologies, such as those that allow for the design of goods for disassembly and the implementation of closed-loop manufacturing processes.
- **Sustainable Supply Chains:** Using blockchain technology, supply chains may be more transparent and traceable, which helps ensure ethical sourcing techniques and reduces the environmental effect of extracting, transporting, and producing raw materials.
- **Green Product Innovation:** Rapid prototyping and product modification are made possible by Industry 4.0, paving the way for creating environmentally friendly goods and long-term fixes. Green design concepts may be included with the help of digital technology, reducing negative effects on the environment at all stages of the product's life cycle.
- **Sustainable Manufacturing Processes:** Optimizing manufacturing using AI and ML means less waste and higher-quality output. Predictive maintenance, made possible by Industry 4.0 tools, helps avoid malfunctions, which cuts down on downtime and waste.
- **Social Responsibility:** The importance of social responsibility cannot be overstated when discussing the integration of sustainability in Industry 4.0. Fair working conditions, a diverse and inclusive workforce, and ethical labor practices are all possible thanks to technological advancements such as AI.
- **Stakeholder Collaboration:** Sustainability in Industry 4.0 can only be achieved through concerted efforts from governments, businesses, academic institutions, and the general public. They may encourage sustainable practices in the industry by working together to create and execute standards, rules, and initiatives.
- **Data Security and Privacy:** Data security and privacy are crucial as digital technologies grow more pervasive in Industry 4.0. Trustworthiness and long-term viability in digital transformation rely on the data's ethical and secure usage.
- **Sustainable Business Models:** Rather than concentrating simply on short-term earnings, companies that embrace sustainability in Industry 4.0 may develop novel business models that put long-term value creation, social impact, and environmental stewardship front and center.

Ultimately, achieving success in sustainability integration in Industry 4.0 is a complex task that calls for a comprehensive strategy. Industries may improve their impact on the environment, society, and the economy by responsibly using digital technologies and coordinating them with sustainable development objectives (Ciliberto et al., 2021; Gebhardt et al., 2022;

Fonseca et al., 2021). When sustainability and Industry 4.0 come together, it creates a powerful potential to solve global problems and pave the road for a more environmentally friendly and economically stable manufacturing sector.

2.2 Role of digital technologies in Industry 4.0

Industry 4.0 relies heavily on digital technologies since they power the revolutionary shifts and improvements that define the fourth industrial revolution. Collectively referred to as the "digital toolbox," these technologies are crucial in evolving production techniques, distribution networks, and business designs (Tripathi and Gupta, 2021). Fig. 14.3 shows examples of important digital technologies and their roles in Industry 4.0.

- **Internet of Things (IoT):** The Internet of Things (IoT) facilitates the collection and sharing of data between physical devices and things over the Internet. Sensors and other Internet of Things (IoT) devices are built into Industry 4.0 equipment, goods, and assembly lines to collect data in real time. Predictive maintenance, process optimization, and improved automation are all made possible by this data-driven strategy, resulting in more productivity and less downtime.

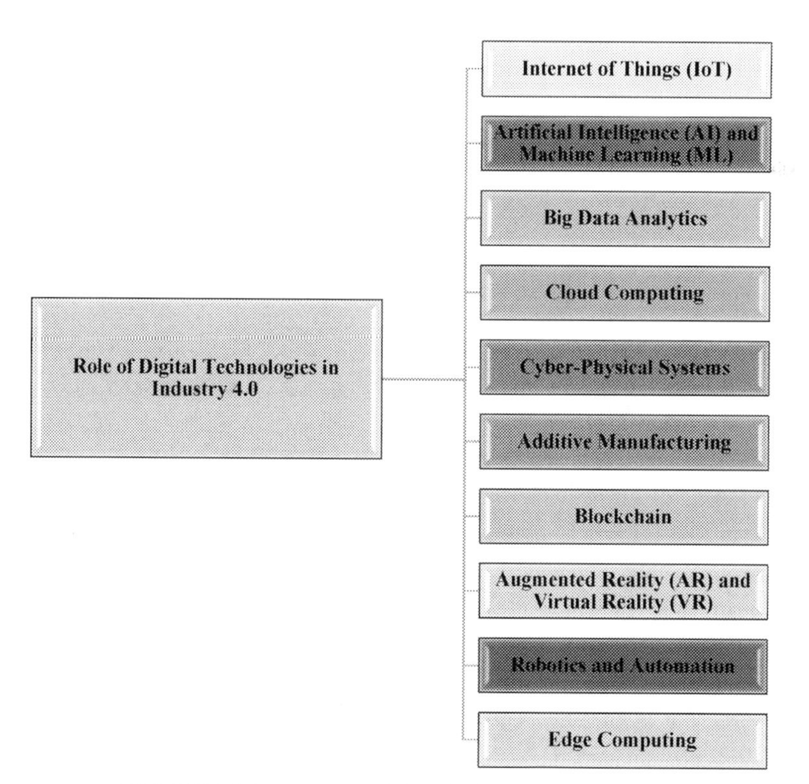

FIGURE 14.3 Role of digital technologies in Industry 4.0.

- **Artificial Intelligence (AI) and Machine Learning (ML):** For robots to mimic human intellect and learn from data patterns, artificial intelligence (AI) and machine learning (ML) are essential components of Industry 4.0. Industries may improve their output and quality thanks to AI-powered analytics that help them conclude data. Improvements in robots and autonomous systems are also driven by AI and ML, which streamline industrial processes and make more complicated jobs possible (Belhadi et al., 2022; Dilyard et al., 2021; Ghobakhloo et al., 2021).
- **Big Data Analytics:** Massive volumes of information are produced by Industry 4.0's many moving parts, such as Internet of Things (IoT) gadgets, manufacturing procedures, and supply chains. By processing and analyzing these data using big data analytics technologies, useful patterns, trends, and correlations are uncovered that may guide strategic decision-making and process improvement. It helps businesses spot waste, plan for future demand, and improve operations.
- **Cloud Computing:** Businesses may store and analyze data on distant servers thanks to cloud computing, which offers scalable and flexible computer resources. With cloud computing, distributed teams can share and collaborate on real-time data, regardless of location, saving time and money while improving productivity in Industry 4.0.
- **Cyber-Physical Systems:** Cyber-physical systems combine digital and physical components to facilitate interaction and collaboration between machines and infrastructures. These technologies provide real-time monitoring, control, and feedback loops to improve industrial responsiveness and efficiency.
- **Additive Manufacturing (3D Printing):** Additive manufacturing is a game-changing process that allows for gradually constructing intricate parts. It revolutionizes conventional manufacturing processes and encourages sustainable practices by facilitating quick prototyping, customization, and decreased material waste.
- **Blockchain:** Transactions are transparent, secure, and verifiable thanks to distributed ledger technology. Industry 4.0 relies on blockchain technology for its secure data sharing, smart contracts, and trackable supply chains, fostering greater stakeholder confidence and encouraging more environmentally friendly procurement methods.
- **Augmented Reality (AR) and Virtual Reality (VR):** In the context of Industry 4.0, augmented and virtual reality technologies improve training, maintenance, and design. They boost output and efficiency by facilitating access to real-time data visualization and scenario simulation.
- **Robotics and Automation:** The concept of "smart factories" at the heart of Industry 4.0 relies heavily on robotics and automation technology. Working in tandem with people, collaborative robots (cobots) streamline repetitive operations, improve accuracy, and lessen risks.
- **Edge Computing:** With edge computing, the processing is done closer to the data source, allowing faster response times and instantaneous decisions. Applications such as autonomous cars and predictive maintenance rely heavily on real-time data processing.

In summary, digital technologies play a game-changing role in Industry 4.0 by increasing output per worker and stimulating creative problem-solving throughout the production and distribution sector. The fourth industrial revolution will usher in a more connected, intelligent, and sustainable age with the help of these technologies, which will allow companies to unleash new opportunities, improve operations, and embrace sustainability (Reyes et al., 2021).

3. Energy efficiency and renewable energy in Industry 4.0

Advanced technologies and digitalization have created new energy efficiency and renewable energy integration possibilities in Industry 4.0. Industry 4.0's smart systems and data-driven decision-making enable energy optimization, greenhouse gas reduction, and sustainable manufacturing. IoT devices and sensors help Industry 4.0 save energy. These smart gadgets track energy, production, and environmental data in real time. These data help firms identify energy-intensive processes, find inefficiencies, and optimize operations to reduce waste and energy use. Predictive analytics using artificial intelligence and machine learning algorithms may estimate energy needs and optimize production schedules, minimizing energy waste. Industry 4.0 makes renewable energy integration easier. IoT-enabled smart grids and sophisticated energy management systems allow enterprises to integrate solar, wind, and other renewable energy sources easily. These systems improve energy distribution, storage, and usage, optimizing renewable energy use and providing a steady energy supply (Jayashree et al., 2022).

Smart and adaptable production processes in Industry 4.0 also save energy. Robotics, automation, and modern control systems provide accurate production control, minimizing waste and energy use. These technologies let manufacturers alter real-time output depending on energy supply and demand. Finally, Industry 4.0 offers industrial sectors a chance to boost energy efficiency and integrate renewable energy. IoT, AI, and smart manufacturing may help companies become more sustainable and eco-friendly. These solutions reduce operating costs, mitigate climate change, and green the industrial landscape.

Renewable energy and energy efficiency are important in making Industry 4.0 sustainable. Reduced energy usage and a shift to renewable energy sources are crucial for companies to avoid environmental consequences and maintain long-term profitability as they use digital technology to enhance operations (Vereycken et al., 2021). Fig. 14.4 gives examples of the important roles that energy efficiency and renewable energy play in Industry 4.0.

- **Energy Efficiency through Digital Optimization:** With the help of IoT and AI, among other Industry 4.0 technologies, industrial facilities can track and analyze their energy use in real time. Industries may find energy-intensive processes, inefficiencies, and opportunities for growth by collecting and analyzing data from a wide range of sensors and systems. Using this data-driven strategy, we can reduce energy use, enhance productivity, and reduce waste.
- **Smart Energy Management:** Smart energy management systems that use renewable energy generation, energy storage, and demand response are made possible by Industry 4.0. These systems balance energy supply and demand, allowing peak power use while minimizing energy waste (AL-Zyadat et al., 2022).
- **Renewable Energy Integration:** With the advent of Industry 4.0, businesses can switch to cleaner energy sources such as solar, wind, and hydro and manage the

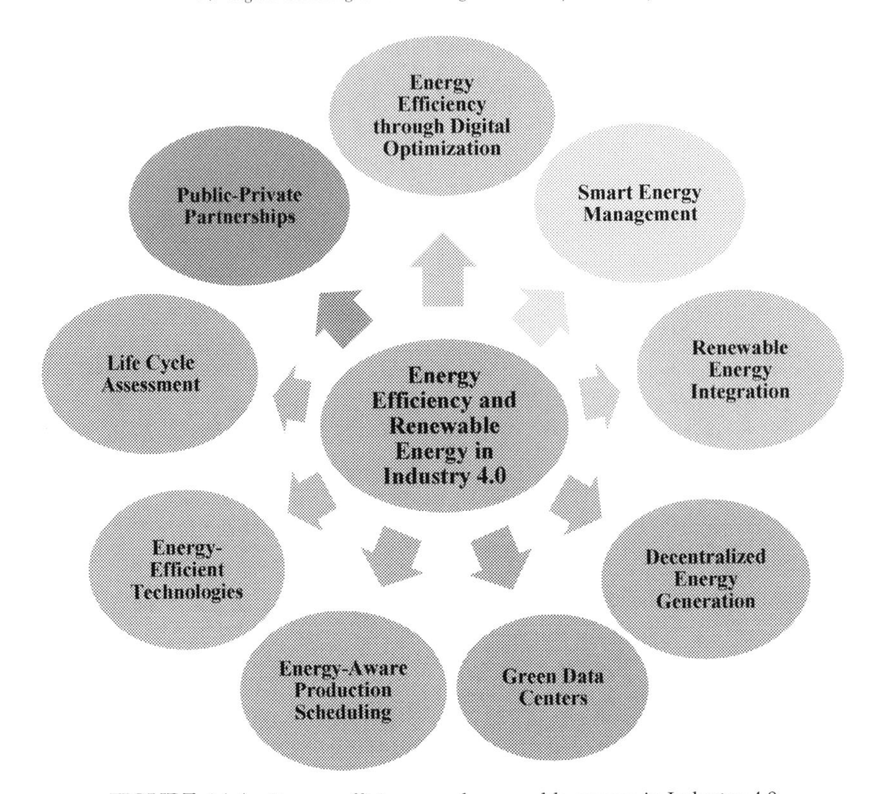

FIGURE 14.4 Energy efficiency and renewable energy in Industry 4.0.

unpredictability of renewable energy sources using IoT sensors and AI-based algorithms that analyze data to optimize energy consumption.

- **Decentralized Energy Generation:** Industry 4.0 has allowed businesses to experiment with microgrids and other forms of distributed energy. Businesses may improve energy security and save costs by producing power locally from renewable sources.
- **Green Data Centers:** The growing need for storing and processing data directly affects the digital transition in Industry 4.0. Using green data centers that run on renewable energy may greatly reduce the environmental impact of computer systems and networks.
- **Energy-Aware Production Scheduling:** Energy prices and availability are included in AI-driven production scheduling, with energy-intensive procedures carried out during low demand or peak renewable energy generation. This strategy optimizes the use of renewable power while minimizing the use of fossil fuels.
- **Energy-Efficient Technologies:** Energy-efficient motors, variable-frequency drives, and LED lighting are just some of the energy-saving technologies that are being promoted by Industry 4.0.
- **Life Cycle Assessment:** With the help of Industry 4.0 technology, it is now possible to conduct thorough analyses of a product's energy and environmental effects across its

entire life cycle, from raw material extraction to final disposal. Companies may use these data to improve their goods' energy efficiency and environmental friendliness.
- **Public–Private Partnerships:** Sustainable energy solutions can be developed, and renewable energy can be adopted more readily in the industrial sector if businesses, governments, and renewable energy sources work together.

In a nutshell, Industry 4.0's sustainability objectives cannot be met without focusing on energy efficiency and renewable energy. Industries may lessen their negative environmental impact and help create a greener future by embracing digital technologies that help them better manage their energy use, incorporate renewable energy sources, and promote eco-friendly activities. Industry 4.0's adoption of renewable and energy-efficient technologies is not only the right thing to do for the planet but also a need for maintaining competitiveness in a world where resources are becoming scarcer by the day (Taddei et al., 2022; Javaid et al., 2022; Delera et al., 2022).

3.1 Circular economy and waste management

The circular economy and waste management are important in promoting environmental responsibility and efficient use of available resources. By maintaining goods, materials, and resources in perpetual cycles of use, reuse, and recycling, the circular economy presents a novel economic model with the potential to uncouple economic development from resource depletion. The circular economy is an alternative to the typical linear "take, make, dispose" model that prioritizes product durability, repairability, and recycling. IoT, AI, and blockchain are just a few examples of digital technologies that are helping to propel the circular economy and improve waste management in the context of Industry 4.0. The Internet of Things (IoT) enables real-time data collecting through sensors implanted in goods and industrial processes, shedding light on product performance, use patterns, and maintenance requirements. This data-driven strategy makes predictive maintenance easier, increasing product durability and decreasing the frequency of unnecessary discards (Kumar et al., 2021b).

AI and ML systems may help improve garbage collection and recycling efforts by analyzing massive datasets. Artificial intelligence (AI) increases material recovery rates and decreases landfill waste by automating trash sorting procedures. Blockchain technology in the waste management industry guarantees the openness and traceability of all transactions. This allows for better tracking of trash, confirmation of proper recycling and disposal methods, and encouragement of environmentally conscious behavior. The environmental and financial advantages of adopting circular economy concepts and modern waste management systems are mutually reinforcing. Industries may save money by reusing and recycling materials, decreasing their waste output, and capitalizing on emerging markets in the remanufacturing and secondhand markets. Additionally, circular economy methods promote cooperation between businesses, government, and consumers, leading to a more robust and sustainable industrial environment. Industry 4.0 can be crucial in building a more sustainable and ecologically responsible future by embracing the circular economy and implementing effective waste management practices (Strandhagen et al., 2022).

3.2 Waste reduction and recycling in Industry 4.0

In Industry 4.0, digital technologies are pivotal in maximizing resource usage and reducing environmental impact, making waste reduction and recycling essential components of sustainability. Industries may benefit from a more circular and environmentally friendly industrial ecosystem by using cutting-edge technology to achieve effective waste reduction and recycling methods (Mabkhot M et al., 2021).

- **Smart Manufacturing and Process Optimization:** The Internet of Things (IoT) and artificial intelligence (AI), two key components of Industry 4.0, make it possible to monitor and analyze production in real time. Manufacturing firms may pinpoint wasteful practices by analyzing production characteristics and resource use data.
- **Product Design for Circularity:** Products designed with Industry 4.0's circularity in mind are encouraged to be easily disassembled, modular, and made from recyclable materials. Design simulations made possible by digital technologies provide longevity, repairability, and deference to recycling and remanufacturing.
- **Predictive Maintenance:** Predictive maintenance powered by AI allows businesses to use their gear and tools more. Preventative maintenance reduces the amount of useable assets that must be thrown away because of minor problems.
- **Closed-Loop Supply Chains:** By increasing visibility and tracking throughout supply chains, digital technologies facilitate the introduction of closed-loop models. The material loop may be closed and waste reduced through reverse logistics, which allows industries to collect and recycle obsolete items.
- **Smart Waste Management:** Garbage collection, sorting, and processing are all enhanced by IoT-enabled waste management systems. Waste is collected more quickly and easily with sensor-equipped "smart bins," recycling rates are increased with the help of artificial intelligence–powered sorting devices (Cezarino et al., 2021).
- **Blockchain for Transparent Recycling:** The recycling process may be tracked in a decentralized and unchangeable ledger made possible by blockchain technology, which also verifies the validity of recycled materials. This encourages honesty and accountability in recycling operations.
- **Waste-to-Energy Solutions:** The integration of waste-to-energy technology, which may turn certain forms of garbage into renewable energy sources, is made possible by Industry 4.0. These measures lessen the burden on landfills and increase energy diversity, which are important components of a sustainable energy system.
- **Collaborative Ecosystems:** Industry 4.0's emphasis on digital connection improves cooperation between businesses, waste management firms, and recycling centers. As a result, recyclables may be traded with one another, resulting in a more robust recycling infrastructure.
- **Consumer Engagement:** The direct connection between businesses and customers made possible by Industry 4.0 helps to educate people about the value of recycling and reducing trash. Information on environmentally friendly consumption patterns and proper waste management may be disseminated through digital channels.
- **Continuous Improvement:** The effectiveness of trash minimization and recycling initiatives may be tracked and evaluated in real time thanks to digital technology. Key

performance indicators allow businesses to monitor their progress and adjust their methods.

Finally, recycling and waste prevention are critical parts of Industry 4.0's commitment to sustainability. Industries may streamline operations, create circular goods, and adopt effective waste management systems using digital technology. Industry 4.0 may help save the environment, preserve resources, and create a more sustainable future if it adopts waste reduction and recycling practices (Goel et al., 2022).

4. Sustainable manufacturing and production

Sustainable manufacturing and production are cornerstones to achieve a more ecologically and socially responsible industrial landscape. Sustainable practices entail utilizing cutting-edge technological innovations and methods that boost ethical and equitable procedures while reducing environmental negative effects. Industries may reduce waste, save precious materials, and boost product reuse and recycling by embracing circular economy ideas and employing resource-efficient production techniques. The carbon footprint of industrial activities may be reduced further and help mitigate climate change by switching to clean and renewable energy sources. A complete, beyond-the-factory-floor approach to sustainability is ensured through responsible supply chain management, which includes ethical sourcing and fair labor standards. In addition, sustainable manufacturing is built on a foundation of care for workers that prioritizes their health, safety, and career advancement. Collaboration in achieving mutual sustainability objectives is facilitated by stakeholder involvement and open lines of communication. Sustainable manufacturing aims to establish quantifiable objectives, track performance, and continually innovate to create a more resilient, resource-efficient, and socially aware industrial environment. The manufacturing and production sectors may be powerful agents of change by adopting sustainable practices that encourage more ethical consumption and output (Grybauskas et al., 2022).

4.1 Sustainable supply chain management

The social, environmental, and economic impacts of supply chain activities are all considered in a sustainable supply chain management strategy. Its primary objective is to establish an environmentally friendly supply chain ecosystem that supports fair labor conditions and is profitable. Responsible sourcing, green procurement, circular economy concepts, and waste reduction programs are all part of sustainable supply chain management. Suppliers that uphold social and environmental norms, such as those guaranteeing decent pay and safe working conditions, are given preference in ethical sourcing. Choosing carbon-neutral items and limiting the usage of harmful compounds are important tenets of green buying. Reducing waste and preserving scarce resources are two of the many benefits of adopting circular economy concepts. Responsible sourcing and manufacturing can only occur when there is full visibility throughout the whole production process. Shared accountability and dedication to sustainability objectives are fostered through supplier/manufacturer/customer/regulator collaboration and participation. Supply networks can adjust to shifting market and environmental situations by continually improving and adapting. Building a more resilient, efficient, and ethically responsible business environment is a primary goal of sustainable supply chain management (Hervas-Oliver et al., 2021).

4.1.1 *Supply chain transparency and traceability*

Traceability and transparency in the supply chain relate to the availability of data about the history, location, and treatment of goods and components from their origin to their final destination. Ethical sourcing, regulatory conformity, and environmental responsibility depend on keeping data open and available. Improvements in transparency, trust, and responsible behavior across the supply chain are all possible thanks to tools such as blockchain and data analytics. A more sustainable and ethical supply chain ecology is fostered through transparent and traceable supply chains, which allow for better decision-making, risk reduction, and customer trust (Benitez et al., 2022).

4.1.2 *Supply chain digitization and collaboration*

The Internet of Things (IoT), artificial intelligence (AI), and cloud computing are all examples of digital technologies that may be used in a supply chain to improve efficiency and effectiveness by streamlining inventory management, demand forecasting, and logistics through real-time data gathering, analysis, and communication. Supply chain partners can more effectively share information, react to changes, and coordinate their activities through digital platforms. Supply chains may save money, provide better customer service, and foster a more robust and adaptable ecosystem by adopting digital tools and coordinating their efforts.

4.1.3 *Green logistics and transportation*

The goal of green logistics and transportation is to reduce the negative effects of transporting products and people on the environment using sustainable methods. Emissions may be reduced by increasing the use of fuel-efficient cars, encouraging the use of other fuels, and adjusting travel patterns. The use of renewable energy in transportation operations and efficient packing and storage practices are all essential components of sustainable logistics. Smart and environmentally friendly logistics management is made possible through GPS, telematics, and data analytics. The logistics and transportation sectors may help reduce the effects of climate change, enhance air quality, and develop a more environmentally friendly transportation network by adopting green practices (Satyro et al., 2022).

5. Environmental monitoring and compliance

Compliance with environmental norms and standards may be ensured by environmental monitoring, which is the systematic collection, analysis, and evaluation of data on the environment. Air quality, water quality, waste management, emissions, and other ecological aspects are only some of the environmental criteria that must be monitored. IoT sensors, satellite images, and remote sensing are just a few examples of cutting-edge technologies enabling remote and real-time environmental monitoring. Verifying an organization's compliance with environmental laws and best practices for doing so helps ensure that negative environmental effects are kept to a minimum and that ecological balance is preserved. Environmental monitoring and compliance must be robust to protect ecosystems, public health, and long-term sustainable development (Chiarini, 2021).

5.1 IoT sensors for environmental monitoring

IoT sensors have a revolutionary impact on environmental monitoring by providing real-time data on a wide range of ecological indicators. Air, water, soil, and animal habitats are just some places where these sensors monitor environmental factors, including temperature, humidity, air quality, water quality, pollution levels, and more. These sensors send their data over the Internet, allowing constant, off-site monitoring. Improved early detection of environmental changes, identification of pollution sources, and informed decision-making for sustainable resource management and conservation are all made possible by environmental monitoring powered by the Internet of Things. Incorporating IoT sensors into environmental monitoring is a game-changer for more proactive and data-driven methods to protect the environment (Somohano-Rodríguez et al., 2022).

5.2 Data-driven environmental compliance

"Data-driven environmental compliance" describes relying on information gathered through modern technology when protecting the environment. Internet of Things (IoT) sensors, satellite images, and remote sensing data may all be used to keep tabs on and evaluate environmental characteristics in real time. By using this data-driven method, firms may see possible compliance concerns sooner, make better, more informed decisions, and pinpoint where to improve. Companies may monitor their environmental progress, establish a baseline for improvement, and devise strategic plans to lessen their negative effects on the environment by evaluating environmental data. Data-driven environmental compliance promotes openness, accountability, and efficient use of available resources, ultimately leading to a more long-term and ethical strategy for protecting the environment (Hassoun et al., 2022b).

5.3 Predictive maintenance for sustainable operations

Predictive maintenance is a data-driven strategy that uses cutting-edge technology such as Internet of Things (IoT) sensors and artificial intelligence (AI) to foresee equipment faults and maximize maintenance efficiency. Predictive maintenance can foresee possible problems with equipment and systems by continually monitoring their state. Reduced downtime, less energy use, and a longer lifetime for equipment are all benefits of this preventative maintenance strategy that helps businesses run more sustainably. Predictive maintenance is a useful tool for encouraging sustainability in various sectors because it allows firms to optimize maintenance schedules and resource utilization, increasing operating efficiency, decreasing waste, and lessening environmental impact (Kumar et al., 2022).

6. Social sustainability and Workforce 4.0

During the revolutionary changes ushered in by the Fourth Industrial Revolution, Workforce 4.0 is committed to fostering a socially responsible and inclusive workplace. Focusing on staff happiness, growth, and agency while adopting cutting-edge innovation is essential. There should be a focus on balancing work and personal life, giving opportunities to learn new skills, encouraging diversity and inclusion, paying fairly, and maintaining a safe and healthy work environment. In the era of Workforce 4.0, firms may foster a highly motivated and engaged workforce, increase productivity, and create a robust and socially aware business ecosystem by emphasizing social sustainability (Bai et al., 2022).

Human-centered methods in Industry 4.0 put employees' health and agency first amidst the technological revolution. They highlight the importance of lifelong learning, employee participation in introducing new technologies, and work–life harmony. Protecting employees' personal information, encouraging a diverse and welcoming workplace, and fostering ethical business practices are all essential. Augmented intelligence encourages human–AI cooperation, and user-centric design results in straightforward software. Social and economic imbalances must be addressed, and socially responsible action must be taken. These strategies, which understand that people are at the center of Industry 4.0, strive to improve people's skills and build a sustainable, egalitarian, and resilient workforce.

The workforce is bolstered by digital technology in several ways. These include the ability to work remotely, access to online learning platforms for skill development, and the availability of effective collaboration tools. In addition to improving decision-making and productivity, data analytics and automation liberate workers from repetitive activities. AR and VR training may be more engaging, which helps students learn and remember more. Staff development is facilitated through monitoring and receiving feedback on performance. Digital means of communication and feedback have been shown to improve employee engagement and well-being. Strategic talent management is made possible by workforce analytics. Personalized services are tailored to each customer. To thrive in the modern digital world, businesses must provide employees with the skills to operate quickly and independently (Zheng et al., 2021).

In the Fourth Industrial Revolution, diversity and inclusion were critical for securing equal access to resources and stimulating creative problem-solving. Everyone benefits when people from different walks of life and experiences work together to solve problems. Employees are more invested and productive in businesses that provide community and emotional security for their workers. The Internet and other digital tools have opened up new channels of communication and cooperation for widely dispersed teams, regardless of their physical location. A more innovative and flexible workforce results from an organization's emphasis on diversity and inclusion. When it comes to succeeding and staying competitive in today's fast-paced, globally networked business environment, diversity, and inclusion are good moral principles and a major competitive advantage (Moldabekova et al., 2021).

6.1 Policy, regulation, and standards

Fostering sustainability in Industry 4.0 through digital technology requires policy, legislation, and standards.

- **Policy:** Sustainable practices in Industry 4.0 may be widely adopted with the help of policies developed by governments and organizations. Policies in this category may include regulations for energy efficiency, funding for circular economy projects, and subsidies for environmentally friendly technology. Digital technology use in sectors may positively affect the environment, resource efficiency, and social responsibility if the appropriate policy frameworks are in place.
- **Regulation:** To ensure that digital technologies in Industry 4.0 are used responsibly, regulations may provide clear standards and criteria. Information privacy and security regulations, technology evaluation criteria, and green production standards are all examples of what this entails. There won't be any detrimental or exploitative uses of digital technology because of the rules put in place to keep them safe and effective.
- **Standards:** Industry 4.0 relies on consistent, interoperable, and transparent digital technologies; establishing industry standards is essential. The concepts of eco-design, methods of calculating one's carbon footprint, and mechanisms for exchanging data may all be included in such standards. Following these guidelines improves the capacity to compare sustainability performance across sectors and encourages cooperation toward shared sustainability objectives (Costa et al., 2022).

Organizations and governments may advance sustainability, responsible innovation, and an inclusive industrial ecosystem by incorporating policy, regulation, and standards into using digital technologies in Industry 4.0. To achieve a more sustainable and resilient future, these governance systems guarantee that digital technologies are used for their beneficial environmental and social impacts.

7. Case studies

- **Case Study 1: Siemens AG—Industrial IoT for Energy Efficiency**

The global company Siemens AG installed an Industrial IoT solution to reduce factory power consumption. The collection and analysis of real-time data on energy use and industrial processes were made possible by linking machines and equipment to the cloud. By analyzing the data, Siemens could pinpoint sources of energy loss and improve efficiency. Consequently, the firm was able to significantly cut its energy consumption, lessen its impact on the environment, and increase the sustainability of its production processes.

- **Case Study 2: Volvo Group—Digital Twins for Circular Economy**

Digital twin technology was used by Volvo Group, a major automaker, to improve circular economy procedures. Volvo could track cars throughout their lifetimes and find places for refurbishment, repair, and remanufacturing if it had digital copies of its goods. Digital twins helped the corporation increase the usefulness of its components, lessen its environmental impact, and save money. Volvo Group was able to save money, lessen its effect on the environment, and increase its dedication to sustainability.

- **Best Practice 1: IBM—Blockchain for Supply Chain Transparency**

IBM used blockchain technology to increase visibility across the food industry's supply chain. They collaborated with major distributors and manufacturers to create a blockchain-based technology that could track the path of food items from farm to fork. Because of this increased transparency, the danger of food fraud was reduced, and ethical sourcing procedures were maintained. A more sustainable and ethical food supply chain resulted from the platform's efforts to educate customers.

- **Best Practice 2: Schneider Electric—Data Analytics for Energy Management**

Schneider Electric, a multinational energy management and automation solutions provider, analyzed building energy use patterns using data analytics. Facility managers could save money and reduce carbon emissions by analyzing energy consumption data in real time with the help of an integrated platform. Schneider Electric's strategy helped its customers save money, cut their carbon footprint, and minimize their energy usage, all while contributing to environmental good.

These examples and recommendations show how Industry 4.0 may be made more environmentally friendly through digital technology. Organizations may improve energy efficiency, spread awareness of circular economy concepts, guarantee supply chain transparency, and maximize resource management using IoT, digital twins, blockchain, and data analytics. There is a strong economic justification for incorporating digital technology into sustainable practices since their adoption helps achieve sustainability goals and provides low-cost alternatives (Khan et al., 2021).

7.1 Sustainable supply chain success stories

- **IKEA—Sustainable Timber Sourcing:** IKEA, a multinational furniture retailer, has made great steps toward more environmentally friendly supply chain methods, especially in procuring lumber. The firm has promised to exclusively use wood from responsible sources going forward. IKEA works with groups such as the Forest Stewardship Council (FSC) to ensure the wood it uses comes from sustainably managed forests, which benefits biodiversity and community involvement.
- **Unilever—Sustainable Palm Oil Sourcing:** Global consumer goods giant Unilever has been an early advocate for responsible palm oil sourcing. The firm has pledged to get all its palm oil from RSPO-verified sustainable sources. This dedication results from reduced deforestation, ecosystem protection, and improved working conditions for palm oil growers.
- **Patagonia—Transparent Supply Chain:** Patagonia, a maker of outdoor clothes and equipment, has committed to supply chain transparency. Customers may learn all they need to know about the company's product's origins and production process before purchasing. Customers have faith in and devotion to Patagonia because of its dedication to transparency in all aspects of business, not just its labor and environmental policies.

- **Tesla—Renewable Energy in Manufacturing:** Tesla, a manufacturer of electric vehicles and clean energy, has adopted renewable energy in its facilities. Solar panels and energy storage systems are used in the company's production facilities to lessen the environmental impact caused by the use of fossil fuels. Tesla's dedication to sustainability is not limited to its final products; it is reflected in every aspect of the company.
- **Nestlé—Sustainable Cocoa Farming:** The international food and drink giant Nestlé has made tremendous efforts to ensure the ethical production of chocolate for its products. The Nestlé Cocoa Plan is an initiative by the firm to aid cocoa farmers by providing them with sustainable agricultural techniques and improved lives through training, technical assistance, and access to resources. Nestlé's program helps with things such as fighting poverty and purely obtaining cocoa.

These examples of achievement show how sustainable supply chain methods may improve corporate, community, and environmental outcomes. These businesses have shown how sustainability can be incorporated into supply chain operations by placing a premium on ethical sourcing, transparency, renewable energy, and social responsibility (Suleiman et al., 2022; Kurniawan et al., 2022b).

7.2 Future outlook and challenges

The potential of digital technology to promote sustainability in the fourth industrial revolution is exciting. There will be even more positive effects on the environment and society as a result of technological progress. A few notable developments are discussed further, as presented in Fig. 14.5.

- **AI and Data Analytics:** Optimization of resource use, prediction of maintenance requirements, and identification of possibilities to increase sustainability will all rely heavily on artificial intelligence and data analytics. These innovations will improve

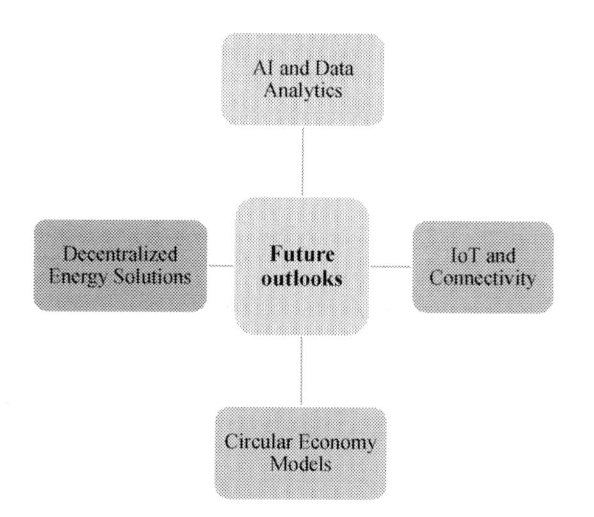

FIGURE 14.5 Future outlooks.

industrial operations' efficiency and sustainability by facilitating data-driven decision-making.

- **IoT and Connectivity:** Connectivity across the supply chain will be driven forward by the IoT, allowing for real-time monitoring and remote management of resources. Transparency and traceability will be improved through Internet of Things–enabled sensors, leading to more ethical supply chains and widespread adoption of circular economy principles.
- **Circular Economy Models:** More advanced versions of the circular economy will emerge as Industry 4.0 develops. This is because digital technology will facilitate product-as-a-service ideas, remanufacturing projects, and recycling programs, all of which contribute to a more sustainable economy by decreasing waste and increasing reusability.
- **Decentralized Energy Solutions:** By incorporating renewable energy sources, smart grids, and energy storage technologies, enterprises can reduce their reliance on centralized power plants in favor of more environmentally friendly, decentralized systems (Murmura et al., 2021).

7.2.1 Challenges

Despite the promising future, digital technologies for fostering sustainability in Industry 4.0 also face significant challenges, as shown in Fig. 14.6.

- **Data Privacy and Security:** Data privacy and security are becoming more important as the usage of data and connections grows. Gaining confidence and keeping private data safe will need diligent cybersecurity precautions and honest data management policies.
- **Technological Divide:** Some sectors or areas may have less access to cutting-edge technology due to the digital transition, which might undermine their attempts to promote sustainability.
- **Skills Gap:** There may be a skills gap in the workforce due to the quick pace of technological change, with some workers unable to keep up with the necessary levels of digital literacy and technical competence.

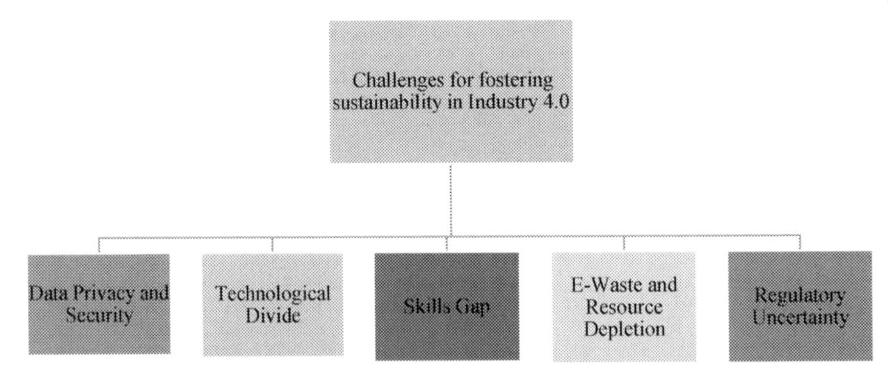

FIGURE 14.6 Challenges for fostering sustainability in Industry 4.0.

- **E-Waste and Resource Depletion:** In the absence of prudent management, the widespread adoption of digital technology risks contributes to the accumulation of electronic waste and the depletion of natural resources. Systems for proper disposal and recycling will need to be developed.
- **Regulatory Uncertainty:** Because digital technologies are rapidly developing, regulatory frameworks may not keep up, leaving businesses unsure of their legal obligations and preventing them from investing in environmentally responsible methods.

To effectively tackle these issues, governments, businesses, and technology providers will need concerted efforts to create a more inclusive and equitable future in Industry 4.0 through digital technologies (Nagy and Lăzăroiu, 2022; Jovanovic et al., 2023; Ogiemwonyi et al., 2023).

7.3 The road to sustainable Industry 5.0

A sustainable future using cutting-edge technology, "Industry 5.0" foresees a more sustainable and socially responsible future for the industrial sector. To improve productivity, well-being, and skill development, Industry 5.0 builds on the foundations of Industry 4.0 by putting people at the heart of technology adoption. With an emphasis on product life extension, remanufacturing, and closed-loop supply chains, the circular economy is woven into the very fabric of Industry 5.0 to reduce waste and preserve scarce resources. Decarbonization and renewable energy are at the forefront of a sustainable Industry 5.0 that uses data analytics and the Internet of Things to manage resources intelligently and track real-time environmental consequences. Social inclusion, diversity, and fair labor standards provide an engaged and empowered workforce, while ethical data management and AI governance guarantee responsible decision-making. Fostering collective action toward shared sustainability objectives and collaborative initiatives across the industry, governments, and civil society creates responsible innovation, transparency, and traceability throughout supply chains. An industrial future that is greener, fairer, and more resilient may be achieved through a dedication to continuous development and the responsible embrace of technology (Trevisan et al., 2023).

Sustainability issues in the Fourth Industrial Revolution (Industry 4.0) can only be solved by a proactive and collaborative strategy that uses technology to improve environmental and social outcomes. Managing the widening digital gap is a significant obstacle since certain areas and businesses may have difficulty accessing and adopting cutting-edge technology. Investment in digital infrastructure, capacity-building programs, and knowledge-sharing activities are all necessary to overcome this obstacle. Trust in digital technology and protecting sensitive information depend on measures to ensure ethical data usage and cybersecurity. Businesses can use circular economy concepts such as product life extension and resource optimization to reduce electronic waste and encourage responsible purchasing practices. For the Fourth Industrial Revolution to be environmentally friendly and socially inclusive, collaborations between governments, industry, and civil society are essential for driving innovation, developing sustainable policies, and fostering collective action. The full potential of Industry 4.0 may be unlocked for a sustainable and resilient future if technical advances are combined with ethical practices and stakeholder involvement (Tsolakis et al., 2023).

8. Conclusion

Finally, digital technologies have emerged as game-changing instruments for promoting sustainability within the context of Industry 4.0. Industry can improve efficiency, lessen its ecological footprint, and increase social responsibility using IoT, AI, data analytics, and blockchain technology. These innovations promote a more sustainable and resilient industrial environment by facilitating open and efficient supply chains, circular economic practices, and energy-saving business procedures. Further, digital technology enables workers by encouraging skill improvement, a healthy work—life balance, and a more welcoming and diverse workplace. Collaborative efforts, ethical data practices, and responsible innovation are necessary to fully realize the sustainability benefits of these technologies. Industry 4.0 may help usher in a more sustainable, egalitarian, and successful future if we carefully embrace digital technology.

Industry 4.0's focus on sustainability through digital technology has good social and environmental effects. Sustainable methods can mitigate climate change and protect biodiversity by decreasing carbon emissions, optimizing resource use, and reducing waste. These innovations support social sustainability by fostering equitable workplaces, improving workers' lives, and giving voice to underrepresented groups in the workplace. Opportunities to solve critical sustainability concerns across sectors are expanding with digital innovation. The full promise of digital technologies has yet to be realized because of obstacles, including data privacy worries, inaccessible technology, and the need to ensure that growth is inclusive and does not leave anybody behind. The potential of digital technology in creating a more sustainable and successful industry cannot be unlocked without collaborative relationships between the public and private sectors and supporting legislative frameworks. 4.0. Businesses and society can pave the way toward a more resilient, fair, and sustainable future by using digital technology responsibly.

References

AL-Zyadat, A., Alsaraireh, J., Al-Husban, D., Al-Shorman, H., Mohammad, A., Alathamneh, F., Al-Hawary, S., 2022. The effect of Industry 4.0 on sustainability of industrial organizations in Jordan. International Journal of Data and Network Science 6 (4), 1437—1446.

Bai, C., Orzes, G., Sarkis, J., February 1, 2022. Exploring the impact of Industry 4.0 technologies on social sustainability through a circular economy approach. Industrial Marketing Management 101, 176—190.

Belhadi, A., Kamble, S., Gunasekaran, A., Mani, V., November 23, 2022. Analyzing the mediating role of organizational ambidexterity and digital business transformation on Industry 4.0 capabilities and sustainable supply chain performance. Supply Chain Management: International Journal 27 (6), 696—711.

Benitez, G.B., Ferreira-Lima, M., Ayala, N.F., Frank, A.G., January 3, 2022. Industry 4.0 technology provision: the moderating role of supply chain partners to support technology providers. Supply Chain Management: International Journal 27 (1), 89—112.

Cezarino, L.O., Liboni, L.B., Oliveira Stefanelli, N., Oliveira, B.G., Stocco, L.C., August 23, 2021. Diving into emerging economies bottleneck: Industry 4.0 and implications for circular economy. Management Decision 59 (8), 1841—1862.

Chiarini, A., November 2021. Industry 4.0 technologies in the manufacturing sector: are we sure they are all relevant for environmental performance? Business Strategy and the Environment 30 (7), 3194—3207.

Ciliberto, C., Szopik-Depczyńska, K., Tarczyńska-Łuniewska, M., Ruggieri, A., Ioppolo, G., November 2021. Enabling the Circular Economy transition: a sustainable lean manufacturing recipe for Industry 4.0. Business Strategy and the Environment 30 (7), 3255–3272.

Costa, A.C., de Mello Santos, V.H., de Oliveira, O.J., August 5, 2022. Towards the revolution and democratization of education: a framework to overcome challenges and explore opportunities through Industry 4.0. Informatics in Education 21 (1), 1–32.

Cricelli, L., Strazzullo, S., July 23, 2021. The economic aspect of digital sustainability: a systematic review. Sustainability 13 (15), 8241.

da Silva, T.H., Sehnem, S., May 10, 2022. The circular economy and Industry 4.0: synergies and challenges. Revista de Gestão 29 (3), 300–313.

Delera, M., Pietrobelli, C., Calza, E., Lavopa, A., April 1, 2022. Does value chain participation facilitate the adoption of Industry 4.0 technologies in developing countries? World Development 152, 105788.

Di Maria, E., De Marchi, V., Galeazzo, A., February 2022. Industry 4.0 technologies and circular economy: the mediating role of supply chain integration. Business Strategy and the Environment 31 (2), 619–632.

Dilyard, J., Zhao, S., You, J.J., September 2021. Digital innovation and Industry 4.0 for global value chain resilience: lessons learned and ways forward. Thunderbird International Business Review 63 (5), 577–584.

Fonseca, L., Amaral, A., Oliveira, J., March 12, 2021. Quality 4.0: the EFQM 2020 model and Industry 4.0 relationships and implications. Sustainability 13 (6), 3107.

Gebhardt, M., Kopyto, M., Birkel, H., Hartmann, E., December 2, 2022. Industry 4.0 technologies as enablers of collaboration in circular supply chains: a systematic literature review. International Journal of Production Research 60 (23), 6967–6995.

Ghobakhloo, M., Iranmanesh, M., Grybauskas, A., Vilkas, M., Petraitė, M., December 2021. Industry 4.0, innovation, and sustainable development: a systematic review and a roadmap to sustainable innovation. Business Strategy and the Environment 30 (8), 4237–4257.

Goel, P., Kumar, R., Banga, H.K., Kaur, S., Kumar, R., Pimenov, D.Y., Giasin, K., April 7, 2022. Deployment of interpretive structural modeling in barriers to Industry 4.0: a case of small and medium enterprises. Journal of Risk and Financial Management 15 (4), 171.

Grybauskas, A., Stefanini, A., Ghobakhloo, M., August 1, 2022. Social sustainability in the age of digitalization: a systematic literature Review on the social implications of Industry 4.0. Technology in Society 70, 101997.

Hassoun, A., Prieto, M.A., Carpena, M., Bouzembrak, Y., Marvin, H.J., Pallares, N., Barba, F.J., Bangar, S.P., Chaudhary, V., Ibrahim, S., Bono, G., 2022a. Exploring the role of green and Industry 4.0 technologies in achieving sustainable development goals in food sectors. Food Research International 20, 112068.

Hassoun, A., Aït-Kaddour, A., Abu-Mahfouz, A.M., Rathod, N.B., Bader, F., Barba, F.J., Biancolillo, A., Cropotova, J., Galanakis, C.M., Jambrak, A.R., Lorenzo, J.M., 2022b. The fourth industrial revolution in the food industry—Part I: Industry 4.0 technologies. Critical Reviews in Food Science and Nutrition 27, 1–7.

Hervas-Oliver, J.L., Gonzalez-Alcaide, G., Rojas-Alvarado, R., Monto-Mompo, S., January 17, 2021. Emerging regional innovation policies for Industry 4.0: analyzing the digital innovation hub program in European regions. Competitiveness Review: An International Business Journal 31 (1), 106–129.

Javaid, M., Haleem, A., Singh, R.P., Suman, R., Gonzalez, E.S., January 1, 2022. Understanding the adoption of Industry 4.0 technologies in improving environmental sustainability. Sustainable Operations and Computers 3, 203–217.

Jayashree, S., Reza, M.N., Malarvizhi, C.A., Gunasekaran, A., Rauf, M.A., May 1, 2022. Testing an adoption model for Industry 4.0 and sustainability: a Malaysian scenario. Sustainable Production and Consumption 31, 313–330.

Jovanovic, V., Kuzlu, M., Cali, U., Utku, D.H., Catak, F.O., Sarp, S., Zohrabi, N., April 9, 2023. Digital twin in Industry 4.0 and beyond applications. In: InDigital Twin Driven Intelligent Systems and Emerging Metaverse. Springer Nature Singapore, Singapore, pp. 155–174.

Khan, I.S., Ahmad, M.O., Majava, J., May 15, 2021. Industry 4.0 and sustainable development: a systematic mapping of triple bottom line, circular economy and sustainable business models perspectives. Journal of Cleaner Production 297, 126655.

Kumar, P., Singh, R.K., Kumar, V., 2021a. Managing supply chains for sustainable operations in the era of Industry 4.0 and circular economy: analysis of barriers. Resources, Conservation and Recycling 164, 105215.

Kumar, R., Banga, H.K., Kumar, R., Singh, S., Singh, S., Scutaru, M.L., Pruncu, C.I., 2021b. Ergonomic evaluation of workstation design using taguchi experimental approach: a case of an automotive industry. International Journal on Interactive Design and Manufacturing 15, 481–498.

Kumar, R., Rani, S., Awadh, M.A., June 3, 2022. Exploring the application sphere of the Internet of things in Industry 4.0: a review, bibliometric and content analysis. Sensors 22 (11), 4276.

Kunkel, S., Matthess, M., Xue, B., Beier, G., July 1, 2022. Industry 4.0 in sustainable supply chain collaboration: insights from an interview study with international buying firms and Chinese suppliers in the electronics industry. Resources, Conservation and Recycling 182, 106274.

Kurniawan, T.A., Othman, M.H., Hwang, G.H., Gikas, P., 2022a. Unlocking digital technologies for waste recycling in Industry 4.0 era: a transformation towards a digitalization-based circular economy in Indonesia. Journal of Cleaner Production 357, 131911.

Kurniawan, T.A., Maiurova, A., Kustikova, M., Bykovskaia, E., Othman, M.H., Goh, H.H., 2022b. Accelerating sustainability transition in St. Petersburg (Russia) through digitalization-based circular economy in waste recycling industry: a strategy to promote carbon neutrality in era of Industry 4.0. Journal of Cleaner Production 363, 132452.

Lepore, D., Micozzi, A., Spigarelli, F., March 2, 2021. Industry 4.0 accelerating sustainable manufacturing in the COVID-19 era: assessing the readiness and responsiveness of Italian regions. Sustainability 13 (5), 2670.

Mabkhot M, M., Ferreira, P., Maffei, A., Podržaj, P., Madziel, M., Antonelli, D., Lanzetta, M., Barata, J., Boffa, E., Finžgar, M., Paśko, Ł., February 27, 2021. Mapping Industry 4.0 enabling technologies into united nations sustainability development goals. Sustainability 13 (5), 2560.

Moldabekova, A., Philipp, R., Reimers, H.E., Alikozhayev, B., 2021. Digital technologies for improving logistics performance of countries. Transport and Telecommunication Journal 22 (2), 207–216.

Mourtzis, D., Angelopoulos, J., Panopoulos, N., August 28, 2022. A literature review of the challenges and opportunities of the transition from Industry 4.0 to society 5.0. Energies 15 (17), 6276.

Mubarak, M.F., Tiwari, S., Petraite, M., Mubarik, M., Raja Mohd Rasi, R.Z., July 20, 2021. How Industry 4.0 technologies and open innovation can improve green innovation performance? Management of Environmental Quality: An International Journal 32 (5), 1007–1022.

Mukhuty, S., Upadhyay, A., Rothwell, H., July 2022. Strategic sustainable development of Industry 4.0 through the lens of social responsibility: the role of human resource practices. Business Strategy and the Environment 31 (5), 2068–2081.

Murmura, F., Bravi, L., Santos, G., January 3, 2021. Sustainable process and product innovation in the eyewear sector: the role of Industry 4.0 enabling technologies. Sustainability 13 (1), 365.

Nagy, M., Lăzăroiu, G., September 28, 2022. Computer vision algorithms, remote sensing data fusion techniques, and mapping and navigation tools in the Industry 4.0-based Slovak automotive sector. Mathematics 10 (19), 3543.

Ogiemwonyi, O., Alam, M.N., Hago, I.E., Azizan, N.A., Hashim, F., Hossain, M.S., May 25, 2023. Green innovation behaviour: impact of Industry 4.0 and open innovation. Heliyon.

Ortega-Gras, J.J., Bueno-Delgado, M.V., Cañavate-Cruzado, G., Garrido-Lova, J., December 9, 2021. Twin transition through the implementation of Industry 4.0 technologies: desk-research analysis and practical use cases in Europe. Sustainability 13 (24), 13601.

Reyes, J., Mula, J., Díaz-Madroñero, M., October 2021. Development of a conceptual model for lean supply chain planning in Industry 4.0: multidimensional analysis for operations management. Production Planning & Control 27, 1–6.

Satyro, W.C., de Almeida, C.M., Pinto Jr Jr, M.J., Contador, J.C., Giannetti, B.F., de Lima, A.F., Fragomeni, M.A., February 20, 2022. Industry 4.0 implementation: the relevance of sustainability and the potential social impact in a developing country. Journal of Cleaner Production 337, 130456.

Sharma, M., Kamble, S., Mani, V., Sehrawat, R., Belhadi, A., Sharma, V., January 25, 2021. Industry 4.0 adoption for sustainability in multi-tier manufacturing supply chain in emerging economies. Journal of Cleaner Production 281, 125013.

Somohano-Rodríguez, F.M., Madrid-Guijarro, A., López-Fernández, J.M., July 4, 2022. Does Industry 4.0 really matter for SME innovation? Journal of Small Business Management 60 (4), 1001–1028.

Strandhagen, J.W., Buer, S.V., Semini, M., Alfnes, E., Strandhagen, J.O., July 27, 2022. Sustainability challenges and how Industry 4.0 technologies can address them: a case study of a shipbuilding supply chain. Production Planning & Control 33 (9–10), 995–1010.

Suleiman, Z., Shaikholla, S., Dikhanbayeva, D., Shehab, E., Turkyilmaz, A., December 31, 2022. Industry 4.0: clustering of concepts and characteristics. Cogent Engineering 9 (1), 2034264.

Taddei, E., Sassanelli, C., Rosa, P., Terzi, S., August 1, 2022. Circular supply chains in the era of Industry 4.0: a systematic literature review. Computers & Industrial Engineering 170, 108268.

Tavera, R.C.A., Ortiz, J.H., Khalaf, O.I., Ríos Prado, A., September 7, 2021. Business intelligence: business evolution after Industry 4.0. Sustainability 13 (18), 10026.

Thiede, S., January 1, 2021. Digital technologies, methods and tools towards sustainable manufacturing: does Industry 4.0 support to reach environmental targets? Procedia CIRP 98, 1–6.

Trevisan, A.H., Lobo, A., Guzzo, D., de Vasconcelos Gomes, L.A., Mascarenhas, J., April 15, 2023. Barriers to employing digital technologies for a circular economy: a multi-level perspective. Journal of Environmental Management 332, 117437.

Tripathi, S., Gupta, M., November 5, 2021. A holistic model for Global Industry 4.0 readiness assessment. Benchmarking: An International Journal 28 (10), 3006–3039.

Tsolakis, N., Schumacher, R., Dora, M., Kumar, M., August 2023. Artificial intelligence and blockchain implementation in supply chains: a pathway to sustainability and data monetization? Annals of Operations Research 327 (1), 157–210.

Vereycken, Y., Ramioul, M., Desiere, S., Bal, M., July 13, 2021. Human resource practices accompanying Industry 4.0 in European manufacturing industry. Journal of Manufacturing Technology Management 32 (5), 1016–1036.

Zhang, C., Chen, Y., Chen, H., Chong, D., June 2021. Industry 4.0 and its implementation: a review. Information Systems Frontiers 7, 1-1.

Zheng, T., Ardolino, M., Bacchetti, A., Perona, M., March 19, 2021. The applications of Industry 4.0 technologies in manufacturing context: a systematic literature review. International Journal of Production Research 59 (6), 1922–1954.

CHAPTER

15

A biobjective mixed integer nonlinear programming model for the cell switch-off problem considering quality of service and energy consumption

Diego Gabriel Rossit[1], *Francisco Luna-Valero*[2,3], *Jesús Galeano-Brajones*[4] *and Javier Carmona-Murillo*[4]

[1]Department of Engineering, INMABB, Universidad Nacional del Sur (UNS)-CONICET, Bahía Blanca, Argentina; [2]Departamento Lenguajes y Ciencias de la Computación, Universidad de Málaga, Málaga, Spain; [3]ITIS Software, Universidad de Málaga, Málaga, Spain; [4]Departamento de Ingeniería de Sistemas Informáticos y Telemáticos, Centro Universitario de Mérida, Universidad de Extremadura, Mérida, Spain

1. Introduction

Wireless communications stand at the forefront of modern technology, playing an increasingly vital role across diverse sectors, from mobile devices to industrial automation. The persistent growth in data traffic demands within cellular networks has been a constant since the inception of telecommunication systems, and this trajectory is expected to continue into the foreseeable future. To address these escalating traffic requirements, both vendors and operators are actively shaping the next generation of mobile networks, commonly known as the sixth generation (6G).

The densification of the network infrastructure, in terms of the number of base stations deployed in the area to be serviced, has become an enabling technology to meet the key

Evolution and Trends of Sustainable Approaches
https://doi.org/10.1016/B978-0-443-21651-0.00010-3

performance indicators expected for these newly developed systems (Lopez-Perez et al., 2015). Nonetheless, a significant challenge emerges during the deployment of these ultra-dense networks (UDNs), particularly during periods of low traffic when the entire system operates at full capacity but remains underutilized. An innovative solution proposed to mitigate this wasteful energy consumption involves selectively deactivating a subset of the network's base stations while keeping a minimum level of quality of service (Luna et al., 2018). This problem, known as the cell switch-off (CSO) problem, is well-recognized as NP-complete (González et al., 2016), due to the exponential expansion of the search space with an increasing number of cells. The proposed objectives are clearly conflicting goals. The larger the number of cells switched on, the larger the bandwidth that can be assigned to a user, but also the higher the power consumption. This problem bears similarities to the facility location problem in capacity planning, which is a well-known NP-hard problem (Cornuéjols et al., 1991). In the facility location problem, one must decide which facilities to open from a predefined set with given locations and how to assign users to these opened facilities to provide a certain service. The decision to open facilities is usually associated with installation costs, and, thus, an efficient optimization process aims at opening the smaller number of facilities in order to reduce these costs but simultaneously provide a proper QoS to users. In the CSO problem, instead of which facilities to open, the decision is which cells to switch on and how to assign the users to the active cells.

In general, given the anticipated scale of UDNs (thousands of cells), attempting to tackle this problem with exact optimization algorithms is deemed impractical. As a result, the predominant focus in the literature centers on the development of heuristics and metaheuristic approaches, and there are relatively few works addressing this problem by means of exact techniques. Exact approaches can serve, however, to validating heuristics on smaller-scale instances and providing a meaningful comparison baseline (Toncovich et al., 2019). Many state-of-the-art commercial exact solvers, while they might not achieve optimality for NP-hard problems, they can provide valuable insights into estimating optimal solutions, a capability absent in heuristic approaches. In this context, our work presents a preliminary approach to address the CSO problem using nonlinear mathematical programming and exact solvers. Our contributions encompass a biobjective mixed-integer nonlinear mathematical formulation of the CSO problem, a biobjective resolution strategy, and computational experiments conducted on realistic instances.

This work is structured as follows. Section 2 presents the target problem, including the network system used, the mathematical formulation, and the main related works. Section 3 presents the computational experimentation and the main results. Finally, Section 4 discusses the main outcomes of this works and outlines the future research lines.

2. An MINLP model for the biobjective CSO problem

In this section we present the biobjective CSO problem, including the network system model used in this work, the mathematical formulation, and the related work.

2.1 Network system model

For developing the mathematical formulation of the CSO problem, first, the telecommunications network system model has to be defined. Thus, this section is devoted to detailing the UDN model used, which was developed in Luna et al. (2018). This procedure is based on an area of 500×500 square meters, which has been discretized using a grid of 100×100 points (also called "pixels" or area elements), each covering 25 m^2 where signal power is assumed to be constant. Regarding propagation conditions, 10 different regions have been defined. To compute the received power at each point in the grid x, i.e., $Prx[dBm]$, the following equation is used

$$Prx[dBm] = Ptx[dBm] + PLoss[dBm]$$

Where Ptx is the transmitted power in dBm to grid point x, and $PLoss$ are the global signal losses depending on the propagation region computed with the following equation

$$PLoss[dBm] = GA + PA$$

where GA are the total gains of both antennas (the reception and transmission antenna), and PA are the transmission losses in space, computed as:

$$PA[dB] = \left(\frac{\lambda}{2\pi d}\right)^K$$

where d is the Euclidean distance to the cell, K is the exponent loss, which ranges randomly in [2.0, 4.0] for each of the 10 different propagation regions.

In this work, no interference among cells that work on similar frequencies is considered. Despite being a simplification from other works from the literature, it is a realistic feature considering the technological advancements that have reduced the problem of interference among cells (Aydin et al., 2017; Usama and Erol, 2019). The signal-plus-noise ratio (SNR) received by user u placed at location x when connected to cell c is computed as:

$$SNR_u = \frac{Prx_{uc}[mW]}{Pn_c[mW]}$$

where Prx_{uc} is the received power by user u from cell c, and $Pn_c[mW]$ is the noise power of cell c, computed as:

$$Pn_c[mW] = -174 + 10\, log_{10} BW_c$$

being BW_c the bandwidth of cell c. In this model BW_c is defined as 10% of the cell operating frequency. Assuming a round robin scheduling, all the users assigned to an active cell receive the same bandwidth, which is calculated as:

$$bw_c[Hz] = \frac{BW_c}{N}$$

where N represents the total number of users connected to a cell. Finally, the capacity of the user u is defined by the Shannon capacity model (Parker, 2017) as:

$$Capacity_u[bps] = bw_c[Hz]log_2(1 + SNR_u)$$

Considering this model for the network in the following section, we present the mathematical formulation of the problem.

2.2 Mathematical formulation

Let C be the set of cells and U be the set of users. As aforementioned, the available bandwidth of a cell is distributed uniformly among the number of users connected to it. Consider that Prx_{uc} is the received power at the location of user u location from cell c, Pn_c is the noise power of cell c, BW_c the total bandwidth of cell c, sv_c be the sum of singular values of cell c, and nt_c be the number of transmissions antennas of cell c. Regarding variables, let s_c be the binary variable that is 1 if cell c is active and 0 otherwise, a_{uc} be the binary variable that is 1 if user u is assigned to cell c and 0 otherwise, bw_c be the bandwidth assigned to each user connected to cell c, and SNR_u be the received signal and noise ratio of user u when connected to a cell.

Then, the mathematical formulation can be described as

$$\min FO_e = \sum_{c \in C} s_c \tag{15.1}$$

$$\max FO_{QoS} = \sum_{c \in C} a_{uc} bw_c \, log_2\left(1 + \frac{(sv_c)^2}{nt_c} SNR_u\right) \tag{15.2}$$

Subject to

$$s_c \leq \sum_{u \in U} a_{uc}, \ \forall \ c \in C \tag{15.3}$$

$$\sum_{u \in U} a_{uc} \leq |U| s_c, \ \forall \ c \in C \tag{15.4}$$

$$SNR_u = \frac{\sum_{u \in U} Prx_{uc} a_{uc}}{Pn_c} \tag{15.5}$$

$$\sum_{u \in U} bw_c a_{uc} \leq BW_c s_c, \ \forall c \in C \tag{15.6}$$

$$bw_c \leq BW_c s_c, \ \forall u \in U, c \in C \tag{15.7}$$

$$\sum_{c \in C} a_{uc} = 1, \forall \ u \in U \tag{15.8}$$

$$a, s \in \mathscr{B}, bw \geq 0$$

Regarding objective functions, Eq. (15.1) aims at minimizing the number of cells that are active, and Eq. (15.2) computes the capacity assign to all the users based on Shannon capacity model (Parker, 2017). Regarding constraints, Constraint (3) forces a cell to be switched off if no user is assigned. Constraint (4) forces users to be assigned only to cells that are active. Constraint (5) computes the signal and noise ratio. Constraint (6) is the sum of bandwidth assigned to users connected to a cell to be equal to the total bandwidth of that cell. Constraint (7) is a symmetry breaking constraint and forces the bandwidth variables to be null if the cell is not switched on. Constraint (8) forces that each user is only assigned to one cell.

2.3 Related work

Turning off the base stations is a recognized technique to reduce energy consumption in 5G and 6G wireless networks (Feng et al., 2017). To meet the anticipated traffic demands and extensive connectivity featured by these new generation networks, network densification, which involves deploying heterogeneous and numerous small base stations, is crucial. This approach enhances the spatial spectrum reuse and, consequently, the overall system capacity, leading to the formation of UDNs. However, the proliferation of cells within UDNs significantly contradicts a fundamental design requirement of 5G/6G systems, which is to operate with a 90% reduction in energy consumption. In response to this challenge, various strategies have been proposed in the literature to tackle the energy consumption issue (Buzzi et al., 2016). Among these strategies, one of the primary approaches is the CSO problem, which involves selectively deactivating a subset of cells in a UDN in period of low demand. This problem involves handling the trade-off between the energy consumption and the quality of service provided to the users (Han et al., 2016). The CSO problem has been approached from various angles, including clustering techniques and its formulation as an optimization problem. As a computationally complex optimization problem, the CSO has been usually addressed with heuristics (Femenias et al., 2020; Lagum et al., 2017) or metaheuristics algorithms (Galeano-Brajones et al., 2023; Luna et al., 2018, 2020). Regarding exact approaches, as the one that is proposed in this chapter, the number of works is more scarce. As far as we are concerned, only a few works deal with this problem using an exact approach. In a dynamic version of the problem, Dolfi et al. (2017) proposed a mixed-integer quadratic programming to address the CSO, which aims at simultaneously minimizing the power consumption and the cells switchings, i.e., variations from/to on/off sates, since the switchings are associated with energy consumptions. The model optimizes this function in a weighted sum considering the minimum required QoS as a constraint. In the experimentation, authors could obtain up to 60% energy savings during off-peak hours, guaranteeing QoS target requirements. Also in a dynamic version of the problem, Ahmed et al. (2021) proposed a quadratic integer programming model, which is linearized by means of McCormick envelopes to address the CSO problem. In this work, the authors considered that some cells are powered by renewable sources of energy and that there is a collaborative working procedure for the cells, i.e., cells that have low demand from users are switched off, and the demand is moved to a central (larger) cell.

Thus, we considered that there is still room in the literature to propose mathematical formulations of this problems. The main contribution of this work is to propose a biobjective mixed-integer nonlinear programming model, which has not been proposed before. The model aims at simultaneously minimizing the number of active cells and maximizing the quality of service, measured through the capacity provided to the users. The model was tested on realistic instances using an exact multiobjective resolution methodology showing it competitiveness to analyze the trade-off among objectives.

As a computationally complex optimization problem, the CSO has been usually addressed with heuristics (Femenias et al., 2020; Lagum et al., 2017) or metaheuristics algorithms (Galeano-Brajones et al., 2023; Luna et al., 2018, 2020). Regarding exact approaches, as the one that is proposed in this chapter, the number of works is more reduced. As far as we are concerned, only a few works deal with this problem using an exact approach. In a dynamic version of the problem, Dolfi et al. (2017) proposed a mixed-integer quadratic programming to address the CSO, which aims at simultaneously minimizing the power consumption and the cells switchings, i.e., variations from/to on/off sates, since the switchings are associated with energy consumptions. The model optimizes this function in a weighted sum considering the minimum required QoS as a constraint. In the experimentation, authors could obtain up to 60% energy savings during off-peak hours, guaranteeing QoS target requirements. Also in a dynamic version of the problem, Ahmed et al. (2021) proposed a quadratic integer programming model, which is linearize by means of McCormick envelopes to address the CSO problem. In this work, the authors considered that some cells are powered by renewable sources of energy and that there is a collaborative working procedure for the cells, i.e., cells that have low demand from users are switched off, and the demand is moved to a central (larger) cell.

Thus, we considered that there is still room in the literature to propose mathematical formulations of this problems. The main contribution of this work is to propose a biobjective mixed-integer non-linear programming model which has not been proposed in the related literature. The model, which aims at simultaneously minimizing the number of active cells and maximizing the capacity provided to the users, was tested on realistic instances using an exact multiobjective resolution methodology showing it competitiveness to analyze the trade-off among objectives.

3. Resolution methodology

It is well known that in biobjective problems with conflicting objectives, enhancing solutions in one optimization criterion inevitably leads to a compromise in the other objective. Consequently, it does not exist as a single, universally optimal (preferred) solution; instead, we have a collection of efficient solutions that explore the trade-offs among the objectives. The set of efficient solutions is the Pareto front, and a solution is efficient if it is not dominated by any other solution. For example, in a multiobjective minimization problem, a solution x'

from the feasible region SP is said to be efficient if there is not any other solution x in SP, which is strictly better than x', i.e.,

$$f_n(x) \leq f_n(x') \text{ for all the objectives } n$$
$$\text{and } f_{n'}(x) < f_{n'}(x') \text{ for a at least one objective } n'$$

where $f_n(x)$ is the evaluation of solution x in objective n. The most prevalent approaches to address multiobjective problems with exact methods include the weighting or weighted-sum method and the ϵ-constraint method. To obtain nondominated solutions and iteratively construct the Pareto front, these methods require multiple executions. However, the weighting method typically demands a higher number of iterations, often yielding repeated solutions. Moreover, it has limitations in reaching solutions situated in nonconvex regions of the Pareto front (Rossit et al., 2017).

Consequently, for this work, we adopted the ϵ-constraint method, which has successfully been applied in several multiobjective linear integer programming models (Mavrotas and Florios, 2013). In biobjective optimization, the ϵ-constraint method involves solving a single-objective optimization problem with one objective while constraining the other objective with an additional constraint. In a minimization (maximization) problem, this additional constraint enforces the objective function to be less than (greater than) a specified parameter ϵ:

Original problem	$\Rightarrow$	*Transformed problem with ϵ-constraint method*
Min $f_1(x)$		Min $f_1(x)$
Min $f_2(x)$		St:
St:		$f_2(x) \leq \epsilon$
$x \in SP$		$x \in SP$

Subsequently, the method derives various nondominated solutions to approximate the Pareto front by conducting multiple runs with diverse ϵ values. The number of runs is known as the number of subproblems in the ϵ-constraint method. However, this method can obtain weakly efficient solutions that have the same value for objective $f_1(x)$ but different values of $f_2(x)$ with respect to the epsilon restriction. Thus, instead of using the original epsilon constraint method developed by Haimes (1971), we used an improved version proposed in Mavrotas (2009) and Mavrotas and Florios (2013), which modified the model as follows:

$$\text{Min } f_1(x) + eps * \frac{s}{r_2}$$

$$\text{Min}$$

$$\text{St}:$$

$$f_2(x) + s = \epsilon$$

$$x \in SP,\ s \geq 0$$

Where s is the slack variable to transform the restricted objective into an equality, r_2 is the range of the second objective in the Pareto front, and eps is a relatively small constant between 10^{-2} and 10^{-4}. By minimizing in the objective function slack variable s, the method avoids obtaining weakly efficient solutions. Also the information of the slack variables is used by the augmented ϵ-constraint method to avoid some runs or subproblems that would end up in repeated solutions (Mavrotas, 2009). There are different approaches to estimate the range of the restricted objective within the Pareto front r_2. The simpler approach is to directly apply single-objective optimization to optimize each objective and construct the pay-off tableau. Another approach is to use lexicographic optimization (Mavrotas, 2009). In biobjective problems, as that addressed in this chapter, lexicographic optimization is based on a two-step procedure to find the extreme solutions. First, single-objective optimization model is solved to find the optimum of one objective. Second, to avoid weakly efficient solutions, in the second step another single-objective optimization problem is solved optimizing the second objective provided that the first objective does not worsen its value. This process can be enhanced using the previous solution to warm start the solver in the second step, which can speed up the resolution process (Rossit et al., 2020).

4. Computational experimentation

In this section, the computational tests are presented, including the description of the set of instances that were used, the implementation details, and the main numerical results.

4.1 Description of instances

The set of instances for the computational experimentation were generated using the procedure developed by Luna et al. (2018). This procedure involves creating scenarios where cells and users are deployed based on an independent Poisson point process (PPP) characterized by a deployment probability. In the case of cells, different types of cells can be deployed. These are femtocells, picocells, and microcells, each increasing in size. Notably, macrocells, which are used in Luna et al. (2018), are not used in this work since the instances are relatively smaller than the instances used in Luna et al. (2018). This is connected to the resolution approach. Whereas Luna et al. (2018) used a metahueristc approach, an exact approach is used in this work.

Regarding users' deployment, the social attractors (SAs) model introduced by Mirahsan et al. (2015) is used for this purpose. This deployment model integrates two factors, α and β, which determine the strength of attraction between cells and SAs and between SAs and users, respectively. As in Luna et al. (2018), these factors have been configured as $\alpha = \beta = 0.25$.

The deployment probability of users is set to $\lambda_u = 1000$ (which corresponds to the low-density scenario in Luna et al. (2018)). Regarding the distinct types of cells, Table 15.1 provides the λ_c, the transmission gain (G_{tx}), and the operating frequency of the antenna (f) of each type of cell.

Based on the system model developed by Luna et al. (2018), five instances of 20×20 grid points are built (in which each 1×1 grid point represents an area of 25 m^2). The information about the number of users and number of cells of each instance is presented in Table 15.2.

TABLE 15.1 Description of different types of Cells.

Type of cell	λ_c	G_{tx}	f[GHz] (= 10%BW_c)
micro1	100	12	3.5
micro2	100	10	5
pico1	500	5	10
pico2	500	7	14
femto1	1000	4	28
Femto2	1000	3	66

4.2 Implementation details

The model was implemented using the Pyomo optimization package (Hart et al., 2011) in a Python environment. Pyomo is a modeling language that allows coding a mathematical programming problem as a computational model with an intuitive syntax. However, in order to solved the model, it requires a solver. To solve the optimization model, we used two different solvers: BARON and DICOPT. Both solvers are used through a connection to the pool of solvers of GAMS v.43. In the case of this work, specific solvers for MINLP problems are required. First, BARON version 23.3.11 was used. BARON, which stands for Branch and Reduce Optimization Navigator, is a solver for solving nonconvex optimization problems to global optimality, including mixed-integer nonlinear problems (Sahinidis, 2021). It uses a pool of techniques including constraint propagation, interval analysis, and advanced branch-and-bound optimization concepts. The solver combines branch-and-bound techniques specialized for nonconvex MINLPs with bound reduction based on both feasibility and optimality reasoning, which improve its performance (D'Ambrosio and Lodi, 2013). In the GAMS environment—as is the case of this chapter, BARON has the flexibility to employ various nonlinear programming (NLP) solvers for subproblems and offers the option of using

TABLE 15.2 Description of instances.

		Cells			
Id	Grid points	Micro	Pico	Femto	Users
20×20–1	20×20	–	–	12	19
20×20–2	20×20	–	–	12	23
20×20–3	20×20	–	6	12	26
20×20–6	20×20	3	6	12	14
20×20–11	20×20	–	6	12	27

CBC, CPLEX, and XPRESS as LP/MILP subsolvers (Kronqvist et al., 2019). It is recognized as one of the most effective solvers for global optimization (D'Ambrosio & Lodi, 2013; Kronqvist et al., 2019)

DICOPT, short for discrete continuous optimizer, is a solver that is grounded in the outer approximation method for MINLP (Viswanathan and Grossmann, 1990). This solver incorporates the use of both the equality relaxation and augmented penalty methods in conjunction with the outer approximation approach. Both of these methods are primarily designed as heuristics for addressing nonconvex MINLP problems. DICOPT is exclusively accessible within the GAMS environment, and it offers the flexibility to utilize any available MILP or NLP solvers of GAMS for tackling subproblems. Although DICOPT can obtain optimal solutions in convex problems, in nonconvex problems, DICOPT works as a heuristic procedure.

Regarding the resolution platform, the experiments were carried out using a personal computer with an Intel Core i5-3570 @ 3.40 GHz processor with 12 GB of RAM and a 64-bit Windows 10 operating system. A time limit of 1800 s was set for each run.

4.3 Numerical results

Each instance was solved with the two solvers. The restricted objective from the augmented ε-constraint method is the number of active cells since this can be easily discretized. This discretization is used for fixing the maximum number of subproblems to be solved.

In Table 15.3, we compare both solvers in terms of the best trade-off solution reached. This compromising solution is that with minimal normalized distance to the ideal vector, i.e., the vector that represents an ideal unattainable solution composed by the best result of each objective (Rossit et al., 2022). From left to right, Table 15.3 reports for each instance, the ideal vector, the computing time for obtaining all the efficient solutions, the percentage difference between the ideal vector and the trade-off solution for each objective (number of active cells δ_e and quality of service δ_{QoS}) and the overall distance Δ computed with Eq. (15.9).

TABLE 15.3 Main results of computational tests.

	Ideal vector		BARON				DICOPT			
Instance	FO_e	FO_{QoS}	Time (sec)	δ_e	δ_{QoS}	Δ	Time (sec)	δ_e	δ_{QoS}	Δ
20×20–1	1	20594	7464.1	66.67%	−26.98%	71.92%	7.4	75.00%	−27.77%	79.98%
20×20–2	1	91497	7782.6	75.00%	−33.15%	82.00%	7.3	75.00%	−33.66%	82.21%
20×20–3	1	871612	5218.4	80.00%	−38.78%	88.90%	9.9	87.50%	−495.19%	502.86%
20×20–6	1	693426	5908.3	85.71%	−9.65%	86.26%	9.1	87.50%	−424.80%	433.72%
20×20–11	1	988296	6341.3	85.71%	0.00%	85.71%	14.0	88.89%	−374.38%	384.79%
Average	–	–	6542.9	78.62%	−21.71%	82.96%	9.5	82.78%	−271.16%	296.71%

$$\Delta = \sqrt{\sum_{o \in \{e, QoS\}} \left(\frac{IV_o - CS_o}{CS_o} \right)^2} \tag{15.9}$$

Where IV_o and CS_o are the values of the ideal vector and the compromising solution for objective o, respectively.

In terms of the results, BARON consistently outperforms DICOPT in finding compromise solutions with smaller distances to the ideal vector, particularly in relation to the capacity provided to users. On average, BARON's solutions are only 21.71% worse than the ideal vector, with the best result being 0% deviation in the case of instance 20×20–11. In contrast, DICOPT's solutions exhibit a larger average deviation of 271.16% from the ideal vector, with its best result achieving a 27.77% deviation in the case of instance 20×20–1.

When considering the objective of the number of active cells, the differences between the two solvers are relatively smaller, but BARON still outperforms DICOPT. On average, BARON's solutions have 78.62% more active cells compared with DICOPT's solutions, which have 82.78% more active cells on average. The overall distance, which is clearly influenced by the performance on the objective of capacity provided to users, favors BARON. This distance is on average 82.96% for the compromising solutions obtained by BARON and 296.71% for the compromising solutions obtained by DICOPT. However, it's worth noting that DICOPT demonstrates impressive speed in computing the entire set of solutions for any instance, taking less than 15 s, while BARON, even when working at its fastest (as seen in the 20×20–3 instance), requires significantly more time, specifically 5218 s. Moreover, in most of the runs, BARON uses the time limit of 1800 s without arriving to a proven optimal solution.

To delve deeper into the results, we present a Pareto front plot for four instances in Fig. 15.1 (with similar trends observed for the only instance, which is not plotted –20×20–3). These plots illustrate a clear trade-off between the objectives. Solutions with only a few active cells result in very low capacity for users, while a larger number of active cells leads to increased user capacity. An important takeaway from these plots is that achieving maximum capacity does not necessarily require all cells to be active. For instance, in the case of instance 20×20–6, the highest capacity is attained when only 13 out of 21 cells are active. This demonstrates that, given a particular traffic demand, it is possible to deactivate 8 cells without compromising the quality of service provided to users, and thus, the proposed mode is able to successfully identify opportunities for energy conservation without sacrificing user satisfaction.

In the comparison of solvers, it is evident that BARON consistently outperforms DICOPT, establishing its superiority in terms of solution quality. This dominance is particularly pronounced in instances 20×20–6 and 20×20–11, where the solutions generated by DICOPT significantly lag behind those obtained by BARON, especially concerning the capacity provided to users. This reaffirms the earlier observations made in the analysis of the results related to compromise solutions (Table 15.3). Even in instances 20×20–1 and 20×20–2, where DICOPT manages to produce Pareto fronts that are closer to those generated by BARON, it is still apparent that BARON dominates DICOPT's solutions in terms of solution quality. This consistent trend underscores the superior performance of BARON in this particular optimization problem.

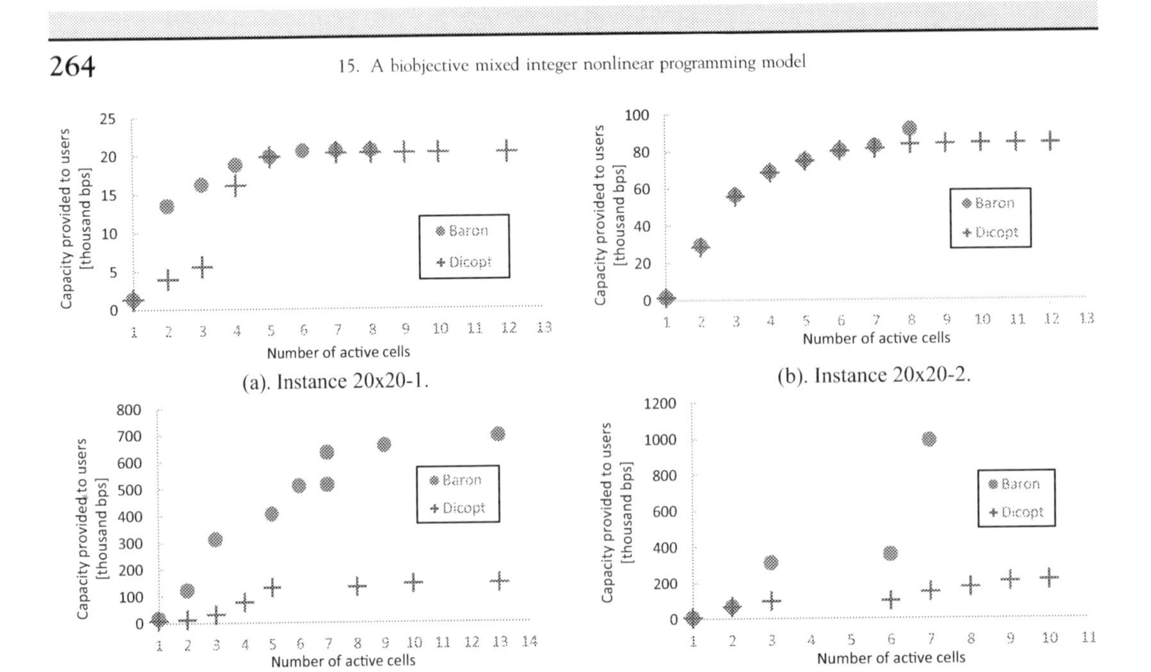

FIGURE 15.1 Pareto fronts obtained on the studied instances. (A) Instance 20×20–1. (B) Instance 20×20–2. (C) Instance 20×20–6. (D). Instance 20×20–11.

5. Conclusions

Telecommunications networks play a pivotal role in our highly interconnected world, necessitating the deployment of numerous cells to maintain a high quality of service for users. However, the quest for faster data transfer rates must be harmonized with the prudent utilization of energy resources. This chapter focuses on the CSO problem, which arises in the context of 5G/6G networks when demand is low, allowing for the deactivation of a subset of cells. This deactivation must be executed properly to avoid compromising the quality of service for the remaining users.

This chapter introduces a biobjective mixed integer nonlinear programming (MINLP) model to address the CSO problem, which aims at simultaneously maximizing the capacity provided to the users (which is a proxy to the quality of service) and minimizing the number of active cells. Preliminary computational experiments conducted on small yet realistic instances demonstrate the effectiveness of our approach in striking a balance between conflicting objectives. Notably, the results reveal that it is possible to reduce energy consumption without impacting the capacity provided to users when certain cells are deactivated. In essence, our work makes a valuable contribution to the ongoing endeavors aimed at managing the intricate interplay between user satisfaction and energy conservation within the telecommunications sector. Furthermore, being an exact approach, our model is suitable for heuristic validation on small instances.

Future research directions will encompass expanding our computational experiments to tackle larger instances and testing other solvers designed for mixed-integer nonlinear programming problems, including commercial solvers such as XPRESS and SBB and open-source solvers such as SCIP, SHOT, and BONMIN.

Acknowledgments

The first author was supported by a scholarship *"Estancias para investigadores de reconocido prestigio en la UMA"* of the Universidad de Málaga of Spain and the research projects 24/J084 and 24/J086 of the Universidad Nacional del Sur of Argentina. This work was partially funded by the MICIU/AEI/10.13039/501100011033 and the "European Union Next Generation EU/PRTR" under grant TED2021-131699B-100.

References

Ahmed, F., Naeem, M., Ejaz, W., Iqbal, M., Anpalagan, A., Haneef, M., 2021. Energy cooperation with sleep mechanism in renewable energy assisted cellular hetnets. Wireless Personal Communications 116, 105–124.

Aydin, O., Jorswieck, E., Aziz, D., Zappone, A., 2017. Energy-spectral efficiency tradeoffs in 5G multi-operator networks with heterogeneous constraints. IEEE Transactions on Wireless Communications 16 (9), 5869–5881.

Buzzi, S., Chih, I., Klein, T., Poor, H., Yang, C., Zappone, A., 2016. A survey of energy-efficient techniques for 5G networks and challenges ahead. IEEE Journal on Selected Areas in Communications 34 (4), 697–709.

Cornuéjols, G., Sridharan, R., Thizy, J., 1991. A comparison of heuristics and relaxations for the capacitated plant location problem. European Journal of Operational Research 50 (3), 280–297.

Dolfi, M., Cavdar, C., Morosi, S., Piunti, P., Zander, J., Del Re, E., 2017. On the trade-off between energy saving and number of switchings in green cellular networks. Transactions on Emerging Telecommunications Technologies 28 (11), e3193.

D'Ambrosio, C., Lodi, A., 2013. Mixed integer nonlinear programming tools: an updated practical overview. Annals of Operations Research 204, 301–320.

Femenias, G., Lassoued, N., Riera, F., 2020. Access point switch ON/OFF strategies for green cell-free massive MIMO networking. IEEE Access 8, 21788–21803.

Feng, M., Mao, S., Jiang, T., 2017. Base station ON-OFF switching in 5G wireless networks: approaches and challenges. IEEE Wireless Communications 24 (4), 46–54.

Galeano-Brajones, J., Luna-Valero, F., Carmona-Murillo, J., Cano, P., Valenzuela, J., 2023. Designing problem-specific operators for solving the Cell Switch-Off problem in ultra-dense 5G networks with hybrid MOEAs. Swarm and Evolutionary Computation 78, 101290.

González, D., Hämäläinen, J., Yanikomeroglu, H., García, M., Senarath, G., 2016. A novel multiobjective cell switch-off framework for cellular networks. IEEE Access 4, 7883–7898.

Haimes, Y., 1971. On a bicriterion formulation of the problems of integrated system identification and system optimization. IEEE Transactions on Systems, Man, and Cybernetics (3), 296–297.

Han, F., Zhao, S., Zhang, L., Wu, J., 2016. Survey of strategies for switching off base stations in heterogeneous networks for greener 5G systems. IEEE Access 4, 4959–4973.

Hart, W., Watson, J., Woodruff, D., 2011. Pyomo: modeling and solving mathematical programs in Python. Mathematical Programming Computation 3 (3), 219–260.

Kronqvist, J., Bernal, D.E., Lundell, A., Grossmann, I., 2019. A review and comparison of solvers for convex MINLP. Optimization and Engineering 20, 397–455.

Lagum, F., Le, Q., Beitelmal, T., Szyszkowicz, S., Yanikomeroglu, H., 2017. Cell switch-off for networks deployed with variable spatial regularity. IEEE Wireless Communications Letters 6 (2), 234–237.

Lopez-Perez, D., Ding, M., Claussen, H., Jafari, A., 2015. Towards 1 Gbps/UE in cellular systems: understanding ultra-dense small cell deployments. IEEE Communications Survey & Tutorials 17 (4), 2078–2101.

Luna, F., Luque, R., Martinez, J., Valenzuela-Valdés, J.F., Padilla, P., 2018. Addressing the 5G cell switch-off problem with a multi-objective cellular genetic algorithm. In: 2018 IEEE 5G World Forum (5GWF). IEEE, pp. 422–426.

Luna, F., Zapata, P., Gonzalez, J., Valenzuela-Valdés, J.F., 2020. Approaching the cell switch-off problem in 5G ultra-dense networks with dynamic multi-objective optimization. Future Generation Computer Systems 110, 876–891.

Mavrotas, G., 2009. Effective implementation of the ε-constraint method in multi-objective mathematical programming problems. Applied Mathematics and Computation 213 (2), 455–465.

Mavrotas, G., Florios, K., 2013. An improved version of the augmented ε-constraint method (AUGMECON2) for finding the exact Pareto set in multi-objective integer programming problems. Applied Mathematics and Computation 219 (18), 9652–9669.

Mirahsan, M., Schoenen, R., Yanikomeroglu, H., 2015. HetHetNets: heterogeneous traffic distribution in heterogeneous wireless cellular networks. IEEE Journal on Selected Areas in Communications 33 (10), 2252–2265.

Parker, M., 2017. Chapter 12 - error-correction coding. Digital Signal Processing 101, 129–147. Newnes. https://doi.org/10.1016/B978-0-12-811453-7.00012-3.

Rossit, D., Tohmé, F., Frutos, M., Broz, D., 2017. An application of the augmented ε-constraint method to design a municipal sorted waste collection system. Decision Science Letters 6 (4), 323–336.

Rossit, D., Toutouh, J., Nesmachnow, S., 2020. Exact and heuristic approaches for multi-objective garbage accumulation points location in real scenarios. Waste Management 105, 467–481.

Rossit, D., Nesmachnow, S., Rossit, D., 2022. A multi objective evolutionary algorithm based on decomposition for a flow shop scheduling problem in the context of Industry 4.0. International Journal of Mathematical, Engineering and Management Sciences 7 (4), 433–454.

Sahinidis, N., 2021. BARON 21.1.13: Global Optimization of Mixed-Integer Nonlinear Programs, User's Manual. Available at: http://www.minlp.com/downloads/docs/baron%20manual.pdf.

Toncovich, A., Rossit, D., Frutos, M., Rossit, D., 2019. Solving a multi-objective manufacturing cell scheduling problem with the consideration of warehouses using a simulated annealing based procedure. International Journal of Industrial Engineering Computations 10 (1), 1–16.

Usama, M., Erol, M., 2019. A survey on recent trends and open issues in energy efficiency of 5G. Sensors 19 (14), 3126.

Viswanathan, J., Grossmann, I., 1990. A combined penalty function and outer-approximation method for MINLP optimization. Computers & Chemical Engineering 14 (7), 769–782.

Index

Note: 'Page numbers followed by *f* indicate figures, *t* indicate tables, and *b* indicate boxes.'

D

E